Hans G. Natke · Geoffrey R. Tomlinson · James T. P. Yao

Safety Evaluation Based on Identification Approaches Related to Time-Variant and Nonlinear Structures

Vieweg International Scientific Book Series

Advanced Lectures in Mathematics,
Gerd Fischer (Ed.)

Advances in Control Systems and Signal Processing,
Irmfried Hartmann (Ed.)

Aspects of Mathematics,
Klas Diederich (Ed.)

Theory and Practice of Applied Geophysics,
Andreas Vogel (Ed.)

Progress in Earthquake Prediction Research,
Andreas Vogel (Ed.)

Notes on Numerical Fluid Mechanics,
H. Hirschel, K. W. Morton, E. M. Murman, M. Pandolfi, A. Rizzi and B. Roux (Eds.)

Methods in Clinical Pharmacology,
Norbert Rietbrock and Barry Woodcock

Advances in Solid State Physics

Advances in Structure Research by Diffraction Methods

Vieweg

Hans G. Natke · Geoffrey R. Tomlinson · James T. P. Yao

Safety Evaluation Based on Identification Approaches Related to Time-Variant and Nonlinear Structures

vieweg

Vieweg is a subsidiary company of the Bertelsmann Publishing Group International.

Printed on acid-free paper

ISBN-13: 978-3-528-06535-5 e-ISBN-13: 978-3-322-89467-0
DOI: 10.1007/978-3-322-89467-0

FOREWORD

Safety evaluation by definition involves many complex factors and thus covers a wide range of topics. In order to focus the content of the workshop the subject matter was specific to the state of the art and the recent developments in nonlinear and time-variant methods employing identification procedures.

Participants in the workshop represented a wide range of expertise. They were selected in order to cover the state of the art of knowledge in fault-detection and damage assessment, system identification, signal processing, mathematical and physical modelling and applications of techniques such as fuzzy logic and neural networks. The emphasis was placed on the exploitation and understanding of nonlinearity arising from structural or material faults. Figure 1 indicates the range of topics covered in the workshop.

Since no unique or general approach yet exists for treating nonlinearity in the field of safety evaluation, many of the topics presented were problem specific. In order to assist the reader in selecting the material of primary interest a matrix of the topics covered by each participant is shown in Table 1. This table relates the authors to the subject matter, providing a guide through the diverse range of topics presented at the workshop.

The structure of the contents of the proceedings, starting with an overview paper on the evaluation of diagnostic methods, follows the manner in which they were presented. The contents can be classified into the following four categories:

- Test design, monitoring of structural behaviour (mainly advanced methods such as intensity measurements, spatially continuous measurements).
- Damage detection and localisation by studying time-variant and nonlinear behaviour.
- Methods of system identification (with respect to time-variant and nonlinear systems) including mathematical solutions of the generally ill-posed problems, treatment of the associated uncertainties, and their effects on structural reliability.
- Procedures for safety evaluation with emphasis on time-variant and/or nonlinear systems (including fuzzy logic and neural nets).

The achievements of the workshop are reflected in the number and diversity of the topics which were succesfully treated, the future directions which have been clearly identified and the areas which require further attention and research. These are discussed in the Summary and Conclusions at the end of the proceedings. It is clear that, as demonstrated by the contents of these proceedings, significant developments have taken place in the field of safety evaluation. However, the ability to model, validate and quantify nonlinear and time-variant problems is still in its infancy and remains to be studied further.

ACKNOWLEDGEMENT

The International Workshop on Safety Evaluation Based on Identification Approaches Related to Time-Variant and Nonlinear Structures was financially supported with a grant from the Volkswagen Foundation, and supplemented with a research award to H. G. Natke and J. T. P. Yao from the Max-Planck Society and Alexander von Humboldt Foundation in the Federal Republic of Germany.

In addition we gratefully acknowledge the support of the following:
Hannoversche Hochschulgemeinschaft, Hannover,
Siemens AG (KWU), Erlangen,
Volkswagen AG, Wolfsburg.

We sincerely appreciate the generous support of these organisations.

In particular, we wish to thank all the participants for their active participation and enthusiastic discussions during this workshop in Lambrecht, Pfalz on 6 - 9 September 1992, and for the preparation of their papers.

H. G. Natke
G. R. Tomlinson
J. T. P. Yao

List of Lecturers

Prof. Dr. *Johann Baumeister*, FB Mathematik, Johann Wolfgang Goethe-Universität Frankfurt
Prof. Dr. *Y. Ben-Haim*, Faculty of Mechanical Engineering, Technion - Israel Institute of Technology, Haifa, Israel
Prof. Dr. *C. Cempel*, Applied Mechanics Institute, Poznan University of Technology
Dr. *P. Cowley*, Rolls-Royce plc, Derby, UK
Prof. Dr.rer.nat. *M. Heckl*, Institut fr Technische Akustik, TU Berlin
Prof. Dr.-Ing. *R. Isermann*, Institut f. Regelungstechnik, TH Darmstadt
Prof. Dr. *L. Jezequel*, Dépt. de Mécanique des Solidesnology, Ecole Centrale de Lyon, Ecully, F
Prof. *M.P. Luong*, Lab. de Mécanique des Solides, Ecole Polytechnique, Palaiseau Cedex, F
Prof. *S.F. Masri*, Dept. of Civil Engineering, University of Southern California, Los Angeles, CA, USA
Prof. Dr.rer.nat. *H.G. Natke*, Curt-Risch-Institut, Universität Hannover
Prof. *T.T. Soong*, Dept. of Civil Engineering, University at Buffalo, USA
Prof. *N. Stubbs*, Dept. of Civil Engineering, Texas A & M University, College Station, TX, USA
Prof. Dr. *G.R. Tomlinson*, Dept. of Engineering, University of Manchester, UK
Prof. *H.S. Tzou*, Dept. of Mech. Eng., Lexington, Kentucky, USA
Dr. *C. Valente*, University "G. D'Annunzio", Pescara, I
Prof. *T. Vinh*, Lab. de Rhéologie et Structures, Institut Superieur des Materiaux et de la Construction Mecanique, Saint-Quen, F
Prof. *J.T.P. Yao*, Dept. of Civil Engineering, Texas A & M University, TX, USA
Prof. *L.A. Zadeh*, Dept. of Electrical Engineering and Computer Sciences, University of California, Berkeley, CA, USA
Prof. *J. Zhang*, Dept. of Engineering Mechanics, Xi'an Jiaotong University, P.R. China

Table of Contents

Evaluation of Diagnostic Methods

Y. Ben-Haim
Faculty of Mechanical Engineering
Technion — Israel Institute of Technology
Haifa, Israel

C. Cempel
Poznan University of Technology
Applied Mechanics Institute
Poznan, Poland

H. G. Natke
University of Hannover
Curt-Risch-Institute
Hannover, Germany

J. T. P. Yao
Texas A& M University
Department of Civil Engineering
College Station, TX, USA

1 Introduction: Systems and Models

In this brief paper we present some comparisons of diagnostic methodologies. Our aims are to enhance our understanding of individual approaches, to indicate the possibilities for combining them and to anticipate possible extensions and new lines of investigation. Our choice of subjects unavoidably reflects our own interests and areas of expertise. We aim not to resolve the conflicts between competing philosophies but rather to collect them under a single roof, "since the human mind has already learned to deal in contradictions" [1, p489].

We deal with systems and with models. By **system** we mean a real, existing object: a bridge or an aircraft wing or a robotic drive mechanism. A **model** is a primarily mathematical description of some aspect of a system. The system in general is more complex than the model. For example, nonlinearities of the system are neglected in a linearized model, or time-variation of system parameters may be frozen in a time-invariant model valid for a specified duration. From a diagnostic point of view some features are usually not significant.

Concerning our subject, we emphasize the distinction between two classes of models: those which describe the dynamic behaviour of the system, and those which describe the uncertainty of the environment within which the system operates and generates signals by installed sensors. We refer to these as **dynamic models** and **uncertainty models**, respectively. For example, the dynamic model of vibration of an elastic mechanical system involves inertia, damping and stiffness forces. The uncertainties which accompany this vibration may, for example, be uncertain inputs to the system such as unknown wind excitations, or uncertain boundary conditions connected with the system, or uncertain faults which can occur in the system.

The line between the system and its dynamic model on the one hand, and the uncertain environment and its model on the other, can usually be drawn at different points. The division between these two types of models is often a matter of convenience. Nevertheless, the distinction between the models is very real: different modelling techniques are employed to represent (1) the dynamics of a system and (2) the uncertainties accompanying those dynamics.

A dynamic model may be erroneous with respect to its structure and its parameters, regardless of the environmental uncertainty described by the model of uncertainty. The uncertainties of a dynamic model can be reduced by system identification (using measurements of the existing system). In consequence the predictions of the model can also be validated to specified reliability for given environmental conditions (loadings, for example). Once a dynamic model has been validated, it can be used for design of diagnostic procedures in order to identify damage existence, its type, location and severity. An uncertainty model can also be either inaccurate or precise, and such models can sometimes be updated and improved by measurements.

Uncertainty-models for fault processes are particularly important in the design and optimization of diagnostic algorithms. Optimization in the presence of uncertainty is contingent upon modelling the uncertain process. Uncertainty models can be combined with dynamic models to predict responses. These predictions, together

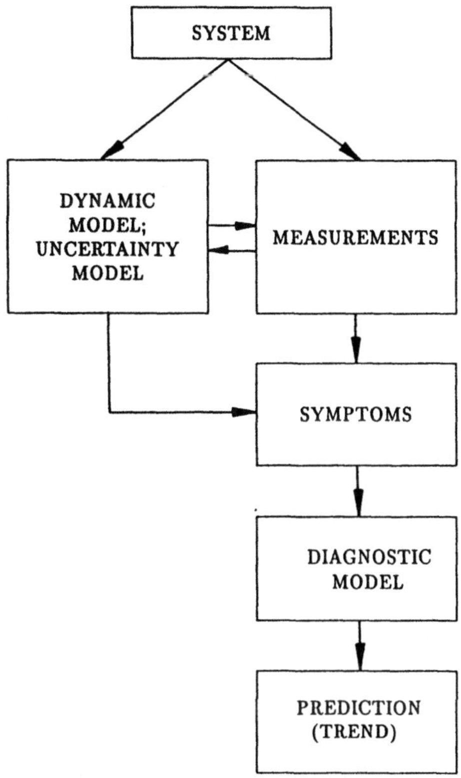

Figure 1 Flow-chart for condition prediction

with the corresponding measurements, form the basis of the diagnosis, as indicated schematically in Fig.**1**. Furthermore, the ability to perform inspection by demand rather than rigid scheduled inspection depends on a knowledge-base consisting of a mathematical model with known confidence and sufficiently small errors and a known model of uncertainty. Inspection by demand is flexible and efficient and more likely to locate trouble spots.

We have discussed the distinction between dynamic- and uncertainty-models. Each of these classes of models can be further subdivided as indicated in Fig.**2**. By "physical models" we mean those based primarily on fundamental physical theories such as elasticity, hydrodynamics and so on. These models are predominantly mathematical. "Heuristic models" are based on linguistic rules which, like rules-of-thumb, phenomenologically describe the behaviour of the subject of the model. "Empirical models" are basically mathematical, like physical models, but are derived empirically rather than from fundamental physical considerations.

We choose a model category — physical, empirical or heuristic — from considera-

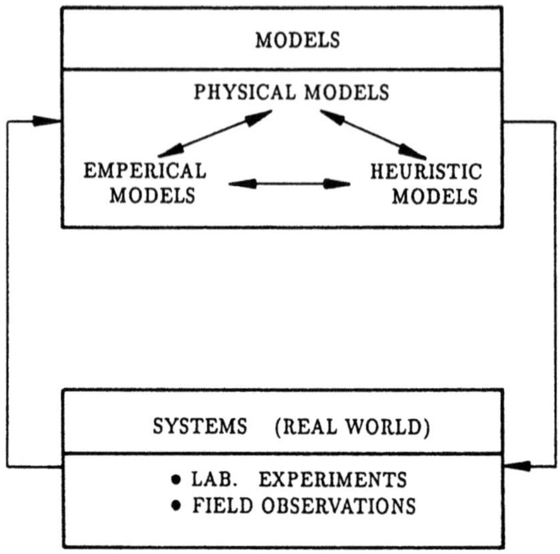

Figure 2 Interaction of systems and their mappings

tions such as robustness, versatility and optimality of the model with respect to the desired application. It is precisely at this point that the difficult decisions arise, since the versatility of one type of model is often accompanied by, say, lack of robustness, while another model-type displays complementary strengths and deficiencies.

Fig.**2** also shows that our models are bilaterally related to the real world. Observations and experiments determine the models, but the models in turn influence the observations and experiments which are performed.

It is appropriate to mention that dynamic models come in great diversity: Continuous, discrete, differential or difference equations, spectral or temporal representations [2] or bond graphs [3] name only a few. Likewise, models of uncertainty can be realized in various alternatives. While by far the most popular approach is based on the theory of probability, it has long been appreciated that probability and uncertainty are not equivalent. Alternatives to probability are fuzzy logic [4] and convex or other set-theoretic methods [5, 6, 7].

Finally we comment that, in order to diagnose a fault/damage, it must result in significant modifications of measurable quantities, here restricted to parameters. Distinguishability of faults by their symptoms or features plays a central role in damage assessment and in the choice of counter-measures. The ability to distinguish between faults, or classes of faults, is a central criterion in evaluating both a diagnostic algorithm and the models upon which it rests.

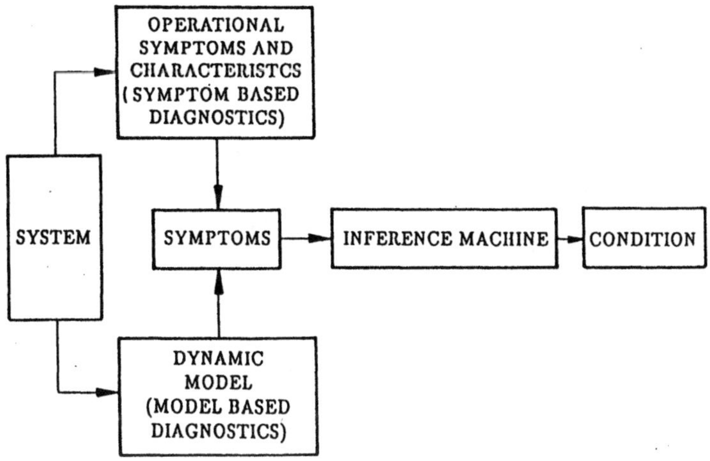

Figure 3 Two way physics-based diagnostics

2 Diagnostic Methods

In this section we will attempt a comparative analysis of the following types of diagnostic methodologies (Tab.**1**).

1. Signature analysis (simple signatures).
2. Extended signature analysis.
3. Diagnostics with physics- or knowlege- based models.
4. Diagnostics with fuzzy logic.
5. Diagnostics with neural nets.

Signature analysis employs simple data-reduction methods. For example, counting the number of peaks of a spectral density, or recording the abscissa of peaks. In general, signature analysis involves scalar information. Multiple features, if employed, are considered independently (Fig.**3**).

Extended signature analysis uses vectorial information and patterns: a set of features can be evaluated and used for decision and assessment. It can be thought of as a logic-filter method (Fig. **3**).

References to the symptom-based diagnostics are e.g. [9], [10], [11], [12].

Fig. **3** shows the skeleton of a flow chart for a physics-based diagnosis. The dynamic model is a structured parametric mathematical model for the system (based explicitly on physical quantities such as mass, stiffness etc.) and either a non-structured or a parametric model for the measurements. The bases for these models

	CATEGORIES OF METHODS	DEFINITIONS	DECISION-MAKING	UNCERTAINTIES	DIAGNOSTIC TASKS
1	(CLASSIC) SIGNATURE ANALYSIS	TIME, FREQUENCY DATA REDUCTION, SCALAR INFORMATION CONTENTS USED INDEPENDENTLY	DETERMINISTIC AND PROBABILISTIC (THRESHOLDS)	VARIANCES, BIAS	DECISION
2	EXTENDED SIGNATURE ANALYSIS (LOGIC FILTERS)	VECTORIAL INFORMATION, PATTERNS, USING A SET OF FEATURES	DETERMINISTIC AND PROBABILISTIC (PATTERNS)	COVARIANCES, BIAS	ANALYSIS, DECISION
3	DIAGNOSIS WITH PHYSICS-BASED MATHEMATICAL METHODS	DYNAMIC MATHEMATICAL MODELS WITH PRIOR SYSTEM KNOWLEDGE	DETERMINISTIC AND PROBABILISTIC (PREDICTED RELIAB.)	DETERMINISTIC (CONVEX SETS), PROBABILISTIC	MODELLING, DECISION, TREND
4	DIAGNOSIS BASED ON AI: FUZZY LOGIC	BEYOND PHYSICS-BASED MODELS	FUZZY LOGIC	FUZZY SETS (PROBABILISTIC, DETERMINISTIC)	DECISION, TREND
5	DIAGNOSIS BASED ON AI: NEURAL NETS	ABILITY OF LEARNING	EMPERICAL	DETERMINISTIC (?)	MODELLING, DECISION, TREND

Table 1 Descriptions of some diagnostic methods

are physical laws and prior knowledge of the system under consideration (method 3 in Table **1**). The diagnosis is implemented, for simple cases often in real time.

The next method falls in the category of artificial intelligence (AI), wherein the diagnostic decision is made with fuzzy logic. This method is characterized by the fact that the systems usually contain more information than can be conveniently quantified and invested in a mathematical, physics-based model.

The last method considered is based on AI using neural nets. Its predominant property is the ready implementation of learning. We should mention, however, that adaptive diagnostic algorithms can be formulated with most of the other methodologies as well.

Table **1** assembles the method and additionally contains some statements on decision making, handling of uncertainties and on diagnostic tasks.

Table **2** presents an attempted assessment. The application of signature analysis is well accepted and widespread. For instance, it is automated and applied in the nuclear power plant industry by a program called SUES [8]. It should be mentioned that the applicability of the extended signature analysis depends strongly on the particular dynamic behaviour of the system observed; it may be possible that the dynamic responses due to the operational forces do not allow extraction of the needed features.

	CATEGORIES OF METHODS	APPLICABILITY	FAULT DISTIN-GUISHABILITY	HANDLING		DIAGNOSIS CONFIDENCE	COSTS	ACCEPTANCE
				ON LINE WITH PC	SIMPLICITY			
1	(CLASSIC) SIGNATURE ANALYSIS	BROAD BUT LIMITED TO SIMPLE SYSTEMS(COMPO-NENTS)	DIFFICULT, ONLY WITH SUFF. PRIOR KNOWLEDGE	YES	HIGH	LOW	LOW	MATURE
2	EXTENDED SIGNATURE ANALYSIS (LOGIC FILTERS)	FOR CRITICAL AND PARTICULAR SYSTEMS	POSSIBLE WITH SUFF.PRIOR KNOWLEDGE	YES	MEDIUM	MEDIUM	MEDIUM	ALMOST MATURE
3	DIAGNOSIS WITH PHYSICS-BASED MATHEMATICAL METHODS	BROAD, FOR COMPLEX SYSTEMS	WITH SUFF.MEA-SURED DATA AND DETAILED MATH.MODEL	YES, RESTRICTED TO DYNAMIC MODELS	LOW	HIGH	HIGH	ALMOST MATURE
4	DIAGNOSIS BASED ON AI : FUZZY LOGIC	POSSIBLE	POSSIBLE	YES	?	?	?	INFANCY
5	DIAGNOSIS BASED ON AI : NEURAL NETS	POSSIBLE	POSSIBLE	YES(?)	?	?	?	INFANCY

Table 2 Assessment of some diagnosis and/or inference methods

Concerning the handling of the methods, the restriction on method no. 3 is explained as follows. A two stage procedure is applied in overall system monitoring. The first stage uses a global, dynamic model for damage detection and localization, while the second stage applies an extended, static model for assessment on the basis of (local) maximum stresses. Trend predictions and studies of counter-measures need the time-variant model, possibly with aid of the static model. The time-varying model will usually be of low dimension (magnitude of 10 degrees of freedom) which is relative easy to up-date and to analyze on a small computer. The detailed static model may have many more degrees of freedom and require large computational capability.

3 Conclusions

3.1 Optimization of Diagnostic Performance

We have discussed a range of alternative approaches to the design of a diagnostic algorithm. We may now ask: how does one compare and evaluate the performance of competing designs or competing design philosophies? This question is fundamental to any rigorous attempt to optimize a diagnostic procedure. Despite the importance of this question, no completely satisfactory answer has yet been found. We may formulate the question in the following more precise manner. Two questions should be considered.

1. How does one evaluate the potential, inherent in the structure of a given system, for distinguishing between distinct classes of faults? In other words, how does one quantify the best fault-diagnosis performance which can be achieved for a given system? We will refer to this inherent potential for distinguishing between classes of failures as **benchmark distinguishability.** The benchmark distinguishability serves as a reference point (or benchmark) against which specific implementable algorithms can be compared. Benchmark distinguishability is a design tool, not a diagnostic tool.

2. How does one evaluate the potential of a specified diagnosis algorithm for distinguishing between distinct classes of faults, when applied to a given system? Alternatively, how does one quantify the performance of a specific algorithm, and compare it to the limiting performance for that system, as expressed by the benchmark distinguishability? The ability of a specific algorithm to distinguish between classes of failures will be referred to as the **algorithmic distinguishability.**

The benchmark distinguishability expresses the limiting or ultimate diagnosis capability inherent in the structure of the system. Thus the algorithmic distinguishability of a specific diagnostic procedure will always be poorer than the benchmark distinguishability. The benchmark capability is a reference point against which the performance of a real algorithm can be compared. If a given algorithm performs far worse than the benchmark, then revision of the algorithm can lead to substantial improvement. On the other hand, if the performance of an algorithm is nearly as good as the benchmark, then further modification of the algorithm will not lead to substantial improvement in the diagnosis. In this way the performance of competing diagnostic designs can be evaluated.

Analyses of benchmark and algorithmic distinguishability have been performed for certain classes of additive failures in linear dynamic systems. See [13, 14].

3.2 Damage, Faults and their Evolution

The usage stage of every system is the stage of acquiring damage. If the damage is "large" it can influence the operational ability resp. the serviceability of the system and it produces some noticable faults. To detect the damage/faults we need to measure some symptoms of it. If the damage/fault as a holistic approach is included in the mathematical model of the system considered, safety state and trend can be predicted and assessed. The damage measure identified from the system and treated as a global quantity and as a local quantity [15] governs the lifetime model behaviour. The goal of the near future research concerning the diagnostics of large systems seems to be elaborate holistic models which include damage measure and evolution.

3.3 Remarks on Fuzzy Logic

Since its debut in 1965, the theory of fuzzy sets and logic as presented by L.A. Zadeh has been expanded and applied to solve many practical problems. Especially, the recent application of fuzzy logic control in combination with neural networks has been highly succesful in numerous commercial and industrial ventures in Japan, and is being followed by other countries including Germany and the USA. These applications have been primarily in automotive and air conditioning industries.

On the other hand, although aerospace, civil and mechanical engineering applications of fuzzy sets and logic were discussed and explored since the early seventies, practical applications in the fields mentioned have not been implemented in any major way. The early application on the control system of a cement manufacturing process in Europe and recent application in a train operating system in Japan might be remotely associated with civil or mechanical systems. However, relatively few aerospace, civil, and mechanical engineers in professional practice are aware of such applications.

The detection, diagnosis, and evaluation of defects and faults in term of damage, safety, and reliability of structural systems are extremely important to aerospace, civil and mechanical engineers. To date, various methods of nondestructive evaluation are available to detect defects and faults. Nevertheless, such resulting data are frequently inexact or incomplete. Moreover, the interpretation of such results is subjective in most cases. In many documented cases, even recognized experts may not agree among themselves on how to properly interpret such results.

The authors believe that fuzzy logic and neural networks provide one rational basis for the orderly assembly and analysis of expertise. With increasing knowledge and more precise data, fuzzy logic and neural networks provide additional tools which may be used to supplement mathematical and other scientific methodologies in solving practical problems. It is indeed important and timely to apply fuzzy logic and neural networks to detect, diagnose, and evaluate defects and faults in terms of damage evaluation and safety assessment of existing structural systems.

4 References

1. H. Adams, 1918, *The Education of Henry Adams,* E. Samuels, ed., Houghton Mifflin Company, Boston.

2. H.G. Natke, 1992, *Einführung in Theorie und Praxis der Zeitreihen- und Modalanalyse,* Friedr. Vieweg, Braunschweig, 3rd edition.

3. R.C. Rosenberg and D.C. Karnopp, *Introduction to Physical System Dynamics,* McGraw-Hill, New York.

4. J.T.P. Yao, 1985, *Safety and Reliability of Existing Structures,* Pitman, Boston.

5. F.C. Schweppe, 1973, *Uncertain Dynamical Systems,* Prentice-Hall, New Jersey.

6. Y. Ben-Haim, 1985, *The Assay of Spatially Random Material,* Kluwer Academic Publishers, Holland.

7. Y. Ben-Haim and I. Elishakoff, 1990, *Convex Models of Uncertainty in Applied Mechanics,* Elsevier, Holland.

8. H.-J. Wehling and R. Warnemünde, 1987, Vibration Monitoring of Light Water Reactors with Advanced Methods and the New Microprocessor-based System SÜS-86, SMORN V, Munich, 12-16 Oct.

9. R.A. Collacott, 1977, *Mechanical Fault Diagnosis and Condition Monitoring,* Chapman and Hall, London.

10. S. Braun (Ed.), 1985, *Mechanical Signature Analysis - Theory and Applications,* Academic Press, New York.

11. C. Cempel, 1991, *Vibroacoustic Condition Monitoring,* Ellis Horwood, UK.

12. H.G. Natke, J.T.P. Yao, 1991, Computer and Test Aided Modelling and its Application to Identify Structural Damage, Computational Stochastic Mechanics, Eds. P.D. Spanos and C.A. Brebbia, Comp. Mech. Publ., Elsevier Appl. Scie., 149-155

13. Y. Ben-Haim, 1986, Convexity analysis: A tool for optimization of malfunction isolation, 25th IEEE Conf. on Decision and Control, Athens, Greece, pp 1570 - 5.

14. Y. Ben-Haim, 1990, Optimizing multi-hypothesis diagnosis of control-actuator failures in linear system, AIAA Journal of Guidance, Control and Dynamics, 13: 744 - 750.

15. C. Cempel, H.G. Natke, 1992, Damage Evolution and Diagnosis in Operating Systems, Proceedings of this International Workshop.

Advantages and Disadvantages of Intensity Measurements

M. Heckl
Technische Universität Berlin
Institut für Technische Akustik

Summary

Appr. 15 years ago very accurate sensors for sound waves became available. Together with modern signal processors they allow the measurment of the energy flow (intensity). Thus it is possible to determine directly sound powers and to separate uncorrelated sources. This makes intensity very useful for machine or product monitoring in a noisy environment. For the localisation of partial sound sources intensity measurements are used quite often although they may give erroneous results in the presence of correlated sources. For sound in structures intensity measurements are still in the laboratory stage - with a few exceptions. They are much more complicated to perform and up to now they can be made only on the surface of a structure. In the paper the basic relations for structural intensity and crude approximations (which generally are used) will be given. Apart from background noise suppression, measurements of structural intensity are used to visualise the energy flow on structures and to find faults such as so-called sound bridges in structures. Because of the approximations that are made in the measurements, the accuracy is not high especially in the low frequency ranges.

1 Introduction

According to the author's limited knowledge, system identification usually is based on the assumption, that the structure of interest can be modelled by a lumped parameter system and that the (linear, time-dependent, or non-linear) parameters can be found from the results of dynamic measurements. There are, however, cases where the parameters (size, flexibility, etc.) of the structure and the frequencies of interest are such, that the motion can only be described by a large number of modes. Such structures would require complicated mathematical models with many unknown parameters. It is then very hard to make all the necessary measurements and to find reliable inverse methods that allow to determine the many unknown parameters that may be required for a reliable identification. In such cases it often is very useful to investigate the energy density and the energy flow in multimodal structures [1] (or gas filled spaces) and to determine some important gross parameters from such measurements. The most simple method of this category is the damping (or absorption measurement) in multimodal structures by measuring the reverberation time. Another method (which also does not involve phase measurements) is the inverse application of the Statistical Energy Analysis (SEA) [1]. In this case a set of energy balance equations is used to relate the energy flow between multimodal structural elements to the energy losses in the elements. This way a model, describing the energy flows is obtained and if it is possible to make a series of measurements on the system, the so-called coupling loss factors that describe the energy flow from one subsystem to another one and/or the loss factors due to material losses can be

[1] A structure is called multi-modal here when the average difference between two modal frequencies is comparable or less than the damping dependent bandwidth of one mode.

determined. The accuracy of the method is not very high [2] because it relies on the equipartition of energy over all modes and because it assumes that the energy density is solely given by the mass density and the mean square velocity. But since the method requires only simple measurements, it can without much effort be used to find out wether the mechanical properties of connecting elements between the substructures (characterised by the coupling loss factor) have changed or wether the inner losses have been altered (e.g. due to crack formation). The most recent methods [2] for solving identification problems in multimodal systems are the application of intensity measurements [3] [4] [5], and the so-called acoustic holography [6] [7] [8]. In both cases the measurement techniques are somewhat complicated (involving also measurements of phases and subtle difference quantities) and time consuming. But in return one obtains rather detailed pictures of the energy flow vectors. From them one can draw a lot of conclusions regarding the location and strength of energy sources or sinks (i. e. spots of high damping). It also is possible to follow the path of the energy flow when subsystems are connected. Because of lack of space only the intensity method is treated in the text. The first part of this discussion deals with intensity in air which has been studied and applied in the past quite extensively. It shows already all the salient features of the intensity concept. In addition intensity measurements in the vicinity of a vibrating structure may help to localise certain mechanical faults that radiate typical sounds (something related to acoustic emission).

2 Intensity

2.1 Sound intensity in gases and liquids

2.1.1 Definition and measurement principle

In media without shear stiffness the acoustic intensity vector $\vec{I}$ is defined as

$$\vec{I} = p(t)\vec{v}(t). \tag{1}$$

Here $p(t)$ is the instantaneous sound pressure and $v(t)$ the instantaneous value of the particle velocity in the sound field. Except for extremely intense sounds (e. g. inside an engine exhaust pipe) $p(t)$ and $v(t)$ are first order linear quantities. Consequently $\vec{I}(t)$ is of second order and does not have in general the property of superposition. As it stands Eq. (1) is a function of time, consisting of a mean value and a fluctuating part which in case of a single frequency sound has twice the frequency. In most applications only the time average or mean value of the intensity is of interest. It shall be denoted here by $\vec{I}$ (without the (t) addendum)

$$\vec{I} = \overline{p(t)\vec{v}(t)}. \tag{2}$$

[2] Computer tomography, sonar, sodar and the rather sophisticated "echoacoustic" methods used in seismology (especially for oil exploration) can also be considered acoustic identification methods; but they are beyond the scope of this paper.

The average is taken over many periods. The quantity I which is also called the "active" intensity, is of great practical importance because it is equal to the magnitude and direction of the energy flow in a sound field. To measure the intensity two small, absolutely identical pressure sensitive microphones are placed in a sound field with a separation d. See Fig. 1. The sound pressures $p_1(t)$ and $p_2(t)$ are related to

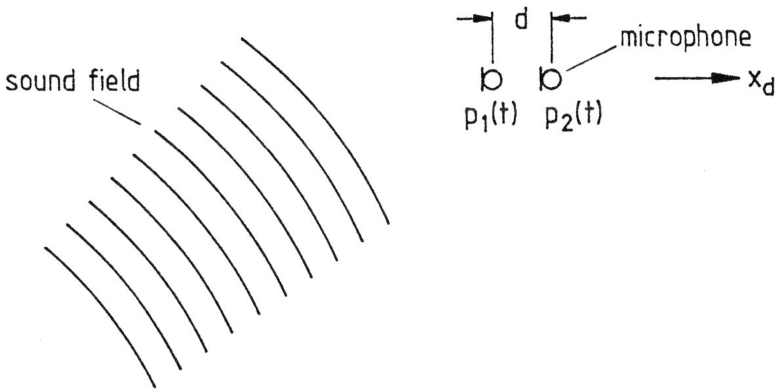

$$\rho \frac{\partial v_d(t)}{\partial t} = - \frac{\partial p(t)}{\partial x_d} \; ; \; \frac{\partial v_d(t)}{\partial t} \approx \frac{-1}{\rho d} \left[p_2(t) - p_1(t) \right]$$

Figure 1 Measurement of sound intensity in gases or fluids

the particle velocity $v_d(t)$ (in the direction of the microphone separation) by the law of conservation of momentum in the acoustic approximation as also shown in Fig. 1. The distance d obviously has to be much smaller than the acoustic wavelength of interest. ρ is the mean density of the medium. Combining the formula in Fig. 1 with Eq. (2) gives after a few manipulations

$$I_d = \frac{1}{2\rho d} \overline{[p_1(t) + p_2(t)] \int_{-\infty}^{t} [p_1(\tau) - p_2(\tau)] d\tau}. \qquad (3)$$

The index d indicates that it is the intensity in the direction of the microphone separation. If the signals $p_1(t)$ and $p_2(t)$ are processed by a two channel FFT-analyser the intensity can be measured rather easily. It turns out [3] that

$$I_d = \frac{-1}{\rho d} I_m \{ \frac{1}{\omega} G(p_1, p_2) \} \qquad (4)$$

with

$$I_m\left\{\frac{1}{\omega}G(p_1,p_2)\right\} = \frac{1}{2\pi T}\int_{\omega_1}^{\omega_2}\frac{1}{\omega}|\check{p}_1(\omega)\check{p}_2(\omega)|\sin(\phi_1-\phi_2)d\omega. \tag{4a}$$

Eq. (4) gives the total time avaraged intensity in the frequency range $\omega_2-\omega_1$. If the single frequency behaviour of the intensity is of interest, the integration from ω_1 to ω_2 is not made. In the equations the following symbols are used:

$I_m(\ldots)$ = imaginary part,
ω = angular frequency,
T = (long) avaraging time,
$\check{p}_1(\omega), \check{p}_2(\omega)$ = spectra of the sound pressures $p_1(t), p_2(t)$,
$\phi_2-\phi_1$ = phase difference between the two pressures at frequency ω,
$G(p_1,p_2)$ = cross power spectral density of p_1 and p_2.

Examples of measured intensity vector fields are shown in Fig. 2 [3, 9].

When intensity measurements are made using the cross-power spectrum density method, one almost automatically gets an "imaginary part" of the intensity which is called here Q_d. It is named "reactive intensity". Its frequency response is given by

$$Q_d = \frac{-1}{2\rho d}\left[\frac{G(p_1,p_1)-G(p_2,p_2)}{\omega}\right]. \tag{5}$$

It can be shown that the divergence of Q_d is related to the potential and kinetic energy densities (or the Lagrangian L). There are also some relations with respect to the curl of the reactive intensity, but up to now no additional insight into the energetics of sound fields were obtained this way.

2.1.2 Applications of the concept of intensity

It is impossible to list here all the applications of the so-called "intensimetry"; therefore only a few examples are given. For further reading the references [5, 10, 11] are recommended. They contain the papers that were given at three conferences that were solely devoted to intensity. One great advantage of the sound intensity is that it allows the direct measurement of acoustic powers (the acoustic Wattmeter [4]). The acoustic power P that is transmitted through an area S is given by

$$P = \int_S I_S dS. \tag{6}$$

Here I_S is the component of the intensity that is normal on the surface element dS. Sound power measurement employing Eq. (6) are used quite frequently to determine

- sound emission from sources; in this case S is a surface enclosing the sound source,

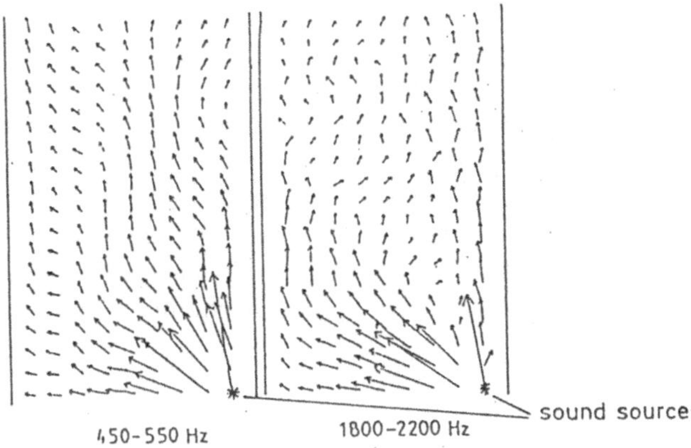

Point source in a duct [1]

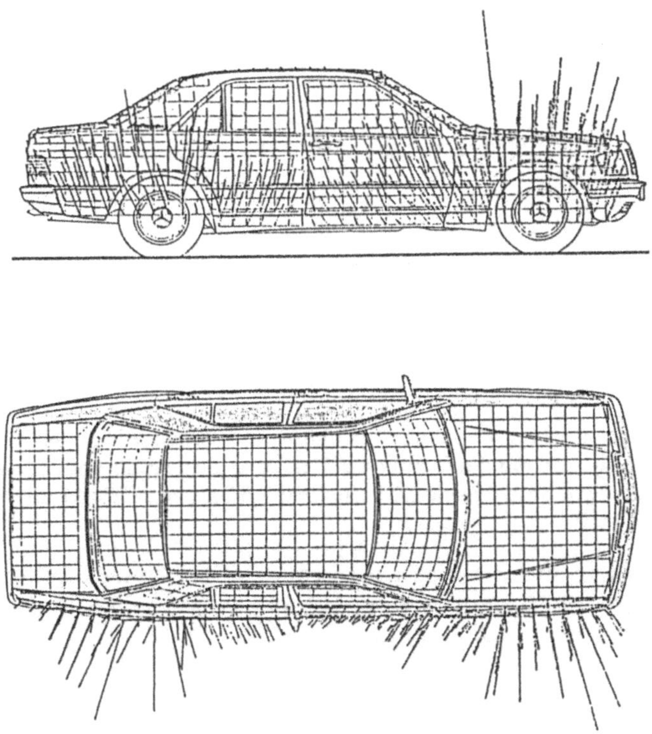

Automobile 888 - 1122 Hz [2]

Figure 2 Examples of intensity vectors

- sound transmission through partitions,
- sound absorption by special materials and constructions, in this case the difference between incoming and reflected sound is measured.

A second advantage is that acoustic power measurements are not influenced by the acoustic nearfield which extends to approximately half a wavelength around a sound source. The rather high sound pressures within the (so-called hydrodynamic) nearfield which have nothing to do with the propagation of acoustic energy are ignored by intensity meters. The third advantage is that unwanted background noise which may interfere with sound power measurements can be greatly reduced by intensity measurements. See Fig. 3. It should be noted, however, that Fig. 3

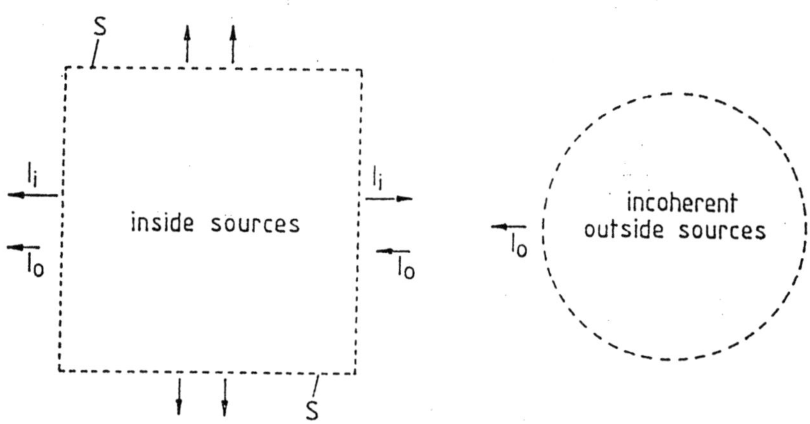

Figure 3 Measurement of sound power through a closed surface in the presence of incoherent outside sources. (If there is neglegible absorption inside S the sound power from outside sources is subtracted on one side and added at another one.)

applies only to uncorrelated sound sources. Source localisation, which quite often is identical with fault detection, is another very popular application of acoustic intensity. It will be discussed in the next section.

2.1.3 Source localisation (fault detection) by measuring intensity

When one looks at plots (see e. g. Fig. 2) of intensity vector fields, one is tempted to use them for source localisation. Clearly there are regions "where the arrows are coming from" and one is tempted to associate them with the origin of the sound; e.g. the engine and the tires of the automobile in Fig. 2. (The exhaust opening does not appear as a seperate source in Fig. 2 because the measurements were made on a "drum" in a closed test cell and therefore the exhaust gases had to be contained in a pipe.) There also may be "sinks" which one is tempted to associate with sound absorbers.

Great successes have been achieved with this intuitive method and many noisy (which quite often is identical with faulty) machine parts, or noise leaks, or unwanted sound bridges have been found. As long as the sources and sinks "seen" by this intuitive method are uncorrelated (i. e. caused by independent mechanisms) reliable results are obtained. There are, however, problems when correlated sources shall be distinguished; e. g. different parts of a machine that is driven by a single motor. As a simple example Fig. 4 shows two neighboring sound sources Q_1 and Q_2. The

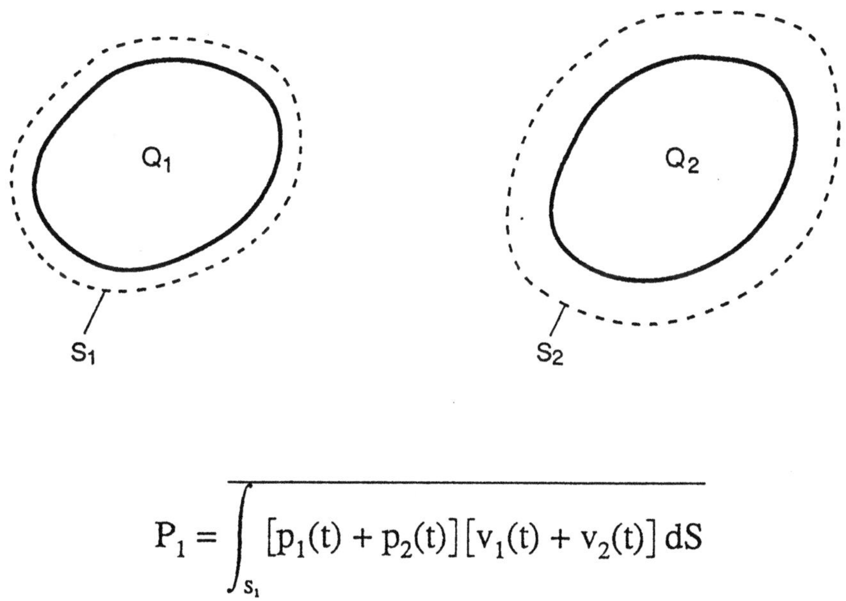

$$P_1 = \overline{\int_{S_1} [p_1(t) + p_2(t)]\,[v_1(t) + v_2(t)]\, dS}$$

$$P_2 = \overline{\int_{S_2} [p_1(t) + p_2(t)]\,[v_1(t) + v_2(t)]\, dS}$$

Cross trems vanish only when Q_1 and Q_2 are uncorrelated

Figure 4 Sound power of neighboring sources Q_1 and Q_2

powers P_1 and P_2 shall be measured on the surfaces S_1 and S_2 enclosing the sources.

If the two sources are uncorrelated, P_1 is identical with the power comming from Q_1 when Q_2 were absent, and vice versa, and the intensity vectors would (more or less) point to the two sources. The story is completely different when Q_1 and Q_2 are highly correlated. In this case power P_1 measured on S_1 depends very much on the sound coming from Q_2 and the intensity vectors can be brought to almost any value by changing Q_2. This way it may happen that P_1 increases tremendously when Q_2 is switched off. It is even possible to dream up [12] situations where the sound transmitted through a surface enclosing a sound source is completely cancelled by additional, coherent sound sources. (This is the goal of the "anti-sound" priciple.) It is surprising that in spite of these very basic problems intensity measurements have been so successful in localising sound sources that are driven by the same mechanism and therefore are highly correlated. The reason seems to be that the interaction effects get averaged out when - as is most often the case - the sound signals contain wide frequency bands and when the sources are seperated by more than half a wave-length. One might even dare the statement: "Since noise control by anti-sound sources over a wide frequency band is very difficult, it is also rather unlikely that two correlated - but more than half a wave length seperated - sound sources cancel each other over a wide frequency band".

2.2 Intensity in elastic bodies

2.2.1 Definition and measurement principle

The sound field in any gas filled room can be considered as a summation of many waves that are criss-crossing the room and that are reflected (scattered) or absorbed at each room surface or other discontiniuty. Similarly the vibrations in and on an elastic structure consist of criss-crossing elastic waves. The difference is that there are two types of waves (shear and longitudinal) which have different (possibly direction-dependent) propagation speeds and that in addition to reflection (scattering) and absorption there is also wave conversion at the discontinuities. It is thus not surprising that there are many similarities between the intensities in gases and solids. This holds already for the definition of the structural intensity which is given by

$$I_i = -\overline{\sigma_{ij} v_j}. \tag{7}$$

Here I_i is the i-th component of the time averaged intensity vector, v_j is the j-th component of the particle velocity vector, and σ_{ij} is the ij-component of the stress tensor (the equivalent of the pressure p in Eq. (2)). The summation convention is employed in Eq. (7). The minus sign is due to a convention. σ_{ij} and v_j are time dependent quantities. The overbar indicates a time average over many periods. Since stresses are hard to measure directly, σ_{ij} in Eq. (7) is replaced in the general (anisotropic material) case by the stress-strain relation

$$\sigma_{ij} = C_{ijkl}\epsilon_{kl} \tag{8}$$

and in the case of an isotropic medium by

$$\sigma_{ij} = G\left[\frac{2\mu}{1-2\mu}\,\varepsilon_{kk}\delta_{ij} + 2\varepsilon_{ij}\right]. \tag{8a}$$

The following notation is used:

$$\begin{aligned}
i,j,k,l &= 1,2,3,\\
C_{ijkl} &= \text{elastic constants},\\
\varepsilon_{kl} &= \text{component of the strain tensor}\\
&\quad \varepsilon_{kl} = \frac{1}{2}\left(\frac{\partial\xi_k}{\partial x_l} + \frac{\partial\xi_l}{\partial x_k}\right),\\
\xi_k &= \text{component of displacement vector},\\
G &= \text{shear modulus},\\
\mu &= \text{Poisson's ratio},\\
\delta_{ij} &= \begin{cases} 1 & \text{for } i=j\\ 0 & \text{for } i\neq j\end{cases}.
\end{aligned}$$

Combination of Eq. (7) and (8a) results in

$$I_i = -G\left[\frac{2\mu}{1-2\mu}\overline{\frac{\partial\xi_j}{\partial t}\left(\frac{\partial\xi_k}{\partial x_k}\right)}\delta_{ij} + \overline{\frac{\partial\xi_i}{\partial x_j}\frac{\partial\xi_j}{\partial t}} + \overline{\frac{\partial\xi_j}{\partial x_i}\frac{\partial\xi_j}{\partial t}}\right]. \tag{8b}$$

This shows that for intensity measurements some material properties have to be known (just as ρ must be known for the evaluation of Eq. (3)). But more importantly it shows that even in an isotropic medium an intensity measurement is a very formidable task when the structure is of arbitrary shape.

For each point it would involve the measurement of the three displacement [3] components ξ and their space derivatives (by using closely spaced sensor pairs) in three directions. Even if all these quantities could be measured (to the author+s knowledge it never was done in the acoustic frequency range) there is still a major difficulty: whenever a sensor is placed inside a solid body (by drilling a hole, etc.) the stress distribution is completely altered and no meaningful results can be obtained. The situation is different for sound in air because microphones have a "hard" surface and therefore scatter much less than discontinuities with a soft surface.

2.2.2 Intensities in thin structures

Since intensity measurements in a body of arbitrary shape seem to be impossible at present, research has concentrated on structures of simple shape. Usually such

[3] In actual measurements in the acoustic frequency range not the displacement is measured but its first or second time derivative, the particle velocity or acceleration, because for these quantities small and reliable sensors are available. The transition from velocity or acceleration to displacement does not pose serious problems; therefore the measurement of the velocity in Eq. (7) causes no additional difficulties.

structures are "thin" (at least one dimension is much smaller than the wavelength) so that measurements on the surface give enough information about the whole body. An additional simplification is due to the fact that most structure surfaces are (outside the excitation region) free of stresses, thus many of the σ_{ij}- terms in Eq. (7) are zero. Results that are obtained this way are [see [11],[13]]

- For a plate of thickness h made of a material with Young's modulus E and Poisson's ratio μ

$$P' = -\int_{-\frac{h}{2}}^{\frac{h}{2}} I_1 dx_2 = P'_{1ip} + P'_{1B} \tag{9}$$

where

$$P'_{1ip} = \frac{Eh}{1-\mu^2}\overline{\left(\frac{\partial\xi_{1M}}{\partial x_1} + \mu\frac{\partial\xi_{3M}}{\partial x_3}\right)\dot{\xi}_{1M}} + G\overline{\left(\frac{\partial\xi_{1M}}{\partial x_3} + \frac{\partial\xi_{3M}}{\partial x_1}\right)\dot{\xi}_{1M}} \tag{9a}$$

is the power per unit width propagating in x_1-direction as in-plane waves; i. e. shear and longitudinal.

$$\begin{aligned} P'_{1B} &= B'\Bigg[-\overline{\frac{\partial}{\partial x_1}\Delta\xi_2\frac{\partial\xi_2}{\partial t}} + \overline{\left(\frac{\partial^2\xi_2}{\partial x_1^2} + \mu\frac{\partial^2\xi_2}{\partial x_3^2}\right)\frac{\partial^2\xi_2}{\partial x_1\partial t}} \\ &+ (1-\mu)\overline{\frac{\partial^2\xi_2}{\partial x_1\partial x_3}\frac{\partial^2\xi_2}{\partial x_3\partial t}} + \rho\frac{h^3}{12}\overline{\frac{\partial^3\xi_2}{\partial x_1\partial t^2}\frac{\partial\xi_2}{\partial t}}\Bigg] \end{aligned} \tag{9b}$$

is the power per unit width propagating in x_1-direction as bending waves. The other quantities used here are

$$\begin{aligned} \xi_{1M}, \xi_{3M} &= \text{in-plane displacement of the plate at its center,} \\ \xi_2 &= \text{bending displacement,} \\ B' = \frac{Eh3}{12(1-\mu^2)} &= \text{bending stiffness.} \end{aligned}$$

The dot $\dot{()}$ indicates a time derivative, the symbol Δ the Laplace operator. For the power in a beam Eqs. (9 – 9b) can also be used. One just has to replace

$$\frac{\partial}{\partial x_3} \to 0; \quad 1-\mu^2 \to 1; \quad \xi_{3M} \to 0.$$

- For a thin shell the power per unit circumferential width in axial direction is

$$P'_z = \frac{Eh}{1-\mu^2}\left[\overline{\frac{\partial \zeta_M}{\partial z}\frac{\partial \zeta_M}{\partial t}} + \mu\overline{\frac{\partial \xi_M}{\partial s}\frac{\partial \xi_M}{\partial t}}\right]$$
$$+Gh\left[\overline{\frac{\partial \zeta_M}{\partial s}\frac{\partial \xi_M}{\partial t}} + \overline{\frac{\partial \xi_M}{\partial z}\frac{\partial \zeta_M}{\partial t}}\right]$$
$$+B'\left[\overline{\left(\frac{\partial^2 \eta}{\partial z^2} + \mu\frac{\partial^2 \eta}{\partial s^2}\right)\frac{\partial^2 \eta}{\partial z \partial t}} - \overline{\left(\frac{\partial \Delta \eta}{\partial z} - \frac{1}{c_L^2}\frac{\partial^3 \eta}{\partial z \partial t^2}\right)\frac{\partial \eta}{\partial t}}\right] \tag{9c}$$
$$+G\frac{h^3}{6}\overline{\frac{\partial^2 \eta}{\partial s \partial z}\frac{\partial^2 \eta}{\partial s \partial t}} + \frac{Eh}{1-\mu^2}\frac{\mu}{a}\left[\overline{\eta\frac{\partial \zeta_M}{\partial t}} + \frac{2h^2}{12a}\overline{\eta\frac{\partial^2 \eta}{\partial z \partial t}}\right]$$
$$+\frac{Gh^3}{12a}\left[\overline{\frac{\partial \zeta_M}{\partial s}\frac{\partial^2 \eta}{\partial s \partial t}} - \overline{\frac{\partial^2 \zeta_M}{\partial s^2}\frac{\partial \eta}{\partial t}}\right].$$

a = radius,
$s = a\vartheta$ = circumferential coordinate,
z = axial coordinate,
ζ_M, ξ_M, η = axial, circumferential, radial displacement in the middle of the shell.

2.2.3 Simplifications and examples

Even for thin structures the expressions (9a – 9c) show that intensity measurements are rather complicated. Therefore it was tried to use the same measurement principle that was developed for air-borne sound measurements and replace the two identical microphones by two identical accelerometers. The quantity that is measured then is [16]

$$I(\omega)_{appr} = -\frac{\sqrt{B'\rho h}}{\omega^2 d} Im\{G(a_1, a_2)\}. \tag{10}$$

Here d is the distance between the sensors, $I(\omega)_{appr}$ is the measured approximate intensity at frequency ω in the direction of the sensor separation, and $Im\{G\ldots\}$ is the imaginary part of the cross-power spectral density. See also Eq. (4). Eq. (10) can be derived from the more general expressions (9 – 9c) if the following approximations are made:

- in-plane waves can be neglected, i. e. $P'_{1ip} = 0$ in Eq. (9a);
- the bending wave power given by (9b) is solely determined by the term

$$\overline{\left(\frac{\partial}{\partial x_1}\Delta \xi_2\right)\frac{\partial \xi_2}{\partial t}} = \omega\sqrt{\frac{\rho h}{B'}}\,\overline{\frac{\partial \xi_2}{\partial x_1}\frac{\partial \xi_2}{\partial t}}; \tag{10a}$$

this in essence means that the bending wave near-fields are neglected, thus correct results within half a wavelength from discontinuities cannot be expected;

- curvature effects are not included.

In spite of these many approximations surprisingly reasonable results can be obtained from Eq. (10). As an example Fig. 5 shows the correct values of the intensity calculated according to Eq. (8b) and the approximate values according to (10). The following can be seen from this figure.

- As one would expect, the normal component I_2 of the intensity is zero outside the excitation region. The approximate value, however, shows some fluctuations. This may be the reason why I_{2appr} hardly ever is measured.

- In some distance from the excitation region the lateral component, i. e. I_1, of the approximate intensity agrees fairly well with the correct one. This is especially true when the plate is thin ($h/\lambda < 0.1$).

- Details of the intensity in and near the excitation region are not given correctly by the approximate intensity.

When the thickness to wavelength ratio becomes larger ($h/\lambda > 0.5$) the values of I and I_{appr} are completely different. But this deficiency can not be blamed fully on the measurement technique, because for $h/\lambda > 0.5$, a structure can no longer be considered as thin (i. e. even Eq. (9a – 9b) are no longer valid) and consequently the intensity on the surface is no longer a good indicator for the energy flow inside the whole structure.

Fig. 6 shows an application in architectural acoustics. The purpose of the measurement was to find out wether intensity measurements give additional information about the flow of structural energy along the flanking wall connecting two rooms [14]. It can be seen that at low frequencies, when the wavelength is almost equal to the height of the room, rather strange results are obtained. At high frequencies, however, a direction of the energy flow from the source room to the rest of the building can be seen fairly clearly. The amplitudes (length of the arrows) are also close to the values that are expected.

Another application of the structural intensity measurement is the localisation of sound bridges [15]. Sound bridges are more or less rigid connections between structural elements that should be isolated from each other. In such measurements mainly the direction of the intensity vector (pointing away from the sound bridge) is of interest; therefore the simple, approximate intensity measurement gives already useful results. Obviously strucural intensity (just as sound intensity in air) can also help to exclude unwanted background noise from uncorrelated neighboring sources. This advantage is used when many samples of a product (e. g. small gear boxes) are tested with respect to their vibration behaviour in a noisy environment.

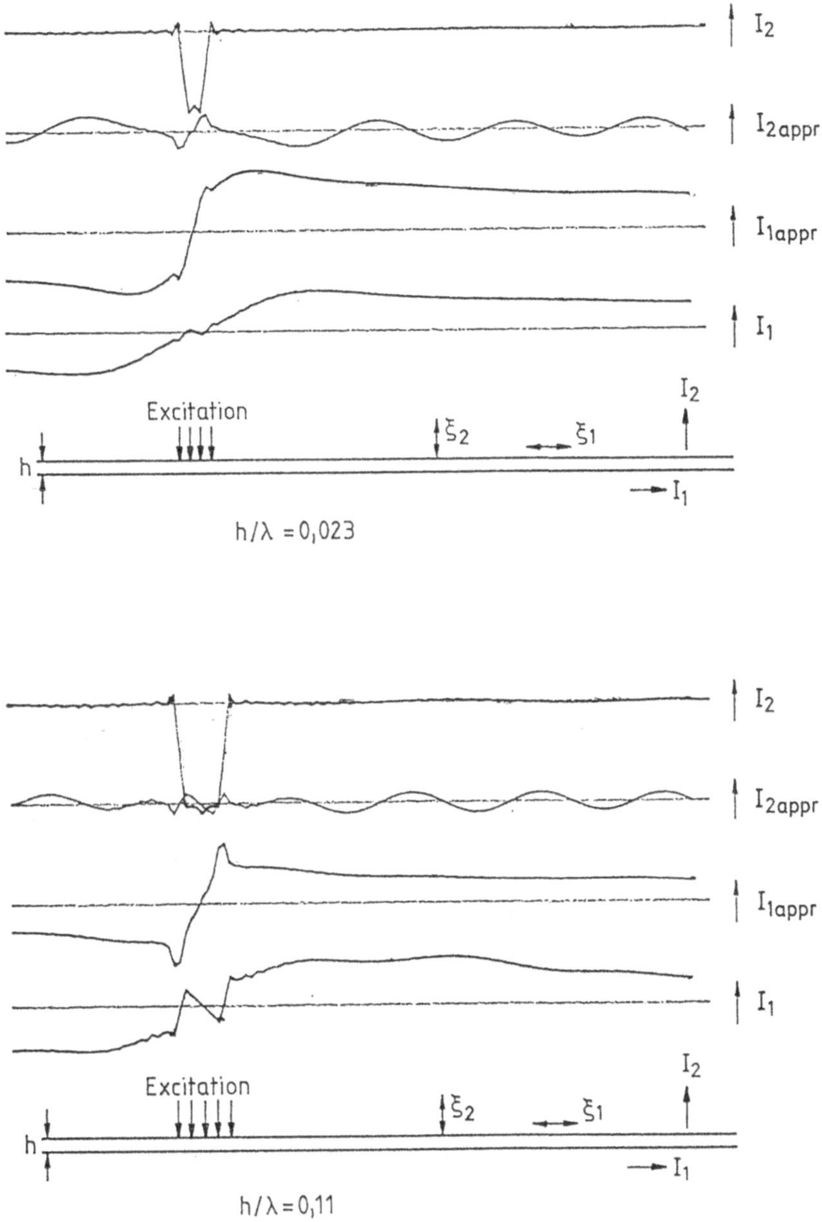

Figure 5 Intensity in a thin plate

I_1, I_2 = correct values

I_{1appr}, I_{2appr} = approximation according to Eq. (10)

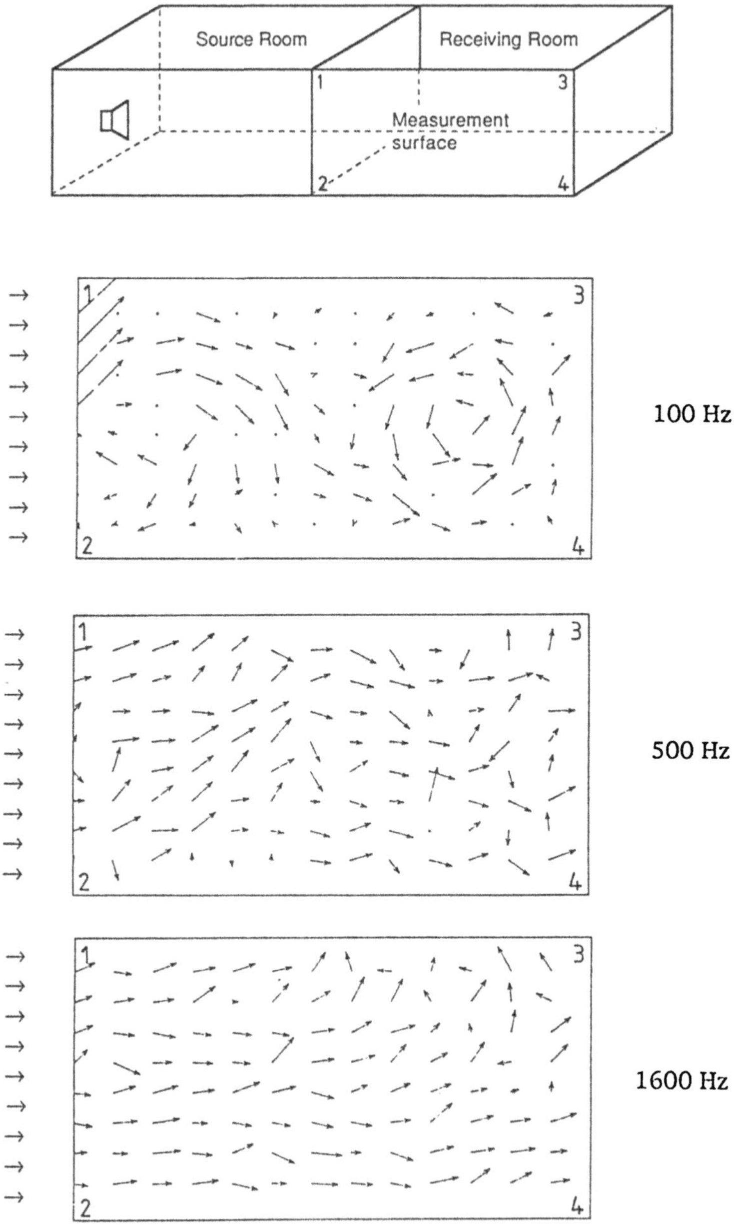

Figure 6 Approximate intensity I_{appr} measured on a wall which is excited at the left hand side.

References

[1] Lyon, R. H.: Statistical energy analysis. M.I.T. Press, Cambridge USA 1975

[2] Lewit, M; Lehmköster, S.: Probleme und Lösungswege in der Energiefluß-analyse, p. 1009-1012. DAGA-92, Fortschritte der Akustik, Bad Honnef: DPG-GmbH.

[3] Fahy, F. R.: Sound Intensity. Elsevier, London 1989

[4] Smith, P.W.; Shultz, T. J.: Malme, C. I.: Intensity measurements in near fields and reverberant spaces. Bolt, Beranek and Newman Inc.; Rep. No 1135 (1964)

[5] Tourret, J.; et al.: First international congress on acoustic intensity. Centre Technique des Industries Mechaniques (CETIM). Senlis, France 1981

[6] Maynard, J. D.; Williams, E. G.; Lee, Y.: Nearfield acoustic holography I. Theory of generalised holography and development of NAH. J. Acoust. Soc. Am. 78 (1985), p. 1395-1413

[7] Pascal, J. C.; Mann III, J. A.; Carniel, X: General computer program for automated measurements in vibroacoustics; p. 69 - 74. FASE-Congress 1992. Swiss Acoustical Society; P. O. Box 251, 8600 Dübendorf, Switzerland

[8] Möser, M.: Ein Beitrag zur Schallquellenrekonstruktion durch akustische Kurzwellen-Holographie. Acustica 61 (1986); 288-300.

[9] Bauer, A.: Bestimmung der Geräuschabstrahlung einzelner Teilquellen einer komplexen Struktur. Dissertation, Technische Universit t Berlin 1987

[10] Tourret, J.; et al.: Second international congress on acoustic intensity (CETIM). Senlis, France 1985

[11] Tourret, J.; et al.: Structural intensity and vibrational energy flow. Third international congress on acoustic intensity (CETIM) 1990

[12] Crighton, D. G.; Dowling, A. P.; Ffowcs-Williams, J. E.; Heckl, M.; Leppington, F.G.: Modern Methods in Analytical Acoustics. Chapter 9.4, Springer, London 1992

[13] Pavic, G.: The influence of curvature on structure-borne acoustical power propagation in a cylindrical shell. Paper D6-6, 12th ICA-Congress 1986 Toronto

[14] Kruppa, P.: Measurement of structural intensity in building constructions. Applied Acoustics 19 (1986), 61-74

[15] Maysenhölder, W.; Schneider, W.: Sound bridge localisation in buildings by structure-borne sound intensity measurements. Acoustica 68 (1989), 258-262.

[16] Noiseux, D. V.: Measurement of power flow in uniform beams and plates. J. Acoust. Soc. Am. 47 (1970), 238-247.

[17] Williams, E. G.: Structural intensity in thin cylindrical shells, J. Acoust. Soc. Am. 89 (1991), 1615-1622.

Measurements of Linear and Nonlinear Structures with Distributed Shell Sensors (Distributed Sensing and Identification)

H. S. Tzou
University of Kentucky
Department of Mechanical Engineering
Lexington, KY, USA

1 Introduction

Structures are distributed and continuous in nature, i.e., structural behaviors are functions of ***spatial*** and ***time*** variables, and they are usually classified as ***continua*** or ***distributed–parameter systems*** (DPS's). Distributed vibration identification and control of distributed parameter systems always represents a challenge, both in theory and practice. Theoretical development for control of distributed parameter systems has been constantly advancing during the past three decades. However, due to practical limitations, e.g., availability of distributed sensors and actuators, real world applications are still relatively limited. Note that conventional sensors and actuators are mostly **discrete** types which measure or control spatially ***discrete*** locations. These devices are ineffective if they are placed at modal nodes or lines. One solution to this problem is to use ***distributed*** sensors and actuators which are spatially distributed, so that they are sensitive to spatially distributed structural behaviors. In this study, a generic deep elastic shell continuum laminated with distributed piezoelectric layers is regarded as a generic DPS from which distributed sensing theory is derived, based on the ***direct piezoelectric effect***. Note that the generic shell can be easily simplified to many other common continua or structures, e.g., plates, cylinders, spheres, etc., by using four system parameters, i.e., two radii and two Lamé parameters [1].

Distributed sensing of flexible structures using synthetic piezoelectric ceramics and/or polymeric polyvinylidene fluoride (PVDF) has been investigated in recent years [2]. Tzou used a PVDF film in a flexible structure [3] and derived a distributed sensing and control theory for a deep shell continuum laminated with distributed sensors and actuators [4]. Lee and Moon proposed a piezoelectric modal sensor design for a plate and applied it to a two–dimensional thin beam [5]. Distributed piezoelectric sensor and actuator have been applied to sensing and control flexible space structures [6,7]. In additional to theoretical and experimental studies, a piezoelectric thin solid finite element was formulated; distributed sensing and active vibration control of a triple layered plate with top and bottom piezoelectric layers was studied [8]. This study is devoted to a development of distributed sensing theory for a generic deep shell continuum laminated with distributed piezoelectric sensor layers. Detailed electromechanics of the distributed piezoelectric sensor are analyzed. Applications of the generic theories to other common geometries are demonstrated in case studies; these include 1) distributed modal voltages and modal identification of a plate, 2) distributed convolving modal sensors for circular rings, and 3) contact measurements of a tactile sensor. Note that the shell could be either linear or nonlinear. However, the emphasis is placed on linear measurements in case studies, although distributed sensing of nonlinear dynamics is discussed.

2 System Definition

In this study, a generic distributed parameter system is an elastic shell continuum, either linear or nonlinear, laminated with a flexible piezoelectric shell layer on its top surface, Figure 1. It is assumed that the piezoelectric layer is much thinner than the elastic shell thickness and its material properties are negligible. The sensor layer can be either fully covered or partially covered. Localization and discretization of the sensor can be defined by shape functions, step–functions, Dirac delta functions, etc [1].

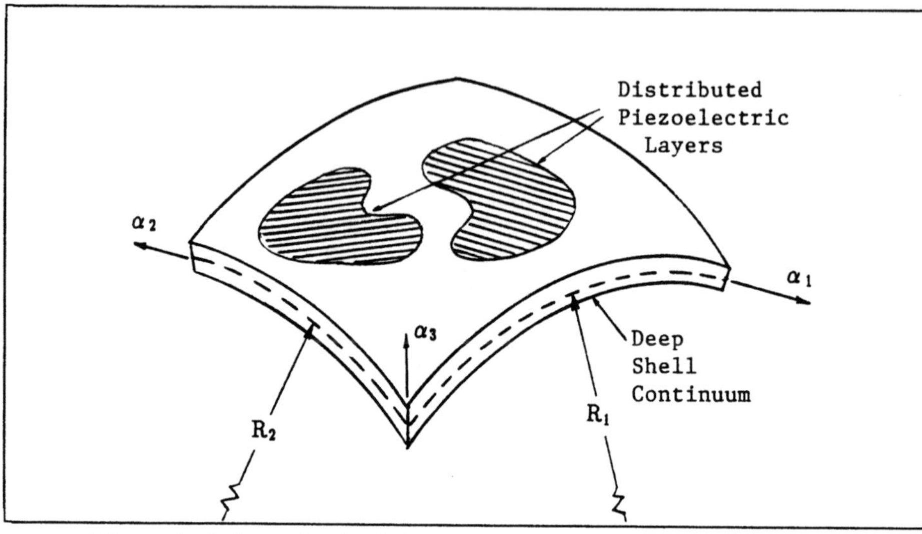

Figure 1 A deep elastic shell with distributed piezoelectric layers.

Equations of motion, in tensor expression, of the elastic shell can be derived by Hamilton's principle [1]:

$$-(N_{11}A_2)_{,1}-(N_{21}A_1)_{,2}-N_{12}(A_{1,2})+N_{22}(A_{2,1})-\frac{1}{R_1}[(M_{11}A_2)_{,1}$$
$$+(M_{21}A_1)_{,2}+M_{12}(A_{1,2})-M_{22}(A_{2,1})]+A_1A_2\,\rho\,h\,\ddot{u}_1=A_1A_2\,F_1\,, \qquad (1)$$

$$-(N_{12}A_2)_{,1}-(N_{22}A_1)_{,2}-N_{21}(A_{2,1})+N_{11}(A_{1,2})-\frac{1}{R_2}[(M_{12}A_2)_{,1}$$
$$+(M_{22}A_1)_{,2}+M_{21}(A_{2,1})-M_{11}(A_{1,2})]+A_1A_2\,\rho\,h\,\ddot{u}_2=A_1A_2\,F_2\,, \qquad (2)$$

$$-\{[(M_{11}A_2)_{,1}+(M_{21}A_1)_{,2}+M_{12}(A_{1,2})-M_{22}(A_{2,1})]/A_1\}_{,1}$$
$$-\{[(M_{12}A_2)_{,1}+(M_{22}A_1)_{,2}+M_{21}(A_{2,1})-M_{11}(A_{1,2})]/A_2\}_{,2}$$
$$+A_1A_2\,[N_{11}/R_1+N_{22}/R_2]+A_1A_2\,\rho\,h\,\ddot{u}_3=A_1A_2\,F_3\,, \qquad (3)$$

where A_1 and A_2 are the Lamé parameters; R_1 and R_2 are radii of two in–plane α_1 and α_2 axes; ρ is the mass density; h is the shell thickness; $\ddot{u}_i$ is the acceleration in the i–th direction; F_i is the externally applied mechanical load. Depending on four system parameters, two lamé parameters and two radii, the system equations can be simplified to many common geometries, e.g., spherical shells, cylindrical shells, conical shells, plates, non–shell type continua, etc. [1,2]. The resultant forces N_{ij} and moments M_{ij} can be defined in terms of generic displacements u_1, u_2, and u_3 on the neutral surface; these forces and moments can be either linear or nonlinear, depending on the strain–displacement relations and stress–strain relations.

3 Distributed Shell Sensors

It is assumed that a distributed piezoelectric sensor is laminated (or embedded) on a shell continuum, Figure 1. The distributed sensor layer is subjected to strain influence from the shell oscillation. Due to the direct piezoelectric effect, the sensor layer generates an electric signal corresponding to the dynamic state of the shell. The strain experienced in the sensor layer can be calculated based on the linear or nonlinear shell theory. It is assumed that the elastic shell follows the Kirchhoff–Love assumptions in the linear responses [1].

Assuming the electric charge is balanced in the piezoelectric sensor layer (an insulator), one can derive an electrical charge equation by using the Gauss theorem:

$$\nabla \{D_i\} = 0 \,, \quad i=1,2,3, \tag{4}$$

where ∇ is a differential operator and D_i is the electric displacement. From the thin shell configuration defined earlier, only the transverse electric field E_3 is considered, so that the strains and dielectric displacement D_3 are independent of the α_3. According to Maxwell's equation, the electric field E_i can be related to the electric potential ϕ by

$$\{E_i\} = -\nabla \phi \,. \tag{5}$$

The electric potential difference (a voltage) across the electrodes of the distributed sensor can be obtained by integrating the electric field over the thickness of the piezoelectric sensor layer, the direct piezoelectric effect, i.e.,

$$\phi = -\int^{h^S} E_3 \, d\alpha_3 = h^S (h_{31} S_{11}^S + h_{32} S_{22}^S - \beta_{33} D_3) \,, \tag{6}$$

where h^S is the sensor thickness; S_{11}^S and S_{22}^S are two normal strains; S_{12}^S is the in–plane shear strain; h_{31}, h_{32}, and h_{36} are the piezoelectric constants in which the first two related to the transverse output due to the in–plane strains and h_{36} the in–plane twisting effect; β_{33} is the impermeability constant. Rearranging the above electric potential equation, one can write the electric displacement D_3 as

$$D_3 = \frac{1}{\beta_{33}} \left(h_{31} S_{11}^S + h_{32} S_{22}^S + h_{36} S_{12}^S - \phi / h^S \right) . \tag{7}$$

Since D_3 is defined as the charge per unit area, one can integrate the electric displacement over the electroded surface S^e to estimate a total surface charge. An open–circuit voltage ϕ^S condition can be obtained by setting the charge zero, i.e.,

$$\phi^S = \frac{h^S}{S^e} \int_{\alpha_1} \int_{\alpha_2} \left[h_{31} S_{11}^S + h_{32} S_{22}^S + h_{36} S_{12}^S \right] A_1 A_2 \, d\alpha_1 d\alpha_2 \,, \tag{8}$$

where S^e is the effective sensor area. Note that if the piezoelectric shell sensor is made of hexagonal symmetrical piezoelectric materials, $h_{31} = h_{32}$ and h_{36} is zero. If the sensor thickness is not constant, the sensor output can be estimated by

$$\phi^S = \frac{1}{S^e} \int_{\alpha_3}\int_{\alpha_1}\int_{\alpha_2} \left[h_{31}S_{11} + h_{32}S_{22} + h_{36}S_{12} \right] A_1A_2 \, d\alpha_1 d\alpha_2 d\alpha_3 \, . \tag{9}$$

Substituting (linear) shell strain expressions into the above equation gives

$$\begin{aligned}
\phi^S = \frac{1}{S^e} \int_{\alpha_3}\int_{\alpha_1}\int_{\alpha_2} \Bigg\{ & h_{31}\Bigg[\left[\frac{1}{A_1} u_{1,1} + \frac{u_2}{A_1A_2} A_{1,2} + \frac{u_3}{R_1}\right] \\
& + d \left[\frac{1}{A_1}\left[\frac{u_1}{R_1} - \frac{1}{A_1} u_{3,1}\right]_{,1} + \frac{1}{A_1A_2}\left[\frac{u_2}{R_2} - \frac{1}{A_2} u_{3,2}\right] A_{1,2}\right]\Bigg] \\
& + h_{32}\Bigg[\left[\frac{1}{A_2} u_{2,2} + \frac{u_1}{A_1A_2} A_{2,1} + \frac{u_3}{R_2}\right] \\
& + d \left[\frac{1}{A_2}\left[\frac{u_2}{R_2} - \frac{1}{A_2} u_{3,2}\right]_{,2} + \frac{1}{A_1A_2}\left[\frac{u_1}{R_1} - \frac{1}{A_1} u_{3,1}\right] A_{2,1}\right]\Bigg] \\
& + h_{36}\Bigg[\left[\frac{A_2}{A_1}\left[\frac{u_2}{A_2}\right]_{,1} + \frac{A_1}{A_2}\left[\frac{u_1}{A_1}\right]_{,2}\right] + d \Bigg[\frac{A_2}{A_1}\left[\left[\frac{u_2}{R_2} - \frac{1}{A_2} u_{3,2}\right]/A_2\right]_{,1} \\
& + \frac{A_1}{A_2}\left[\left[\frac{u_1}{R_1} - \frac{1}{A_1} u_{3,1}\right]/A_1\right]_{,2}\Bigg]\Bigg]\Bigg\} A_1A_2 d\alpha_1 d\alpha_2 d\alpha_3 \, ,
\end{aligned} \tag{10}$$

where u_i is the displacement in the i–th direction and d is the distance measured from the shell neutral surface to the mid–plane of the piezoelectric sensor. Note that the strains include two effects: 1) the membrane effect and 2) the bending effect, leading by d. If the bending oscillation dominates the shell vibration, the membrane stain components can be neglected. Define the total output signal as a summation of two components: 1) a ***membrane sensitivity*** and 2) a ***bending sensitivity***. The generic sensor output can be written as

$$\begin{aligned}
\phi^S = \frac{h^S}{S^e} \int_{\alpha_1}\int_{\alpha_2} \Bigg\{ & h_{31}\left[\frac{1}{A_1} u_{1,1} + \frac{u_2}{A_1A_2} A_{1,2} + \frac{u_3}{R_1}\right] + h_{32}\left[\frac{1}{A_2} u_{2,2} \right. \\
& \left. + \frac{u_1}{A_1A_2} A_{2,1} + \frac{u_3}{R_2}\right] + h_{36}\left[\frac{A_2}{A_1}\left[\frac{u_2}{A_2}\right]_{,1} + \frac{A_1}{A_2}\left[\frac{u_1}{A_1}\right]_{,2}\right] \Bigg\} A_1A_2 d\alpha_1 d\alpha_2 \\
+ \frac{1}{S^e} \int_{\alpha_3}\int_{\alpha_1}\int_{\alpha_2} \Bigg\{ & h_{31}\, d \left[\frac{1}{A_1}\left[\frac{u_1}{R_1} - \frac{1}{A_1} u_{3,1}\right]_{,1} + \frac{1}{A_1A_2}\left[\frac{u_2}{R_2} - \frac{1}{A_2} u_{3,2}\right] A_{1,2}\right] \\
& + h_{32}\, d \left[\frac{1}{A_2}\left[\frac{u_2}{R_2} - \frac{1}{A_2} u_{3,2}\right]_{,2} + \frac{1}{A_1A_2}\left[\frac{u_1}{R_1} - \frac{1}{A_1} u_{3,1}\right] A_{2,1}\right] \\
& + h_{36}\, d \left[\frac{A_2}{A_1}\left[\left(\frac{u_2}{R_2} - \frac{1}{A_2} u_{3,2}\right)/A_2\right]_{,1} + \frac{A_1}{A_2}\left[\left(\frac{u_1}{R_1} - \frac{1}{A_1} u_{3,1}\right)/A_1\right]_{,2}\right]\Bigg\} \\
& \cdot A_1A_2 \, d\alpha_1 d\alpha_2 d\alpha_3 \, .
\end{aligned} \tag{11}$$

Note the the first integration denotes the ***membrane sensitivity*** S^m contributed by the membrane strains and the second integration is the ***bending sensitivity*** S^b contributed by the bending strains (as a function of sensor location and thickness):

$$\phi^s = \frac{h^s}{S^e} \int_{\alpha_1}\int_{\alpha_2} \left[f_1(u_1,u_2,u_3) \right] A_1 A_2 d\alpha_1 d\alpha_2$$

$$+ \frac{1}{S^e} \int_{\alpha_3}\int_{\alpha_1}\int_{\alpha_2} \left[f_2(u_1,u_2,u_3) \right] A_1 A_2 d\alpha_1 d\alpha_2 d\alpha_3$$

$$= S^m + S^b , \tag{12}$$

where $f_1(u_1,u_2,u_3)$ and $f_2(u_1,u_2,u_3)$ denote the terms inside the integration. (Note that these two functions depend on the nonlinear strains in the shell structure in nonlinear measurements.) In addition, each sensitivity is contributed by a transverse oscillation u_3 and two in–plane oscillations u_1 and u_2. Note that for a shell sensor with a uniform thickness, the thickness integration can be carried out as $\int_{\alpha_3}(\cdot)d\alpha_3 = h^s$; the sensor output is an averaged signal over the electroded area.

It is desirable to identify the mode shapes of a shell continuum via the the distributed piezoelectric sensor. Since mode shapes are distinct for the natural modes, the induced potential (spatial) distributions – ***modal voltages*** – should also be unique [1,8,9]. To construct a *modal voltage* of a natural mode, one needs to calculate a number of local point signals $\phi_k^s(\alpha_1^*, \alpha_2^*)$ and graphically connect the local amplitudes together for the natural mode. The ***potential map*** or ***contour*** represent the modal voltage of the natural mode. Note that a number of distributed segmented electrodes is needed to acquire a global voltage distribution; a finite separation of surface electrodes is required to prevent the voltage average since electric charges are freely moving on the electrode. These modal voltages of a plate are demonstrated in a case study presented later.

4 Distributed Modal Filtering – Convolving Modal Sensors

The sensor signal is contributed by all participating structural modes in the above sensor equation. In feedback control of the shell structure, this multiple modal coupling can introduce unstable controlled responses due to observation spillovers. Thus, it is desirable that the distributed sensor is only respond to a specific natural mode, such that the observation spillover prevented [1,5,6]. In this section, distributed convolving modal sensors are proposed.

It is assumed that the sensor thickness is constant and the effective sensor surface can be spatially shaped and the polarity changed. Assume the sensor surface is defined by a shape function $\mathcal{Q}_k(\alpha_1,\alpha_2)$ where the α_2 coordinate is a function of α_1. Thus, the sensor equation becomes

$$\phi^s = \frac{h^s}{S^e} \int_{\alpha_1} \mathcal{Q}_k(\alpha_1)\, \mathcal{P}[\mathcal{Q}_k(\alpha_1)] \cdot \left[f_1(u_1,u_2,u_3) + f_2(u_1,u_2,u_3) \right] A_1 A_2 d\alpha_1 , \tag{13}$$

where $\mathcal{P}[\mathcal{Q}_k(\alpha_1)]$ is a polarity function which defines the polarity changes of the shell sensor. Considering modal orthogonality, one can design a spatially distributed shape function $\mathcal{Q}_k(\alpha_1,\alpha_2)$ which is orthogonal to all $j \neq k$ modes. Thus, this shaped sensor is effective only for the k–th mode, i.e., the k–th piezoelectric modal sensor, such that all other residual modes can be filtered out in the dynamic measurements. This distributed convolving sensors are demonstrated in a ring case presented later.

5 Applications

In this section, three cases are studied: 1) distributed modal voltages of a plate, 2) distributed convolving ring sensors, and 3) contact measurements. The first case is to investigate modal voltages for a number of natural modes of the plate. The second case is to study the membrane and bending sensitivities of distributed convolving modal sensors for ring structures. The third case is to evaluate contact responses of piezoelectric sensors and compared with other sensor responses.

5.1 CASE 1: Distributed Modal Voltages of a Plate

In this section, distributed modal voltages of a zero–curvature shell – a plate are investigated. The physical model is a plexiglas cantilever plate (10cm× 10cm× 0.31cm) with a top distributed piezoelectric PVDF actuator layer (40μm). Figure 2 shows the cantilever plate model with a distributed piezoelectric sensor; this sensor layer is further segmented into twenty five segments in the calculation of voltage distributions. (Note that the distributed piezoelectric sensor can be made either mono–axially oriented, aligned with the x–axis, or bi–axially oriented, effective in both x and y axes.)

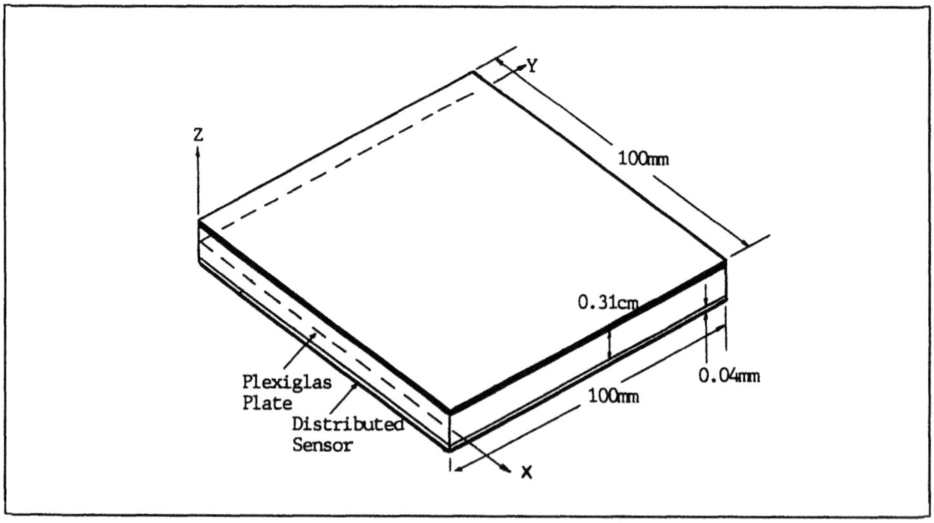

Figure 2 A laminated plate with (segmented) distributed sensors.

As discussed previously, the distributed piezoelectric sensing layer responds to the plate oscillation and generates an electric voltage representing the distributed dynamic state of the plate [8]. Thus, modal voltage distribution of the natural modes represents its unique modal characteristics of the natural modes. Sensor outputs at twenty–five discrete locations of the distributed piezoelectric sensor layer are calculated, and distributed ***modal voltages*** measured by the distributed sensor are plotted for the first three natural modes. The first three plate mode shapes and ***modal voltage*** distributions are illustrated in Figures 3–5.

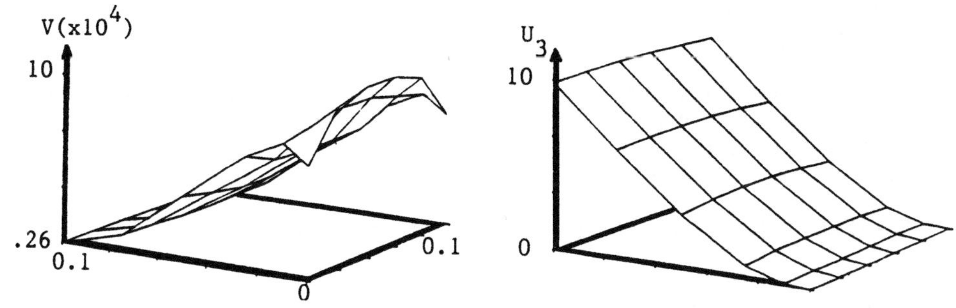

Figure 3 First mode shape and distributed modal voltage.

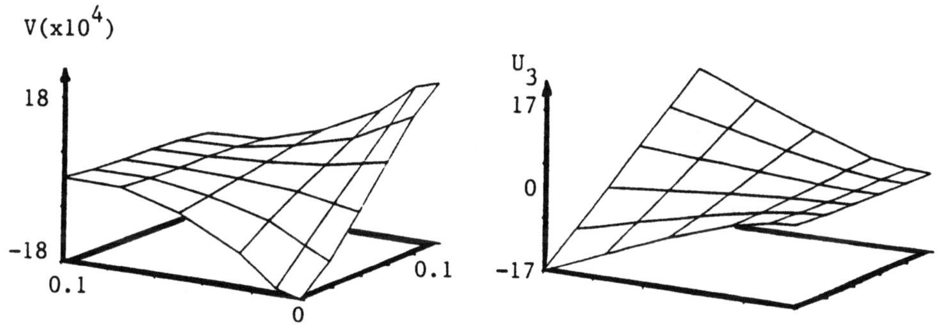

Figure 4 Second mode shape and distributed modal voltage.

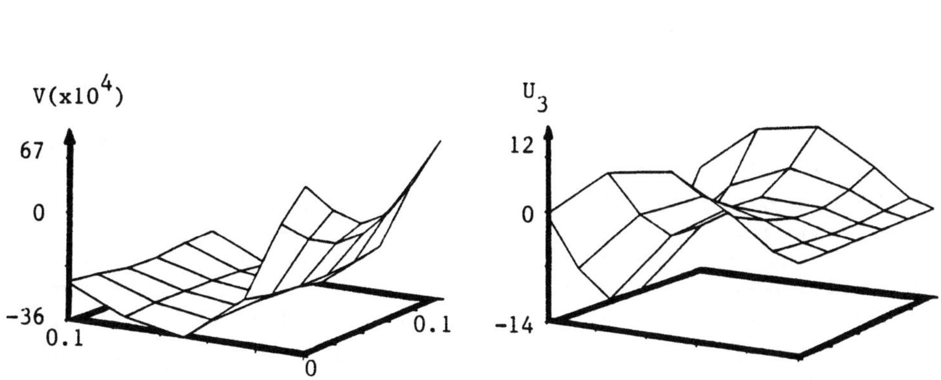

Figure 5 Third mode shape and distributed modal voltage.

It is observed that the first mode is a bending mode, the second a torsion mode, and the third a warping mode. The sensitivities of the modes vary because the tension and compression vary in different modes. Voltage drops at two corners are observed in all three figures; this is because that the strains are (numerically) smaller at these two boundary nodes. It is observed that the maximum positive voltage occurs at the maximum negative strain. The ***modal voltage*** distribution could also be refined if more discrete voltages are calculated such that *modal voltages* of higher natural modes can be observed.

5.2 CASE 2: Distributed Convolving Ring Sensors

In this section, distributed shaped modal sensors based on modal orthogonal functions are designed for a flexible circular ring, and their performance evaluated. Figure 6 shows the flexible circular ring structure. It is assumed that the ring is free–floating with free boundary conditions.

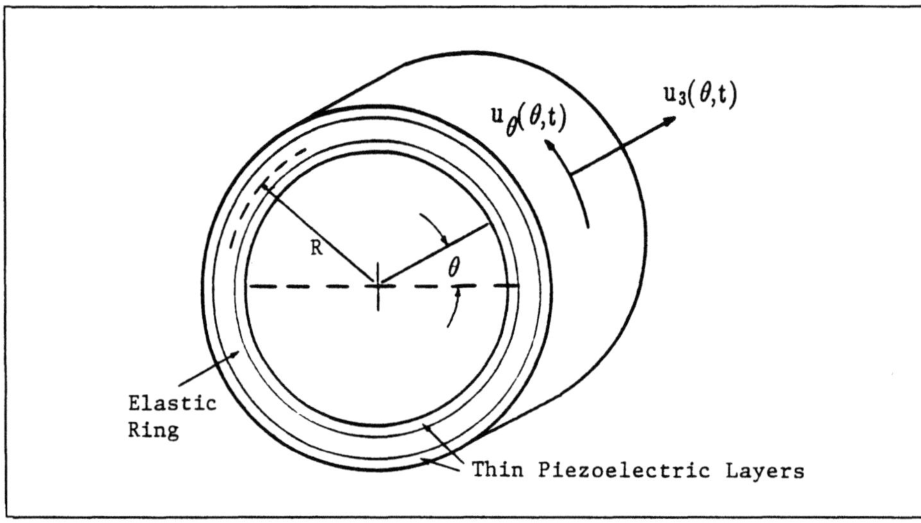

Figure 6 A flexible circular ring.

For a circular ring, Lamé parameters are $A_1 = 1$ and $A_2 = R$; the radii of curvatures are $R_1 = \infty$ and $R_2 = R$. The modal equations in the circumferential and transverse directions, $u_\theta(\theta,t)$ and $u_3(\theta,t)$, of the ring can be derived as a sine and a cosine function, respectively [1]. For a circular ring with free boundary conditions, the first mode $n = 0$ is a breathing mode and the second mode $n = 1$, is a translational rigid–body mode. For $n \geq 2$, sets of a transverse mode and a circumferential mode appear. It should be noted that for a given integer n, there are two component modal frequencies ω_{n_1} and ω_{n_2} in which the former is related to the transverse mode, i.e., transverse oscillation dominates, and the latter to the circumferential mode, i.e., circumferential oscillation dominates [1,10]. For <u>transverse modes</u>, the modal oscillation amplitudes of the transverse and circumferential oscillation components are coupled by a factor: $D_{n_1} \cong -n\, C_{n_1}$. Similarly, for <u>circumferential modes</u>, the ratio between the transverse oscillation component and circumferential component is $C_{n_2} \cong n\, D_{n_2}$. Note that C_{n_2} and D_{n_2}

are the modal amplitudes of circumferential modes and C_{n_1} and D_{n_1} denote the modal amplitudes of transverse modes. In addition, the membrane and bending functions f_1 and f_2 can be derived as

$$f_1(u_1,u_2,u_3) = h_{31}\,[(u_{\theta,\theta} + u_3)/R]\ ; \tag{14}$$

$$f_2(u_1,u_2,u_3) = h_{31}\,d\,[(u_{\theta,\theta} - u_{3,\theta\theta})/R^2]\ . \tag{15}$$

Note that the shell sensor is made of hexagonal symmetrical piezoelectric materials, and the piezoelectric constants $h_{31} = h_{32}$. The first term in the parenthesis is related to the circumferential motion and the second is the transverse motion. Using the modal expansion method, substituting the modal equations and strain expressions into the sensor equation, one can derive the modal sensor signal as a sum of all contributing modes:

$$\phi_k = \frac{h_{31}h^S}{S^e} \sum_{n=0}^{\infty} \eta_n(t) \left\{(nC_{n_2} + D_{n_2}) + \frac{d}{R}(nC_{n_1} + D_{n_1}n^2)\right\}$$
$$\cdot \int_0^{2\pi} \mathcal{Q}_k(\theta)\, \mathcal{P}[\mathcal{Q}_k(\theta)]\cos(n\theta)d\theta\ , \tag{16}$$

where η_n is the modal participation factor and the mode number $n = 1,2,...\infty$. It can be observed that the first part, membrane strains, is primarily contributed by the circumferential modes with amplitudes C_{n_2} and D_{n_2}; the second part, bending strains, is contributed by the transverse modes with amplitudes C_{n_1} and D_{n_1}. The modal amplitudes, either C_{n_1}/D_{n_1} or C_{n_2}/D_{n_2}, are coupled by a constant as discussed previously.

Based on modal orthogonality, appropriate shaped and convolved sensors only respond to the modal amplitudes of selected modes without the spillovers of other residual modes. Thus, observation spillover problem can be prevented. Since the k–th mode shape is orthogonal to all other mode shapes, the shape function $\mathcal{Q}_k(\theta)$ of the distributed ring sensor can be designed as a cosine function, i.e.,

$$\mathcal{Q}_k(\theta) = b\cos(k\theta)\ , \tag{17}$$

where b is a weighting factor. Substituting $\mathcal{Q}_k(\theta)$ into the sensor equation gives

$$\phi_k = \frac{h_{31}h^S\pi b}{S^e_k}\, \mathcal{P}(\cos n\theta)\left((kC_{k_2} + D_{k_2}) + \frac{d}{R}(kC_{k_1} + k^2D_{k_1})\right)\eta_k(t)\ , \tag{18}$$

where $\eta_k(t)$ is the modal participation factor; S^e_k denotes the effective electroded area; k_1 denotes the transverse mode and k_2 the circumferential mode for $n = k$. C_{k_1} and D_{k_1} are respectively the circumferential and transverse modal oscillation amplitudes of the k–th transverse natural mode with a component natural frequency ω_{k_1}. C_{k_2} and D_{k_2} are the circumferential and transverse modal amplitudes of the k–th circumferential natural mode with a component natural frequency ω_{k_2}.

Thus, two modal sensitivities: 1) a ***transverse modal sensitivity*** and 2) a ***circumferential modal sensitivity*** can be defined. (Note that the membrane modal sensitivity used for the generic shell sensor is defined as the circumferential modal sensitivity in ring sensors.) Each of the sensitivities can be further divided into two component sensitivities defined in terms of either the transverse or circumferential oscillation amplitudes. Note that these two amplitudes are coupled by a constant defined earlier. The ***circumferential modal sensitivity*** S_c^t defined by the transverse oscillation amplitude D_{k_2} and the circumferential modal sensitivity S_c^c defined by the circumferential oscillation amplitude C_{k_2} can be respectively expressed as

$$S_t^c = \frac{\phi_k}{D_{k_2}\eta_k(t)} = \frac{h_{31}h^S\pi b}{S^e{}_k}(k^2+1)\,, \tag{19}$$

$$S_c^c = \frac{\phi_k}{C_{k_2}\eta_k(t)} = \frac{h_{31}h^S\pi b}{S^e{}_k}(k+\frac{1}{k})\,. \tag{20}$$

The ***transverse modal sensitivity*** S_t^t defined by the transverse oscillation amplitude D_{k_1} and the transverse modal sensitivity S_c^t defined by the circumferential oscillation amplitude C_{k_1} of the distributed cosine shaped convolving sensor can be written as

$$S_t^t = \frac{\phi_k}{D_{k_1}\eta_k(t)} = \frac{h_{31}h^S\pi b}{S^e{}_k\,R}(k^2-1)\,, \tag{21}$$

$$S_c^t = \frac{\phi_k}{C_{k_1}\eta_k(t)} = \frac{h_{31}h^S\pi b}{S^e{}_k\,R}(k-k^3)\,. \tag{22}$$

These two sensor sensitivities for the distributed convolving ring sensors are to be analyzed. Thus, for bending oscillation where the transverse modes dominate, S^t should be used to estimate the oscillation amplitude. On the other hand, for the circumferential oscillations, S^c is used. For an output signal measured, the voltage can be used to estimate both the transverse and circumferential oscillation amplitudes.

Sensitivity Analysis

To demonstrate the convolving sensor theory derived earlier, a circular ring made of steel and coupled with spatially distributed sensors is studied. The circular ring has a radius of 50 mm, a width of 10 mm, and a thickness of 1 mm. It is assumed that the distributed sensors are made of piezoelectric polymeric polyvinylidene fluoride material with a thickness of 25 μm. Distributed sensor behaviors and parametric studies for transverse and circumferential modes are analyzed and results evaluated. Contributions from the two modal oscillation amplitudes, e.g., an in–plane circumferential component and an out–of–plane transverse component, were analyzed and compared. Variations of the ring and sensor thicknesses are also investigated.

As discussed previously, the first mode, k = 0, is a breathing mode and the second mode, k = 1, is a translational rigid–body mode for a free–floating circular ring. Sets of transverse and circumferential modes, with distinct component natural frequencies, appear when $k \geq 2$. Thus, cosine shaped distributed sensors are primarily designed for $k \geq 2$ and sensitivity analyses of the sensors are also evaluated for $k \geq 2$.

Circumferential modal sensitivity and transverse modal sensitivity of the distributed cosine shaped sensors are calculated and plotted for $k \geq 2$ modes. As discussed previously, there are two component natural modes for each k value, i.e., a circumferential mode and a transverse mode, with distinct natural frequencies. Each mode consists of two modal oscillation amplitudes: 1) an in–plane circumferential oscillation component and 2) an out–of–plane transverse oscillation component. These two oscillation components are coupled by a constant. Thus, for a measured output signal, the voltage can be used to estimate both the transverse oscillation amplitude and circumferential oscillation amplitude for a specific natural mode. Circumferential modal sensitivities, with two component sensitivities, for the same modes are plotted in Figure 7. Transverse modal sensitivities, with two component sensitivities, for natural modes $k = 2 \rightarrow 10$ are plotted in Figure 8. Note that two oscillation components are plotted respectively in each figure.

It is observed, in Figure 8, that for a measured signal, the inferred transverse oscillation amplitude, the bending effect, is larger than the in–plane circumferential oscillation, the membrane effect for the transverse natural modes. This is true because transverse oscillation amplitude has to be much more significant in the transverse natural modes. Figure 7 also leads to similar conclusions for the circumferential natural modes.

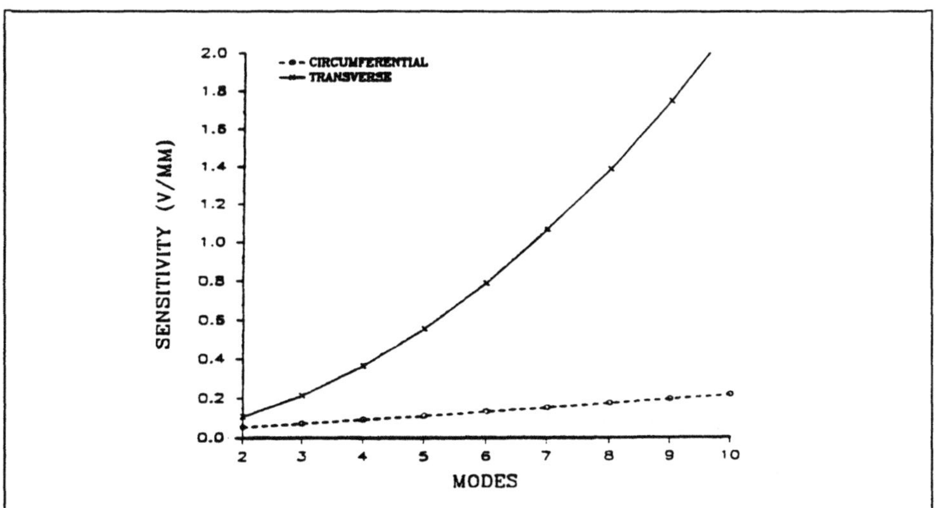

Figure 7 Circumferential modal sensitivities.

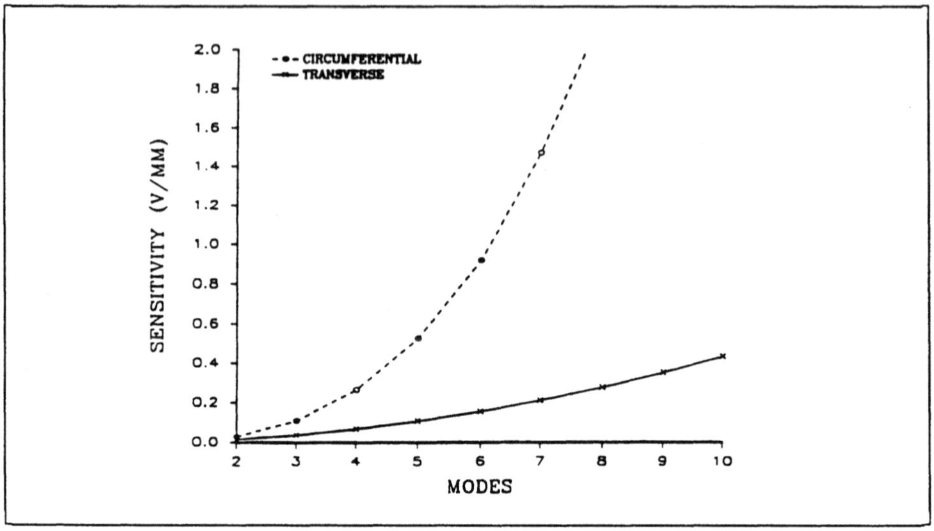

Figure 8 Transverse modal sensitivities.

Circumferential and transverse modal sensitivities are also evaluated when the thickness of the ring and the sensor layer are changed. Analysis results suggested that the transverse sensitivity increases when the ring thickness increases. However, the circumferential sensitivities remain identical because the membrane strains are unchanged regardless the thickness change. Thus, the membrane (circumferential) modal sensitivity is independent of the ring thickness and the transverse modal sensitivity is a linear function of the ring thickness. In general, membrane strains, the circumferential component, in the distributed sensor should be the same regardless the ring thickness. However, bending strains, the transverse component, increase when the ring becomes thicker; higher strains in the sensor layer lead to higher output signals.

5.3 CASE 3: Contact Measurements

A piezoelectric PVDF tactile sensors S–1 made of a thick piezoelectric PVDF polymer (1 mm) is used to evaluate the dynamic contact responses [11]. The polymeric PVDF sensor is composed of three basic components: 1) a protective surface, 2) a piezoelectric polymer – a sensing element, and 3) connectors, equivalent to housing, Figure 9.

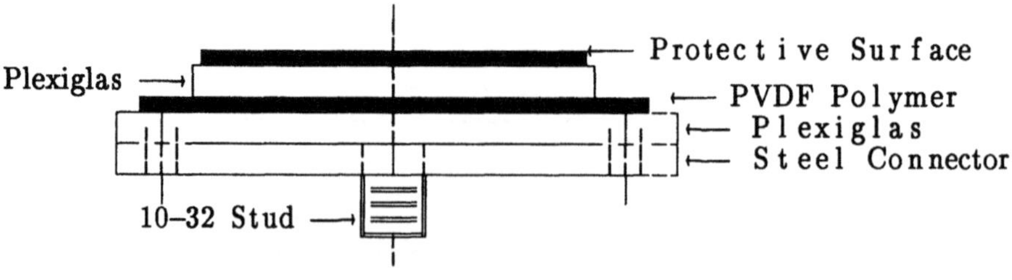

Figure 9 A piezoelectric PVDF tactile sensor.

To evaluate the tactile dynamic response of the polymeric PVDF sensor S–1, three steel balls were used to be dropped at different heights and the outputs from the PVDF polymer were recorded for analysis and comparison. The impact force is proportional to the product of the mass and the contact velocity and the latter is linearly proportional to the square root of drop–height, i.e., $v = \sqrt{2gh}$. Thus, the output from the PVDF system S–1 should be linearly proportional to the ball mass or the square root of drop–height. A reference transducers, a PCB impact force transducer, Figure 10, was subjected to the same excitation in order to compare and evaluate the results.

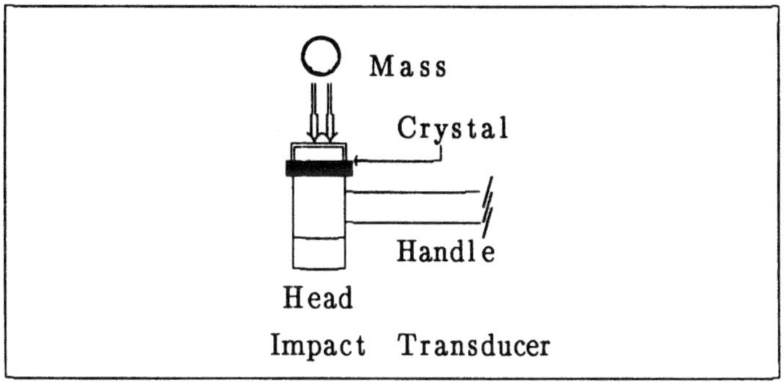

Figure 10 Experimental setup for testing tactile response.

Due to the disturbance of low frequency noise, a Krohn–Hite filter set at high–pass is always used in conjunction with all the experimental set–ups discussed here. Electromechanical dynamics and tactile response of the polymeric PVDF sensor system is evaluated using analytical and experimental techniques.

Tactile Responses

To evaluate the tactile response of the polymeric PVDF sensor S–1, test data obtained for three ball masses and for four different drop heights were recorded and analyzed. An average of seven values was taken for each ball drop from a given height. Two sets of plots are provided. One is the output voltage versus masses, Figure 11, and the other is the output voltage versus square root of drop–heights, Figure 12.

It is observed that the output voltage varies linearly with respect to the dropped masses, Figure 11, and the square root of drop–heights, Figure 12. However, biases in these two figures can be observed, which can be compensated by adjusting a constant DC offset in the measurement instruments. The maximum deviation is calculated to be 1.9% in Figure 11 and 5.3% in Figure 12. These experimental results show that the polymeric PVDF sensor system responds impact loading rather linearly; however, measurement biases should be corrected.

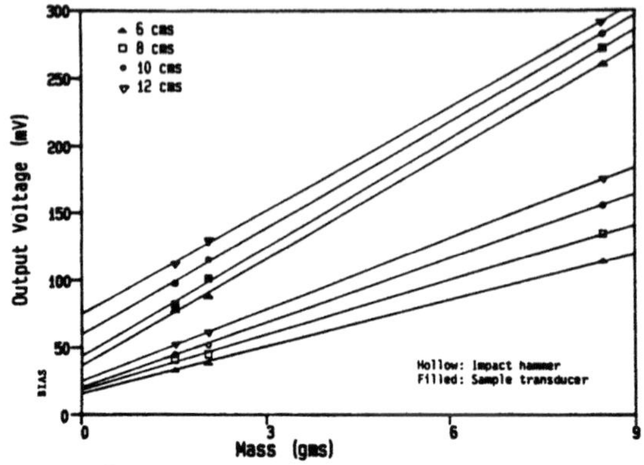

Figure 11 Comparison of the PVDF sensor with an impact force transducer as a function of drop–mass.

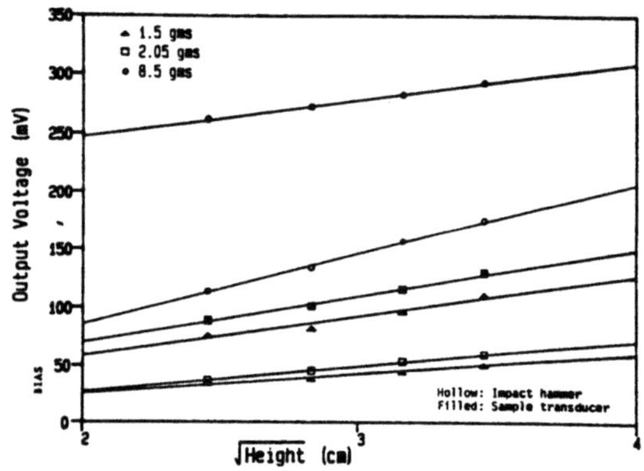

Figure 12 Comparison of the PVDF sensor with an impact force transducer as a function of drop–height.

6 Conclusions

Electromechanics of the distributed piezoelectric shell sensors were studied, and applications demonstrated. A generic sensor equation of a piezoelectric shell sensor laminated on a deep shell continuum was derived, and linear and nonlinear responses discussed. The sensor signal equation shows that the overall output is contributed by two strain components: 1) membrane strains and 2) bending strains. Consequently, two types of sensors can be designed: 1) a membrane sensor and 2) a bending sensor, based on the placement of the distributed shell sensor. In addition, distributed convolving modal sensors can be designed based on the modal orthogonality principle.

The demonstrate the use of the distributed piezoelectric sensor, three cases were studied. The first case was to illustrate the distributed ***modal voltages*** of natural modes. It was observed that the modal voltages are unique and they can be used to distinguish modal characteristics of natural modes. The second case was to illustrate the design of spatially distributed and shaped modal sensors for a circular ring. Transverse and circumferential modal sensitivities of the transverse and circumferential natural modes were derived; their electromechanical behaviors evaluated. The third case was to study the tactile response of a piezoelectric tactile sensor. Experimental results were calibrated with respect to those obtained from an impact sensor.

This study demonstrates the uses of distributed piezoelectric shell sensors in dynamic measurements and identification of continua. Applications of the piezoelectric technology are still wide open and need to be further exploited [2,12].

7 Acknowledgement

This research was supported, in part, by a grant from the Army Research Office (DAAL03–91G–0065) (1991–1994), Technical Monitor: Dr. Gary L. Anderson. A grant from the National Science Foundation (No. RII–8610671) and the Commonwealth of Kentucky (1986–1991) is gratefully acknowledged. A travel grant supported by the Workshop organizer, Prof. H. G. Natke, is very much appreciated.

8 References

[1] H.S. Tzou, 1993, *Piezoelectric Shells: Distributed Sensing and Control of Continua*, Kluwer Academic Pub., Dordrecht/Boston/London, Feb. 1993.

[2] H.S. Tzou and G.L. Anderson, 1992, *Intelligent Structural Systems*, Kluwer Academic Publishers, Dordrecht/ Boston/ London, August 1992.

[3] H.S. Tzou, 1987, "Active Vibration Control of Flexible Structures Via Converse Piezoelectricity", *Developments in Mechanics*, Vol.14c, 1201– 1206.

[4] H.S. Tzou, 1991, "Distributed Modal Identification and Vibration Control of Continua: Theory and Applications," *ASME Journal of Dynamic Systems, Measurements, and Control*, Vol.(113), No.(3), pp.494–499, September 1991.

[5] C.K., Lee and F. Moon, 1990, "Modal Sensors/Actuators," *ASME Journal of Applied Mechanics*, Vol.57, pp.434–441.

[6] S.A. Colins, D.W. Miller, A.H. von Flotow, 1990, "Sensors for Structural Control Applications Using Piezoelectric Polymer Film," MIT SERC Report #12–90.

[7] N. Hagood, W. Chung, and A. von Flotow, 1990, "Modeling of Piezoelectric Actuator Dynamics for Active Structural Control," AIAA Paper No.90–1087, 31st Structures, Structural Dynamics and Materials Conference, Long Beach, CA, April 2–4, 1990.

[8] H.S. Tzou and C.I. Tseng, C.I., 1990, "Distributed Piezoelectric Sensor/Actuator Design for Dynamic Measurement/Control of Distributed Parameter Systems: A Finite Element Approach," *Journal of Sound and Vibration*, Vol.(138), No.(1), pp.17–34.

[9] H.S. Tzou, 1992, "A New Distributed Sensor and Actuator Theory for "Intelligent" Shells," *Journal of Sound & Vibration*, Vol.(153), No.(2), pp.335–350, March 1992.

[10] H.S. Tzou and J. Zhong, 1993, "Sensor Mechanics of Distributed Shell Convolving Sensors Applied to Flexible Rings," *ASME Journal of Vibration & Acoustics*.

[11] H.S. Tzou and S. Pandita, 1987, "A Multi–Purpose Dynamic and Tactile Sensor for Robot Manipulators," *Journal of Robotic Systems*, Vol.(4.6), pp. 719–741, 1987.

[12] H.S. Tzou and T. Fukuda, 1992, *Precision Sensors, Actuators, and Systems*, Kluwer Academic Publishers, Dordrecht/ Boston/ London, December 1992.

Damage Evolution and Diagnosis in Operating Systems

C. Cempel
Poznan University of Technology
Applied Mechanics Institute
Poznan, Poland

H.G. Natke
University of Hannover
Curt-Risch-Institute
Hannover, Germany

1 Introduction

The existing systems in operation, for instance in mechanical, civil, transport, electrical engineering, etc., are usually considered as having only spatial r and fast time t-dynamic coordinates and properties. This description is sufficient if only the design properties are considered, optimized, and the dynamic behaviour is simulated. But when the mechanical properties of real systems in operation are considered, which means taking into account imperfections and damage that have appeared during the manufacturing stage, and when their evolution in operation is considered, one should introduce the concept of life with the lifetime Θ, where the damage evolution can proceed and can be measured. The life properties depend on the environmental conditions as well as on the operational ones. Both can introduce faults and damage into the system, which, if they exist, will have an instantaneous effect or will affect the mechanical system under operation in an evolutionary manner to a certain extent. In consequence, a holistic approach is needed in order to model the total dynamics of the system in operation.

As the further development of the lifetime Θ concept we can introduce the damage concept D (see Cempel, Natke 1990)[1] as the inherent property of the operating system. In the development of this approach the damage may also be a function of two coordinates, $D \equiv D(r, \Theta)$. Additionally, we can assume that damage is cumulative or a monotonous function with respect to Θ and normalized as $0 \leq D \leq 1$.

In this sense, a damage measure defined in this way is an **inherent independent property** of the operating system, described by an appropriate model along with its masses, stiffnesses and dampings. The damage measure will be non-zero for real systems with imperfections at $\Theta = 0$. In such a holistic approach to operating system dynamics, damage is the result of an external action (external energy) and of the internal energy dissipation during system operation. The measure of damage can therefore be based on physical quantities, such as energy expression, or on a corresponding diagnostic symptom (symptom = measurable or observable quantity covariant with the damage development).

2 Energy model of damage

Damage here is understood broadly as the change (worsening) of the operational capability expressed as a change of dynamical behaviour as a consequence of parameter changes (mass, stiffness, damping) within the operating system under consideration.

The concept of damage which we are going to present here will be valid for both cases mentioned above; the global one for the total system and the local one for its elements and its local properties (Fig. 1).

[1] also local or systems age concept

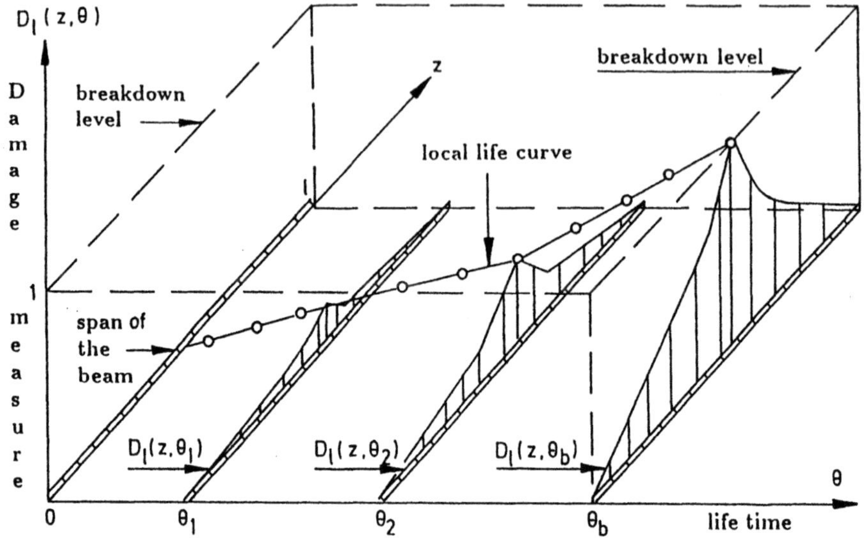

Figure 1 An attempt at an explanation of the local damage concept for a beam under progressive fretting and of the measure $D_l(r,\Theta)$

The operating system can be treated as an open energy transformation system. It is open because it always has its own energy input and output, and at each operation time Θ energy is transformed: producing something or moving something or, for example, changing wind energy into vibration, as in the case of civil engineering structures. This energy is distributed into two flows: applicable energy E_u and dissipated energy E_{tc}. The part V of dissipated energy is radiated outside the system in terms of any physical phenomenon: sound, vibration, light, heat and/or a electromagnetic phenomenon. The remaining part of it, E_t, is dissipated internally in different forms of destructive processes such as fatigue, creep, friction and erosion of different forms. It is obvious that the capacity to dissipate energy in every element of the system is limited, as well as the capacity for energy dissipation by the whole system during operation. Let us denote this limit by E_{tb} as the breakdown energy (of the element or the total system).

In this way we can define the measure of the energy dissipation progress by $e(\Theta)$ as the ratio of currently internally dissipated energy at lifetime Θ, to its breakdown value,

$$e(\Theta) = \frac{E_t(\Theta)}{E_{tb}}, \qquad \text{with :} \quad 0 < e \leq 1. \tag{1}$$

It can be noted from the previous equation that by mathematical definition the initial e value for a new system which has some imperfections, or even is slightly damaged at the beginning of its operational life ($\Theta = 0$), will be non-zero.

Destructive energy (dissipation), lifetime dependent, is measured by the relative function introduced above. It is a direct measure of damage (Cempel, Natke 1990), therefore we define a damage measure[2] as

$$D(\Theta) \equiv e(\Theta). \tag{2}$$

It may be of interest to discuss which are the particular forms of the above damage measure for different wearing processes. Taking into account the type of event and continuous forms of wear separately, one can find (Cempel, Natke 1990; Cempel 1991)

$$D(\Theta) = \begin{cases} \frac{\Theta}{\Theta_b} & \text{– for continuous wear such as friction etc. } (\Theta_b \text{ - breakdown time}) \text{ and for generalized cases of uniform wear} \\ \sum_i \frac{\Theta_i}{\Theta_{bi}} & \text{– for multilevel } (\sigma_i) \text{ continuous wear including creep } (\sigma_i\text{-stress level}) \\ \frac{n}{N_b} & \text{– for simple cyclic wear } (N_b \text{ - breakdown number of cycles}), n\text{-current number of cycles} \\ \sum_i \frac{n_i}{N_{bi}} & \text{– for multilevel } (\sigma_i) \text{ cyclic wear} \\ \left(\frac{Z-y}{C-y}\right)^b & \text{– for single monotonous loading of a large structure, with Z the maximum displacement during loading, C the breakdown displacement and y the yielding displacement of the system.} \end{cases} \tag{3}$$

It can be seen from the above that our damage development measure D is equivalent, for example, to the Odquist-Katchanov law of creep on the one hand, and to the Palmgren-Miner fatigue law for the event type wear on the other hand. Thus the D-measure seems to be simple to interpret and has a good physical (energy) and experimental base.

As has been seen during the course of the discussion, it is a global damage development measure, defined for the total system or its parts, without any spatial dependency by the vector r. There is no contradiction in doing this, so the local measure of damage development $D_l(r,\Theta)$ can be defined:

$$D_l(r,\Theta) = \frac{E_t(r,\Theta)}{E_{tb}(r)} \tag{4}$$

with $E_t(r,\Theta)$ the energy dissipated currently at the local coordinate r, and $E_{tb}(r)$ the local limit of breakdown energy. Of course, if we integrate, or add up, the above-defined local measure over the domain or subdomain in ϑ of our subsystem we will obtain a global measure of damage of the given subsystem

$$D(\Theta) \equiv \frac{1}{\vartheta}\int_{\vartheta} D_l(r,\Theta)d\vartheta\,. \tag{5}$$

[2] also damage development or age

Local measures are appropriate to local problems. For example, some wearing processes, like fatigue or erosion, are local ones. So local models are adequate to these problems, especially when the finite element method (FEM) or the boundary element method (BEM) are used.
The concept of local damage measure can be elucidated easily for one-dimensional systems like beams; it is indicated in Fig. **1** for the progressive fretting wear of a vibrating beam. If we describe the system by its primary mechanical properties, such as mass, stiffness, damping distributions, we may note that the damage measure $D_l(r,\Theta)$ will affect these characteristics in different ways. Moreover, for example, each wearing process will affect them in a different way. In general, we can only write a functional dependency for these parameters as below:

$$\begin{array}{lll} \text{Mass} & - & m\left[r, D_l(r,\Theta)\right] \\ \text{Stiffness} & - & k\left[r, D_l(r,\Theta)\right] \\ \text{Damping} & - & c\left[r, D_l(r,\Theta)\right]. \end{array} \tag{6}$$

It can be seen from the above and from Fig. **1** that the spatial distribution of these parameters, as well as the corresponding damage, are two independent variables. So the damage measure $D_l(r,\Theta)$ in such a holistic approach i.e.: "space-time-life" seems to have *characteristics independent of the operating systems*, as well as of their parts, elements etc. In other words, for the holistic modelling of the system the damage measure of a system should be identified by some proper method.

3 Holistic modelling of systems

When an attempt is made to model dependent on the space r, time t, life-time Θ the behaviour of structures, vehicles, machines etc., one approach is, as already mentioned, to introduce the independence of t and Θ. For illustration of the problem let us consider the example of fatigue wear, which is generated as a result of the high amplitude of vibration. Assuming, for example, the frequency of vibration as 10 Hz and looking for a typical fatigue life behaviour, as for steel, one finds that the first damage effects can occur after 10^3 seconds for a high amplitude of vibration, and after $10^7 \div 10^9$ for small vibration amplitudes (Juvinal 1967). Remembering that in vibration analysis we are interested only in the time span of magnitude of a few tenths of cycles of vibration, we can be almost sure that no negligible error will arise if we separate the lifetime Θ measured in months and years of operation from the dynamic time t measured in some seconds or so (see also Fig. **2**).
Hence, let us introduce the lifetime Θ and the damage characteristics as independent quantities into our model of the dynamics of the operating systems. Trying to keep the formulation as simple as possible, we consider a one-dimensional beam-like system, i.e. a shaft or rod. Adding our damage measure $D_l(r,\Theta)$ to the formulation of Newland (Newland 1989), we can find

$$f(z,t) = m\left[z, D_l(z,\Theta)\right] \frac{\partial^2 y(z,t)}{\partial t^2} + \alpha m\left[z, D_l(z,\Theta)\right] \frac{\partial y(z,t)}{\partial t}$$

$$+ \quad \beta L\left[z, D_l(z,\Theta), \frac{\partial}{\partial z}, \frac{\partial}{\partial t}, y(z,t)\right] + L\left[z, D_l(z,\Theta)\frac{\partial}{\partial z}, y(z,t)\right]. \quad (7)$$

Here, in addition to previous notations, z is the space coordinate and $y(z,t)$ is the dynamic deflection of the system, $L[\cdot]$ is the stiffness operator, β and α are the damping coefficients for the assumed uniform beam.
As we assume linearity of the system, we can solve Equation (7) with the impulse response function $h\left[t, z, D_l(z,\Theta)\right]$, and the system response follows by using the double convolution integral operator to

$$y \equiv y\left[z, t, D(\Theta)\right] \equiv y(z,t,D) = f(z,t) \otimes h\left[t, z, D_l(z,\Theta)\right]. \quad (8)$$

In the above equation the impulse response $h[t, z, D_l(z,\Theta)]$ includes the local damage characteristics. The system response y, being the result of overall integration over the system, includes only the lifetime variable Θ or the global damage D. Hence, the damage measure $D_l(z,\Theta)$, or corresponding characteristics, must be extracted in some way from the system response $y(z,t,D)$.
By applying the modal concept to the system response we can separate the dynamical time from other variables in the form

$$y(z,t,\Theta) = \sum_{j=1}^{\infty} q_j(t) U_j(z,D) \quad (9)$$

where normal modes $U_j(z,D)$ satisfy the orthogonality properties and the equations

$$\left.\begin{array}{rcl} -\omega_j^2(D) m\left[z, D(z,\Theta)\right] U_j(z,D) + L\left[z, D(z,\Theta)\frac{\partial}{\partial z}\right] U_j(z,D) & = & 0 \\ \ddot{q}_j(t) + 2\xi_j(D)\omega_j(D)\dot{q}_j(t) + \omega_j^2(D) q_j(t) - \int_R dz U_j(z,D) f(z,t) & = & 0 \\ 2\xi_j(D)\omega_j(D) - \alpha - \beta\, \omega_j^2(D) & = & 0. \end{array}\right\} \quad (10)$$

In the original course of the solution of the above equations one can find that normal modes $U_j(z,D)$ depend on the wave number $\kappa_j(D)$ and wave length $\lambda(D)$

$$\kappa_j(D) = \frac{\omega_j(D)}{c(D)} = \frac{2\pi}{\lambda_j(D)} \quad (11)$$

which may be marked as

$$U_j(z,D) \equiv U_j\left[z, \kappa_j(D)\right] = U_j\left[z, \lambda_j(D)\right]. \quad (12)$$

Here $c(D)$ is the speed of sound in the material, $\lambda_j(D)$ is the wave length and $\omega_j(D)$ is the i- th eigenfrequency. As we can see, all the characteristic parameters of the system, dampings $\xi_j(D)$, eigenfrequencies $\omega_j(\Theta)$ and the wave numbers $\kappa_j(D)$ and wavelengths $\lambda_j(D)$ now all depend on the damage accumulation D and damage distribution $D_l(z,\Theta)$. But this dependency is not explicit, so the damage development and distribution must be identified by means of special procedures and the use of symptoms. We will consider this problem later on.
As is known, the dynamics of operating systems can be alternatively analysed in a discrete approach. In this case, using a discretization technique (Ritz or Galerkin principle, FEM or BEM), we obtain the matrix form of the equations of motion

$$\left[m_{ij}(D)\right]\left\{\ddot{q}_j(t)\right\} + \left[c_{ij}(D)\right]\left\{\dot{q}_j(t)\right\} + \left[k_{ij}(D)\right]\left\{q_j(t)\right\} = \left\{F_j(t)\right\}. \quad (13)$$

The damage dependent on mass, stiffness and damping was suggested primarily in (Natke, Yao 1987) and (Natke, Cempel 1990), so we can generalize these to our needs as below

$$\begin{aligned} m_{ij}(D) &= m_{ijo}\,[1 - a_{mij}D(\Theta)]^{\alpha_m}\,, \quad 0 \le a_{mij} \le 1 & (14)\\ k_{ij}(D) &= k_{ijo}\,[1 - a_{kij}D(\Theta)]^{\alpha_k}\,, \quad 0 \le a_{kij} \le 1 \\ c_{ij}(D) &= c_{ijo}\,[1 + a_{cij}D(\Theta)]^{\alpha_c}\,, \quad 0 \le a_{cij} \le 1\,, & (15) \end{aligned}$$

where the initial values m_{ijo}, k_{ijo}, c_{ijo} are for lifetime $\Theta = 0$. The exponents α_k, α_c, α_m are positive. Here the indices i, j, in addition to the matrix order, can also play the role of spatial location indices. This statement holds true in particular for the FEM and BEM approaches, where each element/mode has prescribed its own damage (wear) participation factor; of mass a_{mij}, stiffness a_{kij} and damping a_{cij}. Another approach will apply the energy flow and the dissipation within the system. It appears that this approach enables us to locate the areas where damage arises and to investigate it. Looking once more at the definition of the local measure of damage development (4), (Fig. **3** depicting the energy flow through the system) we can propose the following:

$$D_l(r,\Theta) \equiv \frac{E_t(r,\Theta)}{E_{tb}(r)} = \frac{\mathcal{Z}e(r,\Theta)}{E_{tb}(r)} = \frac{\mathcal{Z}}{E_{tb}(r)} \cdot \bar{\sigma}(r,\Theta) \cdot \bar{\epsilon}(r,\Theta)\,, \tag{16}$$

where $\mathcal{Z} = \frac{N_t}{N_i}$ is the energy flow loss coefficient for our system and $e(r,\Theta)$ is the local deformation energy density, $\sigma(r,\Theta)$ stress distribution, and $\epsilon(r,\Theta)$ the respective strain. By looking for the location of maximum energy density by the FEM calculation we can thus assess which areas of our system are the most prone to damage. In addition, assuming the fatigue as the governing wear process in a system and using the Palmgren-Miner law, we can specify the event component of the local damage measure (16) in the form

$$D_l(r,\Theta) = \frac{\bar{\sigma}(r) \cdot \bar{\epsilon}(r)}{e_{tb}(r)} \cdot \sum_i \frac{n_i}{N_b(\sigma_i)}. \tag{17}$$

Here $e_{tb}(r)$ is the critical energy density for a given material, which for the simple uniform case may be calculated as $e_{tb}(r) = e_{tb} = \frac{\sigma_u^2}{E}$ (σ_u ultimate stress, E-Young modulus). The number of fatigue cycles at the stress level σ_i is n_i and $N_b(\sigma_i)$ is the breakdown number of cycles at the stress level σ_i.

If we calculate the value of the above-defined measure for the whole area of the system with some frequency of loading f, we can obtain the evolution at the local fatigue damage development measure $D_l(r,\Theta)$, due to $\Theta = \sum_i f^{-1} \cdot n_i$. The results of these calculations for a simple supported beam excited by white noise at the middle of its span is shown in Fig. **2**. Here, as a first approach, no change in the beam stiffness due to fatigue was assumed. As can be seen from Fig. **2**, the damage is a real local property. What is more important, this characteristic of local damage, $D_l(r,\Theta)$, cannot be inferred from the global one, $D(\Theta)$, or some symptoms $\omega_j(D)$ etc. It should be identified by a special procedure.

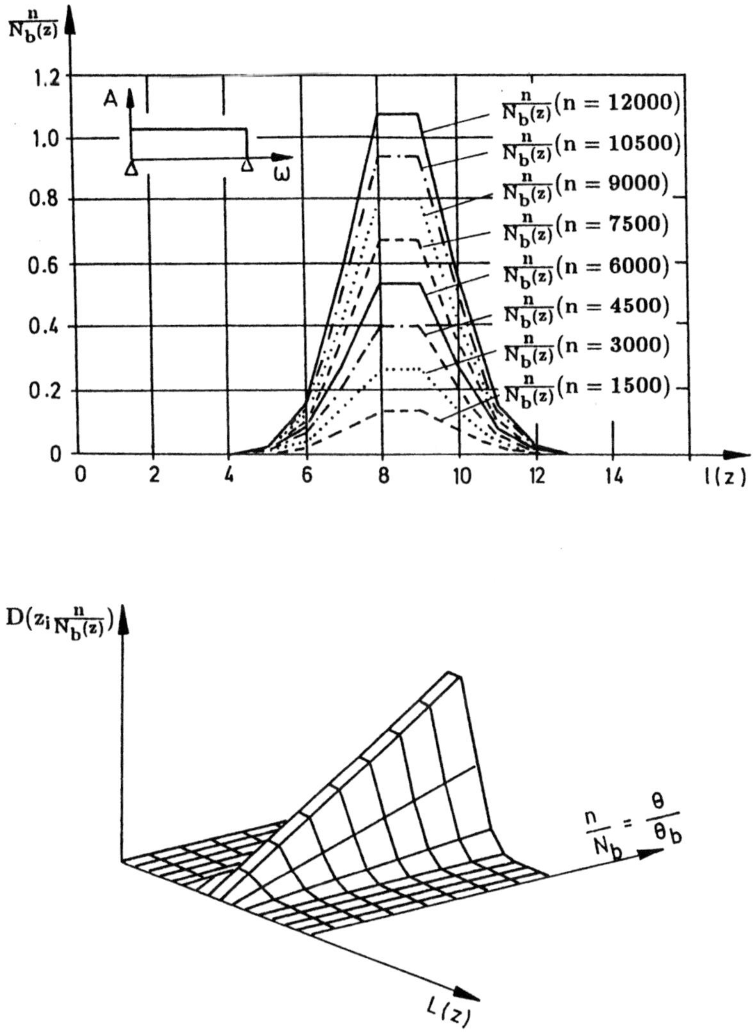

Figure 2 Evolution of damage in a vibrating beam as the result of energy dissipation modelled according to Eq. (17)

4 Damage and symptoms evolution

From the last example it has been seen that the local damage measure increases during the life of the system (Fig. 2). Hence it would be advantageous to explain the mechanism, or to give a model, of damage evolution of this kind. As was stated in the introduction and explained in Section 2, the operating system can be treated as an "open energy transforming system with limited dissipation capacity". This approach

aimed at operating machines was developed under the name "tribovibroacustical machine model" (Cempel 1990). But it is possible to generalize these results to a broad class of system in evolution (Winiwarter, Cempel 1991), and to operating systems of mechanical origin in particular. Hence, let us show briefly (having in mind the damage evolution) the evolution of damage symptoms which we are only able to observe.
Let us consider mechanical systems in operation as structure, machine, equipment

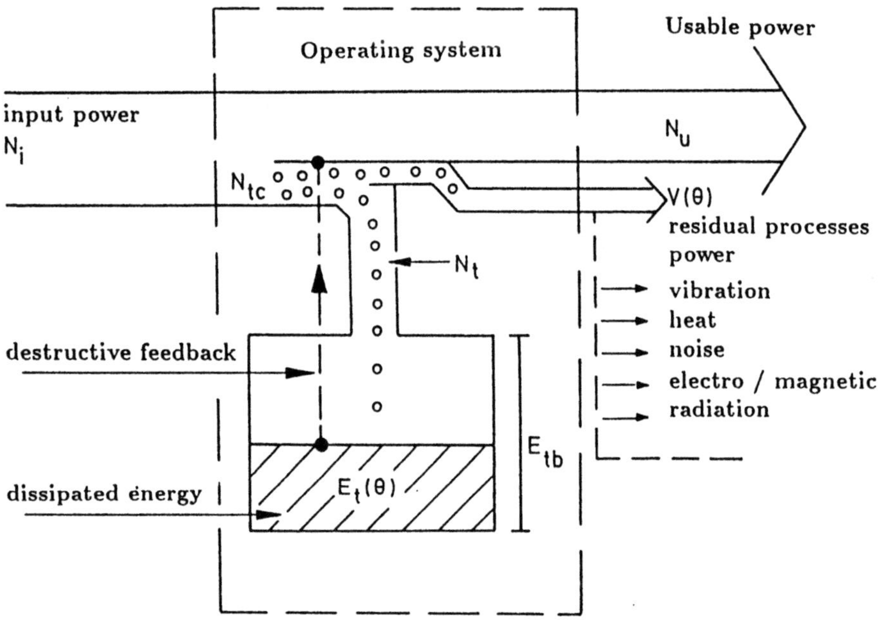

Figure 3 Operating system as an open energy transforming system with limited dissipation capacity and secondary residual processes output $V(\Theta)$

(Cempel, Natke 1990) with an energy flow, and the dissipation of the cumulative type. The input power $N_i(\Theta)$ and output $N_o(\Theta)$, also produce the secondary output $V(\Theta)$ (residual processes), which stimulates the intensity of the wearing process, as depicted in Fig. **3**.

Two relationships are important in creating the damage evolution model.
The limited dissipation capacity

$$E_t(\Theta) = \int_0^\Theta N_t\left[\Theta_o, V(\Theta_o)\right] d\Theta_o \leq E_{tb} \tag{18}$$

and the destructive feedback constitutive relation (see Fig. **3**)

$$dV(\Theta) \equiv \beta dE_t\left[\Theta, V(\Theta)\right]. \tag{19}$$

Starting from this it is possible to obtain and to solve the differential equation for residual output power $V(\Theta)$, colinear with the dissipated power $N_t(\Theta)$, in the form

$$V(\Theta) = V_o \left(1 - \frac{\Theta}{\Theta_b}\right)^{-1}; \quad N_t(\Theta) = N_{to}\left(1 - \frac{\Theta}{\Theta_b}\right)^{-1} \tag{20}$$

with the initial values V_o, N_{to} and the defined system breakdown time

$$\Theta_b \equiv \frac{dE_t(\Theta)}{dN_t(\Theta)}. \tag{21}$$

This is the lifetime $\Theta = \Theta_b$, when the dissipated power tends towards infinity $N_t(\Theta_b) \longrightarrow \infty$ and system breakdown.
By now looking at our previous damage definition in general (1) (and at some particular ones (3)), we can rewrite our evolutionary formula (20) as below

$$V(D) = V_o(1 - D)^{-1}; \quad N_t(D) = N_{to}(1 - D)^{-1}, \tag{22}$$

due to $D \equiv \frac{E_t(\Theta)}{E_{tb}} = \frac{\Theta}{\Theta_b}$. While trying to solve the damage evolution problem we have thus obtained components of dissipated power $V(\Theta), N_t(\Theta)$, as equivalently dependent on damage development $V(D)$, $N_t(D)$. But neither $D(\Theta)$, $V(D)$, $N_t(D)$ can be observed quantitatively to assess the system damage, and the change of its operational properties, parameters etc. We can only observe some symptoms of damage S based on secondary output residual processes $V(D)$ (see Fig. **3**) in terms of measurable quantities of vibration, noise, heat etc.
So it is assumed, that some functional symptoms of the analytical form

$$S(\Theta) = \phi\,[V(\Theta)] \qquad \text{or} \qquad S(D) = \phi\,[V(D)] \tag{23}$$

exist, such as the mapping of $V(D)$ to some measurable quantities, the vibration amplitude, the temperature, etc. Most of the operational systems have a lifetime span of magnitude of tenths of years , and the damage accumulation is usually not purely a deterministic process. Hence, if we observe the symptom $S(D)$ for the set of the systems of the same type that are in operation, we can empirically assess empirically the reliability of the whole set and of the particular operating unit. This can be done on the basis of symptom measurements; i.e. due to the fact that the probability of the observed symptom concerning the operating units exceeds a pre-given value S_p

$$R(S) \equiv P\,(S \geq S_p) = \frac{n(S \geq S_p)}{N_l}, \tag{24}$$

where $n(S \geq S_p)$ is the number of units in a set satisfying this relation, and N_l is the total number of operating units under observation. The particular form of symptom reliability curve $R(S)$ and symptom damage or life curve $S(D)$ depends on the form of the symptom operator (23), and the method of symptom observation. Assuming that the probability density of system observation does not depend on the damage value, (in the first approach), we can obtain the results as in Table **1** (Cempel 1991) and Fig. **4** and Fig. **5**. As is seen from Table **1** and Figs. **4** and **5** all the life

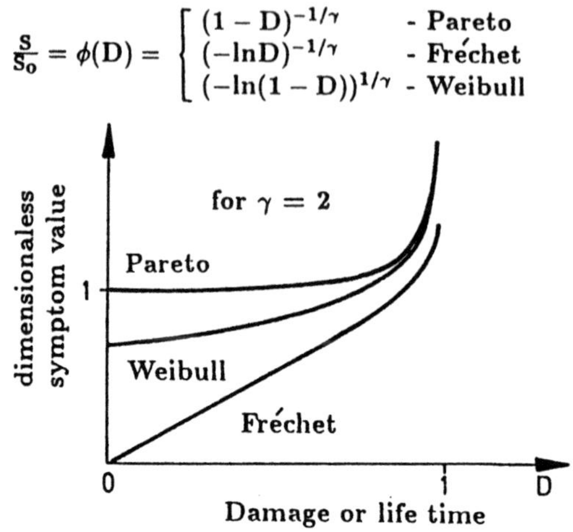

Figure 4 Symptom life curves in correspondence to different symptom operator functions $\phi(V)$ (see Table 1)

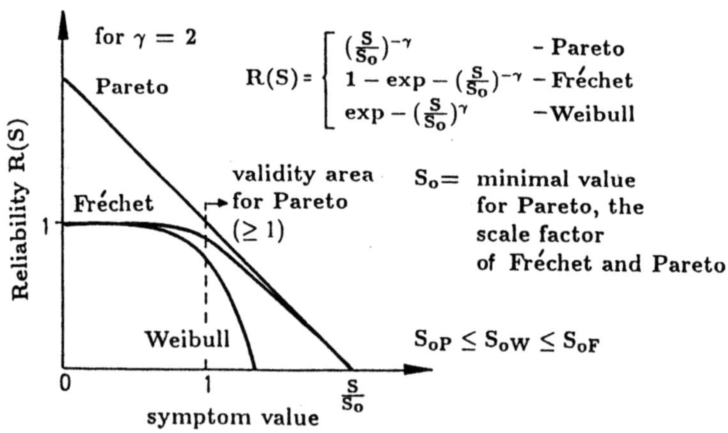

Figure 5 Symptom reliability (cumulative probability distributions) of Pareto, Fréchet and Weibull

curves have asymptotes at $D = 1$, but the highest sensitivity to damage belongs to the Fréchet type life curve. Concerning the reliability, all the curves have right hand skewness, and the Pareto reliability is asymptotic to Fréchet, if $\frac{S}{S_o} > 1$. Here S_o plays the role of the minimum value of the symptom of the Pareto distribution, and the uniform distribution is the Taylor expansion of Pareto distribution near

Symptom operator $V > V_0,\ \gamma > 0$ $\Phi(\frac{V}{V_0}) =$	Symptom life curve $S_0 > 0$, $\frac{S(D)}{S_0} =$	Symptom reliability $R(S) =$	Remarks
$\left(\ln \frac{V}{V_0}\right)^{1/\gamma}$	$[-\ln(1-D)]^{1/\gamma}$	$\exp -(\frac{S}{S_0})^{\gamma}$	**Weibull**, $S \geq 0$
$[-\ln(1-\frac{V_0}{V})]^{-1/\gamma}$	$[-lnD]^{-1/\gamma}$	$1-\exp -(\frac{S}{S_0})^{-\gamma}$	**Fréchet**, $S \geq 0$
$(\frac{V}{V_0})^{1/\gamma}$	$(1-D)^{-1/\gamma}$	$(\frac{S}{S_0})^{-\gamma}$	**Pareto**, $S \geq S_0$ (assymptotics of Fréchet)
$1+\left(1-\frac{V_0}{V}\right)\cdot\frac{1}{\gamma}$	$1+\frac{1}{\gamma}D$	$1+\left(1-\frac{S}{S_0}\right)\cdot\gamma$	**Uniform**, $S \geq S_0$ $S \sim S_0$ (approximation of Pareto)

Table 1 Various symptom operators and the respective symptom life curves and symptom reliabilities for systems in operation

$S = S_0$ distributions. There are also some scaling values $S_o > S_{min}$ for the other two symptom reliability functions.

The exponent γ plays an important role (see Cempel 1990), and for $\gamma < 1$ all three distributions (Weibull, Fréchet, Pareto) have a similar shape. In all the cases of the observation of damage evolution in operating systems we can determine the γ exponent from the symptom reliability curve and use it for determining of the symptom life curve. Moreover, having an empirical assessment of reliability, we can determine a residual damage value or residual time to breakdown, because, when using the formulas of Table 1 and Eq. (22), it is possible to show that the following relationship holds true:

$$R(S) = 1 - D \equiv \Delta D, \text{and also } \Delta D = 1 - \frac{\Theta}{\Theta_b} = \frac{\Delta\Theta}{\Theta_b}. \quad (25)$$

ΔD could be understood here as the residual damage (or life) inherent in the unit when the symptom value S has been observed. This relationship is a suitable basis for operating system diagnosis and prognosis when we are monitoring some symptoms of damage $S(D)$. But what type of symptom should be observed?

5 Damage identification

When looking for damage investigation we should remember that we first need methods of damage detection, localization and assessment, and sometimes each of the enumerated activities can be a hard task. As was said earlier, it is almost impossible to measure the damage development function $D_l(z,\Theta)$ directly, and we need special symptoms, sometimes particular ones for our task. In this respect system identification (theory and practice) is faced with a new challenge due to diagnostic needs. For damage detection it is sufficient to consider the global symptoms, $S(D) = \phi\left[D(z,\Theta)\right]$, and this gives the assessment of the total damage in the system (or the dissipated energy). In mechanical engineering, vibration and temperature symptoms are fully used to monitor the condition of machines. But what kind of symptom may one use for structures? One easy way to observe a system in operation is to measure deflections in representative points during its lifetime. It has already been shown (Cempel, Natke, Ziólkowski 1990) that dimensionless static and dynamic (excited, for example, by random white noise) deflections can be expressed as

$$S_x(\Theta) = \left[1 - a_k D(\Theta)\right]^{-\alpha_k} , \tag{26}$$

if one acknowledges the degradation of stiffness (a_k; $\alpha_k \neq 0$ in Eq. (14)) only. Measuring the deflection of the system can sometimes be no easy task, and thus an alternative is to measure the eigenvibration period of the i-th mode $T_i(D) = \frac{2\pi}{\omega_i(D)}$. The dimensionless symptom for this case, under the same assumption of stiffness degradation only, is

$$S_{T_i}(\Theta) = \left[1 - a_k D(\Theta)\right]^{-\frac{\alpha_k}{2}} \tag{27}$$

and gives a symptom exponent twice greater (less sensitive) than for the deflection: $\gamma_T = \frac{2}{\alpha_k} < \gamma_x = \frac{1}{\alpha k}$.
Hence it may be interesting to demonstrate these relations for a particular system, for example a beam, looking at the change of the stiffness and its influences on the eigenfrequencies and the eigenfunctions. For the one-dimensional model (7) with the stiffness operator degradation $\delta L\left\{z, D_l(z,\Theta), \frac{\partial}{\partial_z}\right\}$ the corrections in the eigenmode shapes and the eigenfrequencies are used for detection purposes and also for localization ones (see: Cawley 1988, Vandiver 1977, Brownjohn 1988, Norton 1986, Cempel, Natke, Ziólkowski 1991). In Cempel, Natke, Ziólkowski 1991, additionally differentiated quantitites (with the example of shear force, without the need of directly performing the differentiation) are used as sensitive symptoms.

6 Damage diagnostics

In general there may be two approaches to the diagnosis of damage in operating system. One is the symptom-based approach, as in machinery diagnostics, and the other is the model-based method. The latter uses FEM, or convex modelling or another analytical approach as a knowledge base (Natke, Yao 1989).

The symptom-based approach is the result of long history measurements during operation, with signficant data containing the type of damage and its location. If we take monotonously growing symptoms $S(D)$, as discussed before (like deflection (26)), and have chosen a proper inspection time interval $\Delta\Theta$, we can assemble the symptom data and elaborate the symptom reliability graph according to relation (24). Assuming a Pareto model of damage evolution in first approximation, we then obtain

$$R(b) = \left(\frac{S}{S_o}\right)^{-\gamma} = 1 - D \equiv \Delta D \tag{28}$$

with $S_o = \text{Min}\,\{S\}$ and γ being the reliability and symptom life curve exponent (see Table 1). With a sample of symptom observations, numbering about 30 , a figure convenient for statistics, the exponent γ can be estimated by applying a regression procedure. All the data are then available for presenting the resulting model of damage evolution, e.g. as a graph as in Fig. 6. With the symptom value S_e measured for

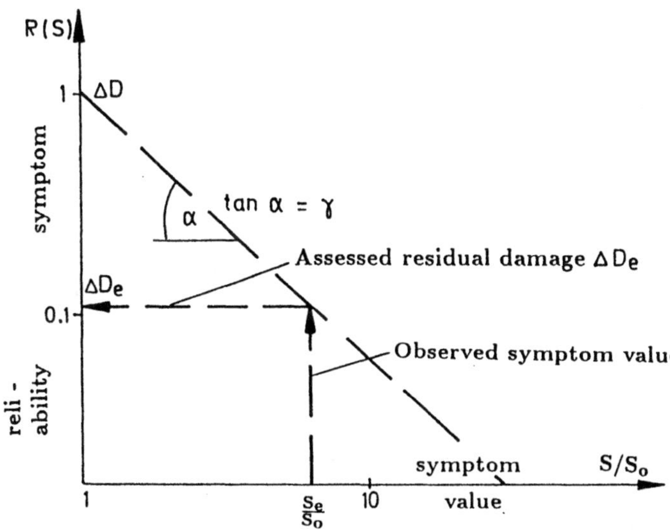

Figure 6 Pareto model of the symptom reliability of an operating system and its use for residual damage ΔD_e assessment

a particular unit, the residual time to breakdown ΔD_e of this unit can be assessed, and it can be used for scheduling the next inspection, maintenance or overhaul (for more information see (Cempel 1990, 1991)).
Additionally, we can deal with model-based diagnostics. Let us remember that the system responses dependent on the excitation function $f(z,t)$, and the system unit impulse response function, depend on the location and the development of damage $h\,[z,t,D_l(z,\Theta)]$. The amplitude $Y_{s,z_e}(D_s)$ will be obtained by using the same excitation during all the tests at the point z_e of the system, and the unit impulse response function $h\,[z,t,D_l(z,\Theta)]$ will be estimated by some identification procedure (Natke

1992), which enables us to build the momentum model of the system, and the model response at the point z_e:

$$\begin{aligned} Y_M\left((z_e, t, D_l(z,\Theta)\right) &= \left[f(z,t) \otimes h\left(z,t,D_l(z,\Theta)\right)\right]_{ze} \\ &= Y_{M,z_e}\left(t, D_l(z,\Theta)\right), \qquad (29) \\ \text{with the amplitude} \quad & \ Y_{M,z_e}\left(D_l(z,\Theta)\right). \end{aligned}$$

During the next inspection we measure the system response amplitude $Y_{s,z_e}(D_s)$, and start from the previous model response. With these we can assess the most probable change of damage function $D_l(z,\Theta)$. This task can be facilitated by using the damage index defined below

$$M(D) \equiv \max_{\Theta,z} \left| \frac{Y_{s,z_e}(D_s)}{Y_{s,z_e}(D_s) - Y_{M,z_e}\left(D_l(z,\Theta)\right)} \right|. \qquad (30)$$

As can be seen, by varying the location of damage z and its development Θ we can assess step by step the maximum probable damage function $D_l(z,\Theta)$. If we have a set of possible damage models designed by the index $i = 1, 2, ...$, instead of a continuous model, we can maximize over this model set

$$M(D) \equiv \max_{\Theta,i} \left| \frac{Y_{s,z_e}(D_s)}{Y_{s,z_e}(D_s) - Y_{M,z_e}\left(D_l(i,\Theta)\right)} \right|. \qquad (31)$$

But this proposal needs verification by simulation and physical modelling, and this will be done in another paper.

The second approach that model-based diagnostics uses for pre-given lifetimes is a corrected mathematical model, for example as written in Eq. (7). This means that these mathematical models have known confidence and sufficiently small errors. Significant deviations in the symptoms leads to residuals of dynamic quantities and parameters which serve for the localization and assessment of the modifications with respect to faults. The prior model, of course updated by measured data (Natke, 1992), should be taken for sensitivity studies and possible fault investigations. Some a priori knowledge is thus available and serves as a pattern. Fig. 7 shows the principles of this extended identification methodology (Natke, Yao 1988). The reader can find more information in (Natke, Yao 1988, Natke 1989, Yao, Natke 1990, 1991, 1992). The introduction of damage measures, if necessary, can be carried out in accordance with the above measures, as already mentioned. However, the models with their properties used, here form the knowledge base for an extended diagnosis. It is possible that the, in general, dynamic (global) models used may not be detailed enough in order to determine, for instance, maximum stresses for assessment (Natke, Yao 1991). The decision can be made deterministically, statistically or by the application of fuzzy logic (Yao, Natke 1992).

7 Conclusions

This paper formulates the first attempt at the holistic approach to operating system dynamics and diagnosis. It is shown that consideration of the space-time-life

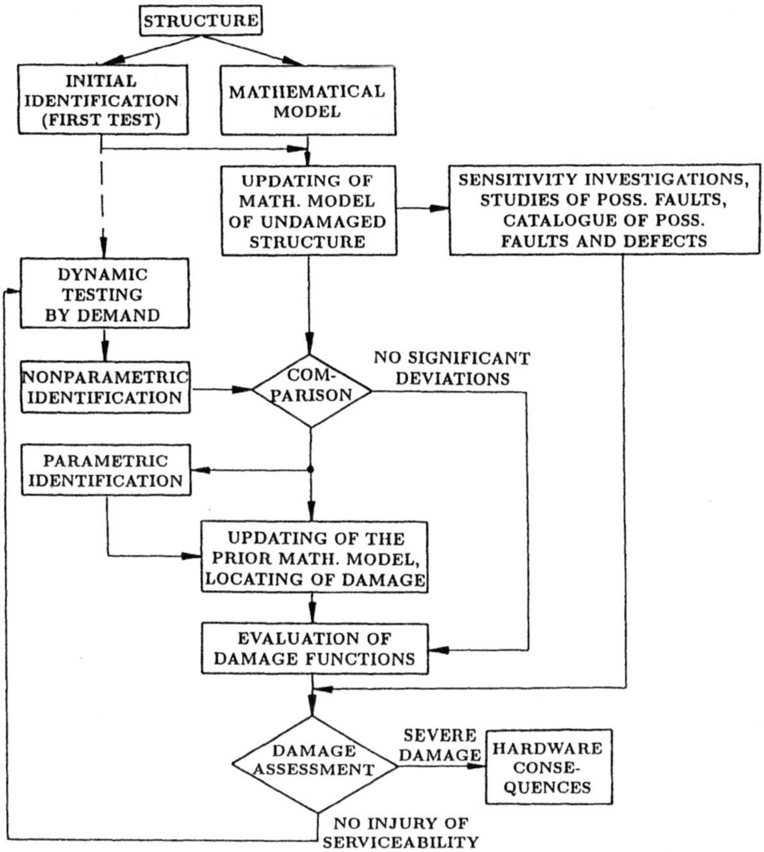

Figure 7 A procedure for structural damage evaluation

behaviour is advantageous and that, additionally, a new quantity should be introduced: *the damage development D* as a global measure, and also the local characteristics of damage. These damage characteristics are energy based. They permit an empirical validation, for example, for the Palmgren-Miner rule. Starting from these the holistic modelling of system dynamics (space-time-lifetime r- t-Θ) was proposed for continuous models as well as for discrete models. On this basis of the energy flow of a system the local method was proposed for the detection of areas of high probability of damage and the evolution of local damage measures.
Because the direct measurement of damage development seems to be imposssible, the general theory of symptom evolution is presented, valid for one operating unit, and for a set of them. It is shown that symptom reliability is governed by Weibull or Fréchet distributions, and can be used further for system residual life or for the potential assessment of a given unit in operation.

Diagnosis of the state and condition of systems in operation is described as two-fold, symptom-based, as in the case of machinery condition monitoring and, it is model-based when the best available knowledge base (a validated mathematicl model) is available. In the former case, if we have proper symptoms of the global damage we can assess the residual time to breakdown or the damage potential. Additionally, a new damage index was proposed in order to facilitate the choice of a proper damage distribution function.

8 References

Cempel, C., Natke, H.G., 1990: Damage measures and evolution in machines and structures; **Curt-Risch-Institute, University of Hannover**, Report CRI-B-1/1990

Cempel, C., 1991: Damage initiation and evolution in operating mechanical systems (Part 1,5 of book in progress "**Model-based fault diagnosis in mechanical systems**", co-authors: Natke, H.G. Ben-Haim, Y.)

Juvinal, R.C., 1967: **Engineering consideration of stress, strain and strength**; Mc Graw Hill Book Co., New York, Chapter 11

Newland, D.E., 1989: **Mechanical vibration analysis and computation**; Longman, London, Chapter 12

Natke, H.G., Yao, J.P.T., 1988: System identification approaches in structural safety, in**Structural Safety Evolution Based on System Identification Approaches**, Natke, H.G., Yao, J.P.T. Editors, Vieweg, Braunschweig.

Cempel, C., 1990: Generalized Trivibroacoustical Model of the Machine in Plant Diagnostic Environment; **Bulletin of the Polish Academy of Sciences**, Technical Sciences, Vol 36, No 1-12, 39-49

Winiwarter, P, Cempel, C., 1992: Life Symptoms, Open Energy Transformation System with Limited Internal Structuring Capacity, sent to: **System Research**

Cempel, C., Natke, H.G. Ziółkowski, H., 1990: Symptoms for Damage Detection in Mechanical Systems; **Curt-Risch-Institute, University of Hannover**, Report RRI-B-2/1990

Cawley, P., Adams, R.D.: 1988 The Mechanics of the Coin-Tap-Method of Non-destructive Testing; **Journal of Sound and Vibration**, 122,(2), 299-316.

Vendiver, J.K.: 1977 Detection of structural failure on fixed platforms by measurement of dynamic response; **Journal of Petroleum Technology**,March 1977, 305-310

Brownjohn, J.M.W., 1988: Assessment of Structural Integrity by dynamic response; **Ph. D Thesis Dept. Civil Engrg**, Bristol Univ. April 1988

Norton, S.J., Testardi, L.R., 1986: Reconstruction of one-dimensional inhomogeneities in elastic modulus and density using acoustic dimensional resonances. **J.Acoust.Soc.Am.** 70(4) April 1986

Armon, D., Ben-Haim, Y., Braun, S., 1991: Crack detection in beams by rank ordering of eigenfrequency shifts. **Safeprocess '91, IFAC/IMACS, Symposium on Fault Detection, Supervision and Safety for Technical Processes, Baden-Baden, Germany**, 79-84

Natke, H.G.,1992: Einführung in Theorie und Praxis der Zeitreihen- und Modalanalyse - Identifikation schwingungsfähiger elastomechanischer Systeme, Vieweg Braunschweig, Wiesbaden, 3rd ed.

Natke, H.G., 1989: Identification Approaches in Damage Detection and Diagnosis, IX. Szola, Diagnostics '89, Rydzyna, 27.8.-1.9.1989, 99-110

Natke, H.G., Yao, J.T.P., 1990: System Identification Methods for Fault Detection and Diagnosis; in: Structural Safety and Reliability (Eds. H.-S. Ang et al.), Proc. of ICOSSAR '89, ASCE New York, 1387-1393

Yao, J.T.P., Natke, H.G., 1991: Computer and Test Aided Modelling and its Application to Identify Structural Damage, in: Computational Stochastic Mechanics, Eds. P.D. Spanos et al., Comp. Mech. Publ. and Elsevier Appl. Scie. London, New York, 149-155

Yao, J.T.P., Natke, H.G., 1992: Reliability of Structures with Active Control, Proc. of FUZZ-IEEE Internat. Conf. on Fuzzy Systems, San Diego CA, USA, March 8-12, 1227-1234

Time-variant Methods in Machinery Diagnostics

W.J. Staszewski G.R. Tomlinson

University of Manchester
Department of Engineering
Manchester, U.K.

1 Introduction

Many signal processing algorithms have been developed recently for mechanical fault detection in machinery diagnostics. Most of those are based on the assumption of stationarity but in many cases fault characteristics appear in non-stationary form e.g. impact excitation. Thus strong non-stationary events can appear in a local time period, eg. one revolution. The analysis of non-stationary signals calls for specific techniques which go beyond the classical Fourier approach. The past ten years have seen major developments in the area of time-variant analysis. Unfortunately, machinery diagnostics have not yet received major benefits from these developments and little attention has been paid to time-dependent methods. These developments can be classified into three major groups [1, 2] : time-dependent models, time-frequency distributions and time-scale methods as indicated in Figure 1.

This paper presents three different time-variant methods currently used by the authors in machinery diagnostics. These are: the moving window procedure, the Wigner-Ville distribution and the wavelet transform. Detailed mathematical discussion is avoided and the reader is directed to the relevant references for more rigorous analysis. In this paper the attention is focused on applications. The structure of the paper is as follows. Section 2 outlines the concept of the moving window procedure, Section 3 describes the Wigner-Ville distribution and Section 4 deals with the wavelet transform. These methods are validated in each case by a series of simulated examples. Finally the methods are applied to experimental data in the form of vibration records taken from a controlled fault (a broken tooth) in a pair of meshing spur gears.

2 The Moving Window Procedure

One solution to the problem of representing non-stationary signals can be given by the concept of time-frequency adaptive methods. A non-stationary signal can be considered as the sum of consecutive quasi-stationary segments for which stationary, classical Fourier-based spectral methods can be applied. As the time blocks are reduced, non-stationary events can be obtained. The most intuitive approach is to apply a moving window $w(t-\tau)$ to the signal $x(t)$, where the time position of the window is given by τ. Then the Fourier transform is applied to the signal. This method is referred to as the short time Fourier transform and is defined as [3]:

$$S(t;\tau) = \int_{-\infty}^{+\infty} x(t)w(t-\tau)e^{-j2\pi ft}dt. \tag{2.1}$$

In practice the squared magnitude of $s(t;\tau)$ called the spectrogram is often used,

$$S_x(t;\tau) = |S(t;\tau)|^2. \tag{2.2}$$

Limitations of this method arise from the Heisenberg uncertainty principle: a good frequency resolution is achieved by means of a large window. This results in poor time resolution. Conversely, a good time resolution implies a small window, which results in a poor frequency resolution.

Several modifications have been proposed for improving the short time Fourier transform method [1]. One method of particular interest is referred to as the moving window procedure [4], which is based on the assumption that local faults in machinery produce impulses in the vibration signals. These signals can be modelled as,

$$x(t) = h_b(t) * \sum_{m=-\infty}^{+\infty} \delta(t - mT) + n(t), \tag{2.3}$$

where $h_b(t)$ is the impulse response function of the system response (bearing, gearbox, shaft, etc.), $n(t)$ is the noise, $\delta(t)$ is the Dirac delta function and T gives the periodicity of the impulses. When the short time window is applied to the signal, the energy spectrum is given by

$$X_x(f;\tau) = \frac{1}{D}\left|\int_{-\infty}^{+\infty} x(t)w(t-\tau)e^{-j2\pi ft}dt\right|^2, \tag{2.4}$$

where $\tau = nD$ $(n = 0, 1, 2, ..., N)$ and D is the width of the window. In contrast to the short time Fourier transform (equation (2.1)) the discrete movement of the window $w(t)$ is applied. The basic idea comes from the nature of the high-frequency impulse produced by the fault. If the width D of the window is comparable to the duration of the impulse, the amplitude level of the energy spectrum (equation (2.4)) is magnified when the impulse exists in the signal. This basic idea is illustrated in Figure 2, where two different positions of the window, correlated and uncorrelated with impulses, produce two different broadband spectra. The energy spectrum associated with the method can be expressed in terms of a chosen frequency bandwidth Ω

$$X_B(k;\tau) = \int_{k\Omega}^{(k+1)\Omega} X_x(f;\tau)df, \tag{2.5}$$

where $k = 1, 2, ..., K$, $K = \Omega_{max}/\Omega$, and Ω_{max} is the maximum frequency of interest. The periodicity of the pulse train $1/T$ can then be found by a frequency analysis of the movement of the window $w(t)$. This is obtained by taking the Fourier transform of the signal described by equation (2.5), giving

$$M_x(k;\xi) = \frac{1}{N}\left|\int_{-\infty}^{+\infty} X_B(k;\tau)e^{-j2\pi\xi\tau}d\tau\right|^2 \tag{2.6}$$

for $k = 1, 2, ..., K$.

Finally the series of equations which form the moving window procedure can be summarized as

$$M_x(k;\xi) = \frac{1}{N}\left|\int_{-\infty}^{+\infty}\left(\int_{k\Omega}^{(k+1)\Omega}\frac{1}{D}\left|\int_{-\infty}^{+\infty} x(t)w(t-\tau)e^{-j2\pi ft}dt\right|^2 df\right)\times\right.$$

$$\times \quad e^{-j2\pi\xi\tau}d\tau\Big|^2, \tag{2.7}$$

where $k = 1, 2, ..., K$, $K = \Omega_{max}/\Omega$, $N = T/D$, Ω_{max} is a maximum frequency of the signal, Ω is a chosen bandwidth of the analysis, N is a number of window positions, T is a number of signal samples and D is a width of the window.

More details about the moving window procedure can be found in [4], where the method was applied to the detection of faults in ball-bearings. In what follows an application of the moving window procedure to the detection of a broken tooth in a spur gear is presented. This is preceded by a validation example using simulated signals.

Applications

Equation (2.7) given in the above section was coded in C, implemented on an HP-9000 series computer and validated by several different simulated signals. An example is given in Figure 3. The series of impulses consisting of two cycles of a cosine wave at 64 Hz, having a cosinusoidal envelope, is analysed. The periodicity of the impulses was equal to 5.4 Hz. Gaussian white noise ($\bar{x} = 0$ and $\sigma_x = 1.0$) was added to the signal. The total number of narrowband spectra is equal to 8 ($K = 8$ in equation (2.7)). The impulses are clearly detected by the moving window procedure by the series of harmonics 1×5.4 Hz, 2×5.4 Hz and 3×5.4 Hz for $k = 1, 2, 3$ (see Figure 3D). This pattern is not observed when the impulses do not exist in the signal (pure Gaussian noise - Figure 3C). Figure 4 shows a similar example but for the lower signal-to-noise ratio. The impulses can be still detected by the harmonics 1×5.4 Hz and 2×5.4 Hz for $k = 2$ (see Figure 4D).

Similar results can be obtained when the method is applied to experimental spur gear vibration data. An input gear with 24 teeth was driven by an electric motor and meshed with 16 teeth of a pinion. A broken tooth fault was simulated by the removal of the facewidth at a given radius of one pinion tooth. Rotational frequency of the pinion was equal to 37.5 Hz which resulted in a meshing frequency equal to 600 Hz. The load applied to the system was equal to 70% of the maximum load. Low-pass filtered vibration signals were sampled with a sampling rate equal to 6.4 kHz. Figure 5 shows one pinion rotation of the acceleration response measured on a pinion bearing housing and the power spectra for different gearbox conditions. Here small advancement of fault means the fault which cannot be detected by any classical Fourier based procedure (compare Figure 5B with Figure 5F). In the case of the advanced fault a pattern of sidebands around the meshing harmonics can be observed in the spectrum (Figure 5D). Figure 6 shows the moving window procedure results for the above signals. The Hamming window with the width equal to 32 samples was used. The window size was comparable to the duration of the impulse produced by the fault. The selected frequency bandwidth Ω was chosen as equal to 200 Hz, which produced the maximum possible total number of narrowband spectra to be 16 ($K = 16$ in equation (2.7)). In all narrowband spectra one predominant component equal to 50 Hz (frequency of the mains power supply) is visible. Harmonics 1×37.5 Hz and 2×37.5 Hz can also be seen in Figure 6B for

$k = 11, 12$ and in Figure 6C for $k = 2$. This predominant component is a symptom of the fault and cannot be observed in Figure 6A.

In summary a broken tooth fault in a spur gear produced an impulse which was detected by narrowband spectra of the moving window procedure. The fault was detected even in the case of the small advancement which was not possible by means of the spectral Fourier analysis.

3 The Wigner-Ville Distribution

The fundamental idea of the time-frequency evolutive methods assumes that the spectral content of a signal is changing in time. This leads to the general time varying spectrum. There exist many approaches to the problem. A unified approach to the time-frequency distribution was given by Cohen [5] and can be defined as

$$C(t;\omega;\phi) = \frac{1}{2\pi}\int\int\int e^{-j(\xi t-\tau\omega-\xi u)}\phi(\xi;\tau)x\left(u+\frac{\tau}{2}\right) \times$$

$$\times\ x^*\left(u-\frac{\tau}{2}\right) du d\tau d\xi, \tag{3.8}$$

where $x(u)$ is the signal and $\phi(\xi;\tau)$ is the kernel function which represents the particular distribution. There exist a lot of different time-frequency distributions represented by the ϕ function. An excellent review is given in [6]. More details can be found also in [2, 7, 8].

If one chooses for the kernel $\phi(\xi;\tau) = 1$, equation (3.8) leads to the Wigner distribution defined as [9]

$$W(t;f) = \int_{-\infty}^{+\infty} x\left(t+\frac{\tau}{2}\right) x^*\left(t-\frac{\tau}{2}\right) e^{-j2\pi f t} d\tau. \tag{3.9}$$

In the frequency domain it can be represented as [9]

$$W(t;f) = \frac{1}{2\pi}\int_{-\infty}^{+\infty} X\left(f+\frac{\theta}{2}\right) X^*\left(f-\frac{\theta}{2}\right) e^{j2\pi f t} d\theta. \tag{3.10}$$

In many cases instead of the signal $x(t)$, its analytic representation is taken,

$$z(t) = x(t) + j\,\hat{x}(t), \tag{3.11}$$

where $\hat{x}(t)$ is the Hilbert transform of the signal $x(t)$. This eliminates interference between positive and negative frequencies in the Wigner distribution which, when the analytic signal is used, is very often called the Wigner-Ville distribution (WVD). There is an extensive list of requirements that should be fulfilled by a time-frequency distribution. These requirements include theoretical and practical properties and are given in the literature together with the theoretical background of the subject, e.g [6, 7, 8, 9, 10, 11]. Most of these properties are derived from the spectral density function by adding time dependence. It has been proved that the WVD is the distribution which fulfils the greatest number of theoretical and practical properties [6].

The WVD has found a lot of successful applications in different areas such as for example acoustics, speech processing, detection and estimation problems, image processing and pattern recognition, biosignal processing, seismology, etc. The first significant contribution in condition monitoring was given by Forrester [12] who applied the method to the study of vibration helicopter gear data. Other applications in this area include [13, 14, 15, 16, 17, 18, 19]. In what follows a few examples are presented. The most important properties and a practical application of the method to the detection of a broken tooth in a spur gear are given.

Applications

The calculation procedure of the WVD distribution is evaluated from equation (3.9) using the FFT algorithm. The analytic signal is calculated using the Hilbert transform also by means of the FFT procedure. Since the WVD is always real-valued, the real part of the WVD is displayed as a contour plot using 25 contour lines. The algorithm was coded in C and implemented on an HP-9000 series computer. Computer simulated examples were used to validate the method and to show the properties of the WVD. The examples are presented below.

The WVD is a bilinear distribution. There exist interference effects between components in the frequency domain due to the bilinearity of the definition (equation (3.9)). Figure 7B shows an example of the WVD of the signal consisting of the sum of two cosine waves (Figure 7A). Two dark solid lines correspond to cosine harmonics. Additionally a dashed line can be observed. This arises from an interference between the given harmonics. Such interferences can be removed by using a so called pseudo Wigner-Ville distribution which is an appropriately smoothed version of the WVD [6]. In general, smoothing is a very time consuming computer procedure and has some limitations, e.g. it causes loss of phase information [20]. Instead of using a smoothing procedure Choi and Williams [21] proposed a distribution which reduces to a large extent the cross terms for multicomponent signals. However this distribution is also very time consuming and exhibits poor energy concentration.

The WVD is well equiped to cope with nonstationary signals. As an example the impulse signal consisting of two cycles of a cosine wave (32 Hz) having a cosinusoidal envelope, is analysed (Figure 7C). A broad band indicated by "X" can be seen in Figure 7D. This corresponds to the impulse signal. A similar pattern with vertical stripes occurs at the same frequency (32 Hz) but 180° earlier. The pattern is produced by the time domain interference between the orginal band and its neighbour in the following rotation since in order to obtain an information about one rotation of the signal two rotations are used in the analysis. . This is due to the fact that the discrete version of the WVD is periodic with π instead of 2π.

Figure 7F shows the WVD result of the signal consisting of a cosine wave (32 Hz) and the impulse described above (Figure 7E). A vertical dark solid line at 32 Hz corresponds to the cosine wave. The horizontal broad band and its interference pattern similar to the pattern obtained in the above example can also be seen. In addition a pattern of curved lines can be observed due to the interference between the harmonic wave and the impulse.

The WVD is a highly nonlocal distribution. It reflects properties of the signal that the signal has at other times. As an example if there is noise in the signal for a small finite time of the signal that noise will appear at other times. That is why the WVD is often called "the noisy distribution". Figure 7G shows the signal consisting of two cosine waves and Gaussian noise. The WVD of the signal is presented in Figure 7H. The result can be compared with Figure 7B where the noise does not exist in the signal. This example shows that any noise in the signal makes the analysis much more difficult regarding interpretation.

The above examples exhibit typical patterns which might occur in the WVD analysis of gearbox vibration data. Gearbox vibration signals described in section 2 were used for the WVD analysis. Noise was minimised in the signal by using a time domain averaging procedure. The signal was additionally low-pass filtered with the cut-off frequency equal to 1.6 kHz. The frequency resolution of the analysis was equal to 6.25 Hz. The analytic signal was used in the analysis. Figures 8 shows the WVD results. Two solid lines due to the meshing harmonics (600 Hz and 1200 Hz) can be seen in the contour plots. A lot of other vertical dashed lines can be observed due to interferences (Figure 8A). Figure 8B exhibits a typical pattern of the horizontal broad band and curved lines due to the impulse caused by a broken tooth. If one compares Figure 8B with Figure 7F strong similarities can be observed. In the case of the small advancement of fault curved lines start to appear on the left side of the dark solid line (Figure 7C).

In summary a broken tooth fault in a spur gear was detected by visual observation of the WVD contour plots. Horizontal broad bands and curved lines in contour plots of the WVD were the symptom features of the fault.

4 The Wavelet Transform

The main interest of a time-frequency approach to the study of non-stationary signals is made on time variations of the spectral characteristics. A different approach can be given if non-stationary signals are considered as a superposition of a number of components which are more or less localized. Thus the representation of the vibration signal $x(t)$ can be given in a discrete form expression [1]

$$x(t) = \sum_{n=-\infty}^{+\infty} \sum_{k=-\infty}^{+\infty} c_{nk} x_{nk}(t), \tag{4.12}$$

or in the continuous form as

$$x(t) = \int_{-\infty}^{+\infty} \int_{-\infty}^{+\infty} g(t;\nu)\psi(t;\tau;\nu)d\tau d\nu. \tag{4.13}$$

In both cases the decomposition is based on a priori chosen functions $x_{nk}(t)$ or $g(t;\nu)$. In this approach ν or k stands for a scale parameter. Thus instead of a time-frequency representation we deal with a time-scale representation. The first work in this area goes back to Gabor [22] and Hoelström [23] who used shifted versions of Gaussian discrete and continuous functions respectively in the time and frequency domain. If time/frequency shift is replaced by a dilation or compression of scale, the time-scale decomposition leads directly to the wavelet transform defined as [24]

$$W(a;b) = \frac{1}{\sqrt{a}} \int_{-\infty}^{+\infty} x(t) g\left(\frac{t-b}{a}\right) dt, \tag{4.14}$$

where $g(t)$ is a function called the analysing wavelet, b is the parameter of translation in time and a is a dilation. In terms of the Fourier transform the wavelet transform can be written as [24]

$$W(a;b) = \sqrt{a} \int_{-\infty}^{+\infty} X(\omega) G(a\omega) e^{j\omega b} d\omega. \tag{4.15}$$

There are a number of different complex and real-valued functions used as analysing wavelets. The minimal requirements for the analysing wavelet g(t) are [25]:

1. admissibility condition

$$\int_{-\infty}^{+\infty} \frac{|G(\omega)|^2}{|\omega|} d\omega < \infty \tag{4.16}$$

2. finite energy

$$\int_{-\infty}^{+\infty} |g(t)|^2 dt < \infty. \tag{4.17}$$

In practice the smoothness of g and G is also required. In many cases a so called progressive wavelet function is used being a complex-valued function that satisfies the admissibility condition and does not have any negative frequencies. Equations (4.14) and (4.15) allow us to interpret the wavelet transform as a filter of constant relative frequency resolution $\Delta f/f$. This transform was originated by Morlet [26] in 1980 and it has been evolving rapidly since that time. The theoretical background of the transform can be found in [25, 27, 28, 29]. The wavelet transform has found many applications in different areas, such as sound processing, seismology, image processing, study of fractals, biosignal analysis, fluid mechanics, etc. As far as the authors are aware, it has not been applied as a fault detection procedure in machinery diagnostics. In what follows some properties of the wavelet transform are validated and pointed out using simulated examples followed by the gearbox data analysis.

Applications

The wavelet transform was performed by generating the family of all dilated and rotated wavelets in the frequency domain and multiplying $G(a\omega)$ and $X(f)$ before transforming the results back to the time domain using the FFT algorithm. The Morlet wavelet function (Figure 9) defined as [25]

$$g(t) = e^{-j\omega_o t} e^{-|t|^2/2)}, \tag{4.18}$$

where $\omega_o = 1.75\pi$ was chosen as an analysing wavelet since it is well adopted to the problem of locating abrupt changes in a signal [25, 30, 31]. The Morlet analysing wavelet function is progressive for chosen value of ω_o [25]. Thus the wavelet transform is complex-valued and is presented as a contour plot of its modulus (25 contour lines) and its phase (5 contour lines). In order to increase the visibility of the phase picture, if the modulus of the wavelet transform is less than 3% of the maximum value, the phase value is not represented on the picture (it is set to zero). For the same reason a smaller number of contour lines of the phase is used in the analysis. Logarithmic scale axis representation was chosen to enhance the small-scale coefficients. In order to validate the algorithm and show some useful properties of the wavelet transform, simulated examples were used with the following properties, the sampling frequency of the simulated signals was equal to 512 Hz, the sampling frequency of the analyzing wavelet function was equal to 87.5 Hz and the minimum value of the scale parameter was chosen to be 0.04. The values were sufficient to avoid numerical errors and aliasing [32]. Figure 10 shows the simulated signals chosen for the analysis. These are: sum of two sine waves, impulse (Dirac) delta function, sum of two sine waves and an impulse and the signal exhibiting the sharp transition between frequencies (a chirp).

Figures 11A,B show the wavelet transform of the sum of two sine waves (8.25 Hz and 27.5 Hz). Two harmonics are represented by two horizontal dark bands in the modulus and two horizontal bands with vertical parallel stripes in the phase. The values of the phase change from 0 to 2π. When the phase reaches 2π, it is wrapped around to the value 0. Thus these vertical stripes where the values of the phase drop abruptly to zero are clearly visible and play an important role in the interpretation.

The wavelet transform of the impulse signal is shown in Figures 11C,D. The increase of the modulus around the point where the impulse exists and the convergence of the straight lines of the constant phase towards the same point can be clearly seen. This is contrasted with parallel phase lines in Figure 11B. This example shows that the wavelet transform can detect and localize short-lived high-frequency components.

The wavelet transform of the sum of two sine waves and an impulse can be seen in Figures 11E,F. The pattern is a superposition of the results obtained in the two previous examples. Again the impulse is detected by a peak value of the modulus and the convergence of lines of constant phase.

Figures 12A,B show the wavelet transform of the chirp signal (17.5 Hz and 32 Hz). Both frequencies are represented by horizontal dark bands in the modulus. The scale transition in the modulus can be observed. Similar transition and the convergence of the straight lines of the constant phase can also be seen. This example shows that the wavelet transform can detect a discontinuity in the signal.

Figures 12C,D shows the wavelet transform of the sum of two sine waves with Gaussian noise. Two horizontal dark bands representing the harmonics can still be observed in the modulus but the phase function is more difficult to interpret (compare Figures 12C,D with Figures 11A,B). This shows that any noise in the signal can make the analysis much more difficult for the interpretation.

The same analysis was applied to gearbox vibration data. Again the same vibration signals discussed in section 2 were used. In order to minimise the noise, a time domain averaging procedure was employed. The sampling frequency of the signal was equal to 6.4 kHz and the sampling frequency of the analyzing wavelet function was equal to 35 Hz. The minimum value of the scale parameter was chosen to be 0.1. All values were sufficient to avoid numerical errors and aliasing [32]. Figures 13A,B show the wavelet transform of "the normal condition" signal, ie. no fault condition. Two horizontal bands in the modulus and two horizontal bands with parallel vertical stripes representing two meshing harmonics (compare Figures 11A,B with Figures 13A,B) can be seen. Figures 13C,D and 13E,F exhibit impulse pattern characteristics similar to those features observed before in the simulated signals. The features (peak values in the modulus and the convergence of straight lines of the constant phase), indicated by the star, are the symptom of impulses caused by a broken tooth using the wavelet analysis.

5 Concluding Remarks

Three different time-variant methods have been presented. Most of the existing work in the area of time-variant fault detection methods is of an exploratory character, especially the wavelet transform which is a very recent mathematical tool and has not previously been used in machinery diagnostics. The theory of time-scale representations is evolving at a very fast pace. It is still too early to compare the effectiveness of these methods. The interpretation of the results seems to be one of the most important and difficult problems. The visual observation and information gathered from the pictures is sufficient to detect a problem of an advanced fault but sensitivity problems might occur if the advancement of a fault is small. However, even in this situation the broken tooth in a spur gear was detected by all three methods presented which was not possible by means of the classical Fourier analysis. One solution regarding interpretation of the results could employ classification methods based on pattern recognition, neural networks, etc. In summary the following points are highlighted.

The moving window procedure is simple to apply. It improves the signal-to-noise ratio and does not require any initial time domain averaging procedure. However the window size requires a priori knowledge of the impulses caused by the fault. Additionally, fault localization and fault separation is not possible.

The Wigner-Ville distribution reflects properties of the vibration signal which are not available by means of the moving window procedure. It gives the possibility of a fault localization in time. The distribution is always real-valued so it does not cause any representation problems common to other complex time-frequency distributions. However the Wigner-Ville distribution produces spurious effects which creates difficulties in the interpretation. It is also rather time consuming in terms of computer calculations and additionally requires initial time domain averaging.

The wavelet transform is less time consuming in terms of computer calculations. The choice of the optimal wavelet function to the given fault remains a problem. The procedure also requires initial time domain averaging. The result is represented by the complex function thus different forms of a presentation are possible.

The time-variant methods presented and applied certainly open some new possibilities in machinery diagnostics. However, in order to fully establish the usefulness and extract the maximum benefits from such methods further investigations are required.

Acknowledgment

The authors would like to acknowledge VSEL Barrow-in-Furness, Cumbria, both for the technical assistance and providing the experimental data used in the analysis.

References

[1] Flandrin P. Some aspects of non-stationary signal processing with emphasis on time-frequency and time-scale methods. In Combes J M, Grossmann A, and Tchamitchian Ph, editors, *Wavelets. Time-Frequency Methods and Phase Space.* Springer-Verlag, Berlin, 1989.

[2] Priestley M B. *Non-linear and Non-stationary Time Series Analysis.* Academic Press, London, 1988.

[3] Allen J B. Short term spectral analysis, synthesis, and modification by discrete fourier transform. *IEEE Trans. on Acoust., Speech and Signal Proc.*, ASSP-25(3):235–238, 1977.

[4] Kim Y H, Lim B D, and Cheoung W S. Fault detection in a ball bearing system using a moving window. *Mechanical Systems and Signal Processing*, 5(6):461–473, 1991.

[5] Cohen L. Generalized phase-space distribution functions. *J. Math. Phys.*, 7(5):781–786, 1966.

[6] Cohen L. Time-frequency distributions - a review. *Proceedings of the IEEE*, 77(7):941–981, 1989.

[7] Boashash B. Time-frequency signal analysis. In S Haykin, editor, *Advances in Spectrum Analysis and Array Processing. vol I.*, Advanced Reference Series. Engineering. Prentice Hall, Englewood Cliffs, New Jersey, 1991.

[8] Boashash B, editor. *Time-Frequency Signal Analysis. Methods and Applications.* Longman Cheshire, Melbourne, 1992.

[9] Claasen T A C M and Mecklenbrauker W F G. The wigner distribution - a tool for time-frequency analysis. part 1: Continuous time signals. *Philips J. Res.*, 35(3):217–250, 1980.

[10] Claasen T A C M and Mecklenbrauker W F G. The wigner distribution - a tool for time-frequency analysis. part 2: Discret time signals. *Philips J. Res.*, 35(4/5):276–300, 1980.

[11] Claasen T A C M and Mecklenbrauker W F G. The wigner distribution - a tool for time-frequency analysis. part 3: Relations with other time-frequency transformations. *Philips J. Res.*, 35(6):372–389, 1980.

[12] Forrester B D. Use of the wigner-ville distribution in helicopter fault detection. In *Proc. ASSPA 89*, pages 77–82, 1989.

[13] Forrester B D. Analysis of gear vibration in the time-frequency domain. In *Proceedings of the 44th Meeting of the Mechanical Failures Prevention Group of the Vibration Institute*, pages 225–239, Virgina Beach, Virginia, 1990.

[14] Forrester B D. Time-frequency domain analysis of helicopter transmission vibration. Technical Report 180, Aeronautical Research Laboratory, ARL Propulsion, Melbourne, 1990.

[15] Forrester B D. Time-frequency analysis in machine fault detection. In Boashash B, editor, *Time-Frequency Signal Analysis.* Longman Cheshire, Melbourne, 1992.

[16] McFadden P D. Time-frequency domain analysis of vibration signals for machinery diagnostics: (1) introduction to the wigner-ville distribution. Technical Report Report No OUEL 1859/90, University of Oxford. Department of Engineering Science., Oxford, 1990.

[17] McFadden P D. Time-frequency domain analysis of vibration signals for machinery diagnostics: (2) the weighted wigner-ville distribution. Technical Report Report No OUEL 1891/91, University of Oxford. Department of Engineering Science., Oxford, 1991.

[18] Meng Q and Qu L. Rotating machinery fault diagnosis using wigner distribution. *Mechanical Systems and Signal Processing*, 5(3):155–166, 1991.

[19] McFadden P D and Wang W J. Analysis of gear vibration signatures by the weighted wigner-ville distribution. In *Proceedings of the Fifth International Conference on Vibrations in Rotating Machinery.* University of Bath, IMechE, September 1992.

[20] Garudadri H, Beddoes M P, Benguerel A P, and Gilbert J H V. On computing the smoothed wigner distribution. In *Proc of IEEE ICASSP 87*, pages 1521–1524, 1987.

[21] Choi H I and Williams W J. Improved time-frequency representation of multicomponent signals using exponential kernels. *IEEE Trans. on Acoust., Speech and Signal Proc.*, ASSP-37(6):862–871, 1989.

[22] Gabor D. Theory of communication. *J. Inst. Electr. Eng.*, 93(3):429–457, 1946.

[23] Helstrom C W. An expansion of a signal in gaussian elementary signals. *IEEE Trans. on Inf. Theory*, IT-12:81–82, 1966.

[24] Grossmann A, Kronland-Martinet R, and Morlet J. Reading and understanding continuous wavelet transform. In Combes J M, Grossmann A, and Tchamitchian Ph, editors, *Wavelets. Time-Frequency Methods and Phase Space*. Springer-Verlag, Berlin, 1989.

[25] Kronland-Martinet R and Grossmann A. Application of time-frequency and time-scale methods (wavelet transform) to the analysis, synthesis and transformation of natural sounds. In Piccialli A Deppoli G and Roads C, editors, *Representation of Musical Signals*. The MIT Press, Cambridge, Massachusetts, 1991.

[26] Morlet J. Sampling theory and wave propagation. In *Proc. 51st Ann. Meeting Soc. Explor. Geophys.*, Los Angeles, 1980.

[27] Chui Ch K. *An Introduction to Wavelets*. Wavelets Analysis and its Applications. vol 1. Academic Press, Inc., Boston, 1992.

[28] Chui Ch K, editor. *Wavelets: A Tutorial in Theory and Applications*. Wavelets Analysis and its Applications. vol 2. Academic Press, Inc., Boston, 1992.

[29] Daubechies I. The wavelet transform - a method for time-frequency localization. In Haykin S, editor, *Advances in Spectrum Analysis and Array Processing. vol I.*, Advanced Reference Series. Engineering. Prentice Hall, Englewood Cliffs, New Jersey, 1991.

[30] Kronland-Martinet R and Morlet J. Analysis of sound through wavelet transforms. *International Journal of Pattern Recognition and Artificial Intelligence*, 1(2):273–302, 1987.

[31] Tuteur F B. Wavelet transformations in signal detection. In Combes J M, Grossmann A, and Tchamitchian Ph, editors, *Wavelets. Time-Frequency Methods and Phase Space*. Springer-Verlag, Berlin, 1989.

[32] Staszewski W J and Tomlinson G R. Application of the wavelet transform to fault detection in a spur gear. In preparation.

Nonlinear Mechanical Systems, Sinusoidal Excitation Signal and Modal Analysis: Theory

L. Jezequel
Ecole Centrale de Lyon
Ecully, France

C.-H. Lamarque
ENTPE
Vaulx-en-Velin, France

1 Introduction

In this study, we shall show that this normal transformation is naturally linked to the notion of nonlinear modes.The nonlinear modes appear as an extansion of the linear modes but, in opposition to the latter, they are associated with complex scheduling laws with periods dependant on the amplitudes. Rosenberg in [16], [17] was the first author to have studied them in the case of a system with two degrees of freedom. Moreover, in the context of a general study on the forced response of systems with several degrees of freedom, Szemplinska-Stupnicka [19], [20] proposed the generalization of the Ritz-Galerkin method, which consists of utilizing a natural normal mode dependant on the movement's amplitude. The author clearly showed that it allowed one to find with precision the resonance amplitudes. The mode shapes utilized corresponded to an approximation of the non- linear modes defined by Rosenberg calculated with the help of the harmonic balance principle. However, it has been shown in [19], [20] that it was not possible to use a superposition principle for the normal non-linear modes analogous to that often used in a linear context. This paper deals with the generalization of superposition technique to the nonlinear case. We consider multidegree-of-freedom mechanical systems with analytical nonlinearities. First, we introduce normal form theory: we recall what is the Jordan normal form of an analytical system of ordinary differential equations and how to compute this normal form and the associated normal transform. Then we show how to compute nonlinear normal modes in the case of forced vibrations of the system from the nonlinear modes obtained in the case of free vibrations. Here we study primary resonance as an example. By using the previous computations, the level of decoupling, the true physical orders of responses and forcings and by slightly improving the former theoretical presentation, we are able to build blocks of general masses, general nonlinear frequencies, and nonlinear modes.The amplitudes of the response look like a generalized superposition type. The numerical part of this work can be realized because of an algebraic nonlinear resonant equation derived when we look for approximated periodical solutions of the normalized system, and the normal transform achieved from the computations of the free case. In order to achieve the computations, we could use Maple computer algebra.

2 Normal Form Theory

We consider mechanical systems with analytical nonlinearities. The idea of the normal transform comes from Poincare [1]. It has been taken up by many authors (Birkhoff [2], Arnold [3](3), Moser [4], Iooss and Joseph [5] ...). It consists of carrying out a change in variable allowing one to pass from one system of ordinary differential equations with non-linearities to a simpler system, possibly linear. Brjuno [6], [7] studies the existence and stability of these normal forms and of analytical forms of ordinary differential equations.The normalization of order 1 corresponds to the diagonalization of the linear part, or at least to putting it into its Jordan form.

The analysis of the intrinsic values of the lower part permits a first study of the system's behavior, distinguishing between critical and non critical intrinsic values according to Hsu [10], [9]. We have shown in previous papers [11] [12] [13], [14] how to compute the normal form and the normal transform corresponding to the change of variables by using a induction procedure. We can too compute directly these two quantities [8], [15]. The main point is to choose before starting the computations the "right" linear part of the system we consider: we mean that detuning are useful, and yield to a special treatment.If we compute this transform in an approximated way, by using successive homogeneous changes of variables according to increasing degrees of nonlinearities, the procedure consists in using a well adapted basis of a product of identical spaces of polynomials. We obtain an linear system written with the help of such a basis. The kernel of the linear system gives resonant non linearities of the concerned order. We suppress these resonant terms and we can solve the linear system: so, we have the normal transform D_k at the concerned order k. It is determined except for an element of the kernel. We take it to be equal to zero. We do not describe more longer these computations here: they are described in [13]. Effective computations are achieved from order 2 to the order K choosen before computing.

3 Equations of the System

We extend the nonlinear superposition technique to the case of Jordan blocks of dimension higher than 3, in the study of the primary resonance. We consider the following general system:

$$M\frac{d^2X}{dt^2}+C\frac{dX}{dt}+KX=F(X) \tag{1}$$

and the associated harmonic forced problem:

$$M\frac{d^2X}{dt^2}+C\frac{dX}{dt}+KX=F(X)+\cos(\omega t) \tag{2}$$

wherein M, C, K are the mass matrix, damping matrix are stiffness matrix, F a nonlinear vector with polynomial components, and f the vector of amplitudes of the periodic forcing of frequency ω. We assume that we can write again the two last equations as:

$$\frac{dY}{dt}=AY+f(Y) \tag{3}$$

$$\frac{dY}{dt}=AY+f(Y)+\cos(\omega t)g \tag{4}$$

with a new variable Y.

4 Choice of a Linear Part

We transform the linear part A into its diagonal or more generally into its "approximated" associated Jordan form by using a detuning α and using one of the two following procedures: from the above Jordan form:

$$J(\Lambda) = \begin{pmatrix} J(\lambda_1,k_1) & 0 & \dots & 0 & \dots & 0 \\ \dots & \dots & \dots & \dots & \dots & \dots \\ 0 & \dots & J(\lambda_s,k_s) & 0 & \dots & 0 \\ \dots & \dots & \dots & \dots & \dots & \dots \\ 0 & \dots & 0 & J(\bar{\lambda_1},k_1) & \dots & 0 \\ \dots & \dots & \dots & \dots & \dots & \dots \\ 0 & \dots & \dots & 0 & J(\bar{\lambda_s},k_s) & \end{pmatrix} \tag{5}$$

we introduce either: a)

$$J(i\omega) = \begin{pmatrix} \mathcal{J} & 0 \\ 0 & \bar{\mathcal{J}} \end{pmatrix} \tag{6}$$

with:

$$\mathcal{J} = \begin{pmatrix} J(-i\omega, k_1+\ldots+k_l) & 0 & \dots & \dots & 0 \\ \dots & \dots & \dots & \dots & \dots \\ 0 & \dots & J(\lambda_{l+1},k_{l+1}) & \dots & 0 \\ \dots & \dots & \dots & \dots & \dots \\ 0 & \dots & \dots & 0 & J(\lambda_s,k_s) \end{pmatrix} \tag{7}$$

and after a choice we define the detuning by the equation:

$$\alpha = \mathcal{J}_{\surd} + \mathcal{I}_{\surd} \tag{8}$$

whith:

$$\mathcal{J}_{\surd} = \begin{pmatrix} \mathcal{P} & 0 & 0 & 0 \\ 0 & 0 & 0 & 0 \\ 0 & 0 & \bar{\mathcal{P}} & 0 \\ 0 & 0 & 0 & 0 \end{pmatrix} \tag{9}$$

$$\mathcal{I}_{\surd} = \begin{pmatrix} i\omega I_{k_1+\ldots+k_l} & 0 & 0 & 0 \\ 0 & 0 & 0 & 0 \\ 0 & 0 & -i\omega I_{k_1+\ldots+k_l} & 0 \\ 0 & 0 & 0 & 0 \end{pmatrix} \tag{10}$$

$$\mathcal{P} = \begin{pmatrix} J(\lambda_1,k_1) & \dots & 0 \\ \dots & \dots & \dots \\ 0 & \dots & J(\lambda_l,k_l) \end{pmatrix} \tag{11}$$

or we define the detuning by the procedure b): we consider:

$$J(i\omega) = \begin{pmatrix} \mathcal{J}_{\sqcup} & 0 \\ 0 & \bar{\mathcal{J}}_{\sqcup} \end{pmatrix} \tag{12}$$

with:

$$\mathcal{J}_{\sqcup} = \begin{pmatrix} J(-i\omega, k_1) & \dots & 0 \\ \dots & \dots & \dots \\ 0 & \dots & J(-i\omega, k_s) \end{pmatrix} \tag{13}$$

then we obtain the detuning matrix of small parameters and parameters of physical orders one:

$$\alpha = J(\Lambda) + \begin{pmatrix} i\omega I_{k_1+\dots+k_s} & \\ & -i\omega I_{k_1+\dots+k_s} \end{pmatrix} \tag{14}$$

We decide to start the computations from this second choice of detuning to generalize the nonlinear superposition technique. We use a linear change of variable in order to write:

$$\frac{dZ}{dt} = J(i\omega)Z + \alpha Z + NL(Z) \tag{15}$$

and:

$$\frac{dZ}{dt} = J(i\omega)Z + \alpha Z + NL(Z) + \frac{1}{2}(f_1 + f_2) \tag{16}$$

$$\frac{df_1}{dt} = -i\omega f_1 \tag{17}$$

$$\frac{df_2}{dt} = i\omega f_2 \tag{18}$$

wherein $f_1 = e^{-i\omega t}$, $f_2 = e^{i\omega t}$. We have nonlinear terms in $NL(Z)$ and also with αZ. F_0 contains the small amplitudes of the forcings. All the coefficients may be complex. We use another linear change of variable of the form:

$$Z = \begin{pmatrix} Z_1 \\ Z_2 \end{pmatrix} \tag{19}$$

$$J = \begin{pmatrix} J_1 & 0 \\ 0 & J_2 \end{pmatrix} \tag{20}$$

with J_1 and J_2 conjugates, and then:

$$Z_1^\star = \begin{pmatrix} Z_{1k} \\ f_1 \\ Z_{1n-k} \end{pmatrix} \tag{21}$$

$$J_1^\star = \begin{pmatrix} & l_1 & \\ J_{1k} & \dots & 0 \\ & l_k & \\ & -i\omega & \\ 0 & & J_{1n-k} \end{pmatrix} \tag{22}$$

wherein the $l_1, \ldots, l_k$ introduce the components of the forcing term. We separate two different behavior of the eigenvalues:

$$Z_1 = \begin{pmatrix} Z_{1k} \\ Z_{1n-k} \end{pmatrix} \quad \begin{matrix} \text{eigenvalue} - i\omega \\ \text{eigenvalue far from} - i\omega \end{matrix} \tag{23}$$

Now we can rewrite the starting system with the variable $Z^\star = (Z_1, f_1, Z_2, f_2)$ in the following form:

$$\frac{dZ_1^\star}{dt} = J_1^\star Z_1^\star + \alpha_1^\star Z_1^\star + NL_1^\star(Z^\star, f_1, f_2) \tag{24}$$

5 Normal Form Calculations

After the normalization up to order K of the corresponding free system:

$$\frac{dZ_1}{dt} = J_1 Z_1 + \alpha Z_1 + NL_1(Z) \tag{25}$$

we obtain in normal coordinates $U = (U_1, U_2)$ the system:

$$\frac{dU_1}{dt} = J_1 U_1 + \alpha_1 U_1 + R_2(U) + \ldots + R_K(U) \tag{26}$$

The normalization up to order K of the forced system leads to:

$$\frac{dU_1^\star}{dt} = J_1^\star U_1^\star + \alpha_1^\star U_1^\star + R_2(U^\star) + \ldots + R_K(U^\star) \tag{27}$$

We consider only the first k equations (these are the only non trivial ones) and we write them again:

$$\frac{dV_{1k}}{dt} = J_{1k} V_{1k} + \alpha_{1k} V_{1k} + R_{2k}(U) + \ldots + R_{Kk}(U) \tag{28}$$

for the free behavior and for the forced one we have:

$$\frac{dV_{1k}^\star}{dt} = J_{1k}^\star V_{1k}^\star + \alpha_{1k}^\star V_{1k}^\star + R_{2k}(U^\star) + \ldots + R_{Kk}(U^\star) \tag{29}$$

We denotes as follows:

$$J_{1k}^\star = \begin{pmatrix} J_{1k} & \begin{matrix} l_1 \\ \ldots \\ l_k \end{matrix} \end{pmatrix} \tag{30}$$

α_{1k} is defined by difference (see above). It is very easy to see that for $j = 2,,k$ we have:

$$R_{jk}^\star(U^\star) = Q_j(R_j(U + S_1 f_1 + S_2 f_2)) \tag{31}$$

with Q_j the projection onto the space of polynomials of U, f_1, f_2 such as each monomial: $U^M f_1^{m_1} f_2^{m_2}$ with M n-tuple of positive integers, m_1, m_2 positive integers too verify :

$$(M, m_1, m_2).(-i\omega, \dots, -i\omega, i\omega, \dots, \omega) + i\omega = 0 \tag{32}$$

N denotes the ordered eigenvalues of J different from $i\omega$ or $-i\omega$, $U + S_1 f_1 + S_2 f_2$ the change of variable bringing out a normalization ut to infinity of terms possessing a α. Now we introduce with V_{1k0} a complex vector:

$$V_{1k} = e^{-i\omega t} V_{1k0} \tag{33}$$

into the resonance equation which becomes:

$$(J_{1k} + i\omega I_k)V_{1k} + \alpha_k V_{1k} + f_1 L_{1k} + \sum_{j=2}^{K} Q_j(R_j(U + S_1 f_1 + S_2 f_2)) = 0 \tag{34}$$

and we write :

$$(J_{1k} + i\omega I_k)V_{1k0} + \alpha_k V_{1k0} + f_1(0) L_{1k} + \sum_{j=2}^{K} R^\star_{jk}(0) = 0 \tag{35}$$

with:

$$L_{1k} = \begin{pmatrix} l_1 \\ \dots \\ l_k \end{pmatrix} \tag{36}$$

The amplitude can be written in a similar way by using the projector P_j onto the space which splits the complete space when added to the projection space of Q_j: for a free behavior:

$$Z_{1k} = V_{1k} + T_{2k}(U) + \dots + T_{Kk}(U) \tag{37}$$

with the normal transform $T_{2k}, \dots, T_{Kk}$ and in forced behavior:

$$Z^\star_{1k} = V^\star_{1k} + f_1 S_{1k} + \sum_{j=2}^{K} T_{jk}(U^\star + S_1 f_1 + S_2 f_2) \tag{38}$$

From this we derive:

$$V_{1k} = -(J_{1k} + i\omega I_k + \alpha_k I_k)^{-1}(f_1 L_{1k} + \sum_{j=2}^{K} Q_j(R_j(U^\star + S_1 f_1 + S_2 f_2)) \tag{39}$$

then:

$$\begin{aligned} Z^\star_{1k} &= -(J_{1k} + i\omega I_k + \alpha_k I_k)^{-1} f_1 L_{1k} + S_{1k} f_1 - \\ &(J_{1k} + i\omega I_k + \alpha_k I_k)^{-1} \sum_{j=2}^{K} Q_j(R_j(U^\star + S_1 f_1 + S_2 f_2)) \\ &+ \sum_{j=2}^{K} P_j(T_{jk}(U^\star + S_1 f_1 + S_2 f_2)) \end{aligned} \tag{40}$$

wherein:

$$U^\star = \begin{pmatrix} U^\star_{1k} \\ f_1 \\ U^\star_{1n-k} \\ Z^\star_{2k} \\ f_2 \\ Z^\star_{2n-k} \end{pmatrix} \tag{41}$$

Above, we distinguish between analytical orders and physical order of each monomial (for example, in the study of primary resonance, normal coordinates will have amplitude of physical order 1 and harmonic excitation will be of physical order 2, twice are of analytical order 1). So when it will be useful we shall operate truncations because of physical orders and or analytical orders. This will be done without a priori hypothesis before starting the computations. Resonant terms in the equation of Z_{1k} exist which possess only components of U without f_1 or f_2. We use the equation obtained for the $2n$ equations after the normalization at order K

$$\frac{dU_1^\star}{dt} = J_1^\star U_1^\star + \alpha_1^\star U_1^\star + f_1 L_1^\star + {}^1R_2(U^\star, f_1, f_2) + \ldots + {}^1R_K(U^\star, f_1, f_2) \tag{42}$$

$$\frac{dU_2^\star}{dt} = J_2^\star U_2^\star + \alpha_2^\star U_2^\star + f_2 L_2^\star + {}^2R_2(U^\star, f_1, f_2) + \ldots + {}^2R_K(U^\star, f_1, f_2) \tag{43}$$

In order to look for periodic solutions, we set with $U_{10}^\star$ and $U_{20}^\star$ complex vectors:

$$U_1^\star = e^{-i\omega t} U_{10}^\star \tag{44}$$

$$U_2^\star = e^{i\omega t} U_{20}^\star \tag{45}$$

Remarks: if ${}^1R_j(Z)$ is resonant, let $Z_1^{M_1} Z_2^{M_2}$ be a resonant monomial of ${}^1R_j(Z)$ with the m-uples of integers M_1 and M_2; the resonance is defined for $N = (M_1, M_2)$ by the equation:

$$(\Lambda, (-i\omega, \ldots, -i\omega, i\omega, , i\omega)) + i\omega = 0 \tag{46}$$

so $Z_1^{M_1} Z_2^{M_2}$ is of main harmonic ω and the term:

$$\frac{Z_1^{M_1} Z_2^{M_2}}{-(\Lambda, (-i\omega, \ldots, -i\omega, i\omega, , i\omega)) + i\omega} \tag{47}$$

belongs to the normal transform $\mathcal{T}_1$ obtained up to a certain order and associated to the equations of eigenvalue $-i\omega$. In other words, as we consider only a primary resonance,

$$\frac{Z_1^{M_1} Z_2^{M_2}}{2i\omega} \tag{48}$$

belongs to $\mathcal{T}_1$. In fact, all the resonant terms have this property, so we have:

$$\frac{{}^1R_j(Z_2, Z_1)}{2i\omega} \tag{49}$$

belongs to$\mathcal{T}_1$. To summarize, we can assert that the main harmonic of $\mathcal{T}_1$ is:

$$\frac{{}^2R_j(Z_1, Z_2)}{2i\omega} \tag{50}$$

since ${}^1R_j(Z_1, Z_2)$ and ${}^2R_j(Z_1, Z_2)$ are conjugate. Back to the starting coordinates Y, we have:

$$Y = P \begin{pmatrix} Z_1 + \frac{1}{2i\omega}\sum_{j=2}^{K} {}^2R_j(Z) \\ Z_2 - \frac{1}{2i\omega}\sum_{j=2}^{K} {}^1R_j(Z) \end{pmatrix} \tag{51}$$

In the forced case, the terms with Z_1, Z_2 above will belong to the resonant terms: they do not possess either f_1 or f_2 but in the resonant equation, we find:

$$(J_1 + i\omega I_1 + \alpha_1 I_1)Z_1 + f_1 L_1 + \sum_{j=2}^{K} {}^1R_j(Z) + \mathcal{R}_1 \tag{52}$$

with $\mathcal{R}_1$ which denotes the resonant terms of main harmonic with f_1 or f_2 of degree $\leq K$, and a simpler expression when we consider primary resonance:

$$(J_1 + i\omega I_1 + \alpha_1 I_1)Z_1 + f_1 L_1 + \sum_{j=2}^{K} {}^1R_j(Z_1 + f_1 S_1, Z_2 + f_2 S_2) = 0 \tag{53}$$

We can use new coordinates of main harmonic which diagonalize the system:

$$\hat{Z}_1 = Z_1 + f_1 S_1 \tag{54}$$

$$\hat{Z}_2 = Z_2 + f_2 S_2 \tag{55}$$

With these coordinates, we obtain the following resonance equation:

$$(J_1 + i\omega I_1 + \alpha_1 I_1)\hat{Z}_1 + f_1(L_1 - (J_1 + i\omega I_1 + \alpha_1 I_1)S_1) + \sum_{j=2}^{K} {}^1R_j(\hat{Z}_1, \hat{Z}_2) = 0 \tag{56}$$

A half expression of the amplitude is (the other is its conjugate):

$$\hat{Z}_1 + \frac{1}{2i\omega}\sum_{j=2}^{K} {}^2R_j(\hat{Z}_1, \hat{Z}_2) \tag{57}$$

Back to Y coordinates, we have the complete equality:

$$Y = P \begin{pmatrix} \hat{Z}_1 + \frac{1}{2i\omega}\sum_{j=2}^{K} {}^2R_j(\hat{Z}_1, \hat{Z}_2) \\ \hat{Z}_2 - \frac{1}{2i\omega}\sum_{j=2}^{K} {}^1R_j(\hat{Z}_1, \hat{Z}_2) \end{pmatrix} \tag{58}$$

What we see in this case of primary resonance implies in the general case that it exists a finite number of complex matrices C_i and, by jointing conjugate terms, real matrices M_i such as the amplitude denoted by A_y for Y are written as:

$$A_y = \sum_i M_i(\hat{Z}_1, \hat{Z}_2)(C_i \begin{pmatrix} \hat{Z}_1 \\ \hat{Z}_2 \end{pmatrix} + \bar{C}_i \begin{pmatrix} \hat{Z}_1 \\ \hat{Z}_2 \end{pmatrix}) \tag{59}$$

with $M_i(\hat{Z}_1, \hat{Z}_2)$ real matrices which terms depend on polynomials of the components of $\hat{Z}_1, \hat{Z}_2$ .Then the following is the beginning of a generalization of the modal superposition technique to the nonlinear case. Up to an analytical order K we write:

$$A_Y = \sum_i M_i()(C_i \begin{pmatrix} b_1 \\ b_2 \end{pmatrix} + \bar{C}_i \begin{pmatrix} b_2 \\ b_1 \end{pmatrix}) \tag{60}$$

with:

$$b_1 = (J_1 + i\omega I_1 + \alpha_1 I_1)^{-1}[(J_1 + i\omega I_1 + \alpha_1 I_1)S_1 - L_1]f_1 \tag{61}$$

$$b_2 = \bar{b_1} = (J_2 - i\omega I_2 + \alpha_2 I_2)^{-1}[(J_2 - i\omega I_2 + \alpha_2 I_2)S_2 - L_2]f_2 \tag{62}$$

The next step deals with distinguishing the factorization with $M_i, C_i, \bar{C}_i$ in the free and forced case. It is better to study a particular case where g is of order K:

$$\begin{aligned} A_Y = & \sum_i [M_i(Z_1, Z_2) + \frac{\partial M_i(Z_1, Z_2)}{\partial Z_1, Z_2} \begin{pmatrix} S_1 f_1 \\ S_2 f_2 \end{pmatrix}](C_i \begin{pmatrix} Z_1 \\ Z_2 \end{pmatrix} + \bar{C}_i \begin{pmatrix} Z_2 \\ Z_1 \end{pmatrix}) \\ & + \sum_i M_i(Z_1, Z_2)(C_i \begin{pmatrix} S_1 f_1 \\ S_2 f_2 \end{pmatrix} + \bar{C}_i \begin{pmatrix} S_2 f_2 \\ S_1 f_1 \end{pmatrix} \end{aligned} \tag{63}$$

6 Normal Transform

Let us explain the transformation of:

$$\frac{d^2 X}{dt^2} + C_0 \frac{dX}{dt} + K_0 X = G_0(X) \tag{64}$$

into the equation :

$$\frac{dY}{dt} = AY + \mathcal{F}(Y) \tag{65}$$

We write first:

$$\frac{d}{dt}(\frac{dX}{dt}) = -C_0 \frac{dX}{dt} - K_0 X + G_0(X) \tag{66}$$

$$\frac{d}{dt}(X) = \frac{dX}{dt} \tag{67}$$

with the matrix and vector:

$$A_0 = \begin{pmatrix} -C_0 & -K_0 \\ I & 0 \end{pmatrix} \tag{68}$$

$$Y_0 = \begin{pmatrix} \frac{dX}{dt} \\ X \end{pmatrix} \tag{69}$$

Then we reach Y with:

$$Y = \begin{pmatrix} \frac{dx_1}{dt} \\ x_1 \\ \ldots \\ \ldots \frac{dx_n}{dt} \\ x_n \end{pmatrix} \tag{70}$$

and to the associated matrix A with a linear change of variables:

$$Y_0 = \mathcal{S}Y \tag{71}$$

So we obtain the two equations:

$$\frac{dY_0}{dt} = A_0 Y_0 + \mathcal{F}_0(Y_0) \tag{72}$$

$$\frac{dY}{dt} = \mathcal{S}^{-1} A_0 \mathcal{S} Y + \mathcal{S}\mathcal{F}_0(\mathcal{S}Y_0) \tag{73}$$

The amplitude linked to Y_0 denoted by A_{Y_0} of main harmonic can be written as:

$$A_{Y_0} = \mathcal{S}A_Y \tag{74}$$

We go back to the coordinates because of the n last components of Y_0 :

$$Y_0 = \begin{pmatrix} \frac{dX}{dt} \\ X \end{pmatrix} \tag{75}$$

7 Approximate Solutions

Let N be the littler integer such as $N \geq 2$ and each equation corresponding to a Jordan block has no nonlinear terms strictly below the analytical degree N. We are going to use truncation of "high order" terms.We assume that each component of g has the same order near N: it is easy to see that the equations of two conjugate Jordan blocks are decoupled from the equations of two others conjugate Jordan blocks. The different parameters are developped according to a common scale of a "small" parameter ϵ which determine the physical orders (and not only analytical order):

$$g_j = g_{j0} + O(n_j + 1) \quad j = 1, \ldots, n \tag{76}$$

The decoupling of the equations of "large parameters" will happen at a common maximal order: let n_j be the order of forcing terms, N_j the minimal analytical order of normal equation number j. Let $C = [C_{ij}(\epsilon)]$ be the development of the damping matrix:

$$C = \epsilon C[0] + \epsilon^2 C[1] + \dots \tag{77}$$

We are able to decouple the system at least up to order N. (We could do better with for example: $C = \epsilon^N C[N-1] + O(N+1)\dots$). Now, it is sufficient to obtain a "modal synthesis" for the modes concerned with the two interesting conjugate Jordan blocks. Because of this decoupling, we go on the computations only for a space of size $2k$ obtained with k modal coordinates corresponding to eigenvalues near the frequency of excitation. Let the following equations be the normal form of the equations associated with $-i\omega$:

$$\frac{dY_{1k}}{dt} = J_{1k}(-i\omega)Y_{1k} + \alpha_{1k}Y_{1k} + f_1 L_1 + R_{2k}(Y, f_1, f_2) + \dots + R_{Kk}(Y, f_1, f_2) \tag{78}$$

Because of the coupling of order N, we can assert that if the forcings are of physical order 2, the detunings are of physical order 1 for the k above equations and their k conjugates and of physical order 0 for the $2(n-k)$ other ones. So the amplitudes of these last Y_j are of physical order 2 at least. We can write:

$$\frac{dY_{1k}}{dt} = J_{1k}(-i\omega)Y_{1k} + \alpha_{1k}Y_{1k} + f_1 L_1 + R_{Nk}(Y, f_1, f_2) + \dots + R_{Kk}(Y, f_1, f_2) \tag{79}$$

and more precisely at the physical order $N+1$:

$$\begin{aligned} \frac{dY_{1k}}{dt} &= J_{1k}(-i\omega)Y_{1k} + \alpha_{1k}Y_{1k} + f_1 L_{1k} \\ &+ R^P_{Nk}(Y_{1k}, 0, \dots, 0, Y_{2k}, 0, \dots, 0, f_1, f_2) \end{aligned} \tag{80}$$

wherein $R^P_{Nk}(Y_{1k}, 0, \dots, 0, Y_{2k}, 0, \dots, 0, f_1, f_2)$ denotes the nonlinear of physical order $N+1$ (every other nonlinear term possesses at least a part of order 2 , so has at least order $N+1$; further, if this term is neither f_1 nor f_2 then the term contains at least an Y_j and its conjugate. So the physical order is at least $N+3$). We can write because of the transform:

$$Z_k = \mathcal{P}_k[P \begin{pmatrix} Y_{1k} \\ 0 \\ Y_{2k} \\ 0 \end{pmatrix} + \mathcal{T}_N(Y_{1k}, 0, \dots, 0, Y_{2k}, 0, \dots, 0, f_1, f_2)] \tag{81}$$

In order to get rid of terms of order $< N$ which can appear in the transform, it is necessary to be in one of the following cases: either $N = 3$ or $N > 3$ but the nonlinear terms of the transform all have the physical order $N-1$. We may write at order N a formula of the type:

$$\mathcal{P}_k[\sum_i M_i(\hat{Z}_1, \hat{Z}_2)(C_i \begin{pmatrix} b_1 \\ b_2 \end{pmatrix} + \bar{C}_i \begin{pmatrix} b_2 \\ b_1 \end{pmatrix})] \tag{82}$$

It is sufficient to use the projector $\mathcal{P}_k$ onto the concerned space of dimension k.

8 Generalization of Nonlinear Superposition

We introduce generalized masses and generalized frequencies:

$$\frac{dY_{1k}}{dt} = J_{1k}(-i\omega)Y_{1k} + \alpha_{1k}Y_{1k} + f_1 L_{1k}$$
$$+^1R_{Nk}(Y_{1k}, Y_{2k}, f_1, f_2) + \ldots + {}^1R_{NK}(Y_{1k}, Y_{2k}, f_1, f_2) + O(K+1) \tag{83}$$

$$\frac{dY_{2k}}{dt} = J_{2k}(i\omega)Y_{2k} + \alpha_{2k}Y_{2k} + f_2 L_{2k}$$
$$+^2R_{Nk}(Y_{1k}, Y_{2k}, f_1, f_2) + \ldots + {}^2R_{NK}(Y_{1k}, Y_{2k}, f_1, f_2) + O(K+1) \tag{84}$$

We truncate at order N:

$$Z_k = \mathcal{P}_k(P \begin{pmatrix} Y_{1k} \\ 0 \\ Y_{2k} \\ 0 \end{pmatrix} + \mathcal{T}_N(Y_{1k}, 0, Y_{2k}, 0, f_1, f_2)) \tag{85}$$

So we derive the following:

$$Z = \sum_i M_i(Y_{1k}, 0, Y_{2k}, 0)[(C_i \begin{pmatrix} b_{1s} \\ b_{2s} \end{pmatrix} + \bar{C}_i \begin{pmatrix} b_{2s} \\ b_{1s} \end{pmatrix})] \tag{86}$$

with:

$$b_{1s} = \begin{pmatrix} (J_{1k} + i\omega I_{1k} + \alpha_{1k}I_{1k})^{-1}[(J_{1k} + i\omega I_{1k} + \alpha_{1k}I_{1k})S_{1k} - L_{1k}]f_1 \\ 0 \end{pmatrix} \tag{87}$$

$$b_{2s} = \begin{pmatrix} (J_{2k} - i\omega I_{2k} + \alpha_{2k}I_{2k})^{-1}[(J_{2k} - i\omega I_{2k} + \alpha_{2k}I_{2k})S_{2k} - L_{2k}]f_2 \\ 0 \end{pmatrix} \tag{88}$$

We go back to initial coordinates in the subspace of size $2k$:

$$Y_0 = \mathcal{S}Z \tag{89}$$

where Z is build from:

$$Y = \begin{pmatrix} Y_{1k} \\ 0 \\ Y_{2k} \\ 0 \end{pmatrix} \tag{90}$$

The approximate response $X(t)$ restricted with the projector $\mathcal{P}_n$, giving the n last components of a vector, is:

$$X(t) = \mathcal{P}_n(\mathcal{S}Z) \tag{91}$$

We denote $|| X |$ the vector of amplitudes of first harmonic. $|| X |$ is given by the vector whose components are (one-to-one) the modulus of the components of the vector:

$$\mathcal{P}_n(\mathcal{S} \sum_i M_i(Y_{1k}, 0, Y_{2k}, 0)[(C_i \begin{pmatrix} b_{1s0} \\ b_{2s0} \end{pmatrix} + \bar{C}_i \begin{pmatrix} b_{2s0} \\ b_{1s0} \end{pmatrix})] \tag{92}$$

with:

$$b_{1s0} = \begin{pmatrix} (J_{1k} + i\omega I_{1k} + \alpha_{1k} I_{1k})^{-1}[(J_{1k} + i\omega I_{1k} + \alpha_{1k} I_{1k}) S_{1k} - L_{1k}]g \\ 0 \end{pmatrix} \quad (93)$$

$$b_{2s0} = \begin{pmatrix} (J_{2k} - i\omega I_{2k} + \alpha_{2k} I_{2k})^{-1}[(J_{2k} - i\omega I_{2k} + \alpha_{2k} I_{2k}) S_{2k} - L_{2k}]g \\ 0 \end{pmatrix} \quad (94)$$

Generalized frequencies come from the quantities:

$$(C_i \begin{pmatrix} b_{1s0} \\ b_{2s0} \end{pmatrix} + \bar{C}_i \begin{pmatrix} b_{2s0} \\ b_{1s0} \end{pmatrix}) \quad (95)$$

The nonlinear modes brought out from the free behavior are hidden in the matrices $M_i(Y_{1k}, Y_{2k0})$.The generalized mass matrix is given by $\mathcal{S}$ and a part of $\sum_i M_i(Y_{1k}, Y_{2k0})$. For example we write the resonance up to the physical order $N+1$, with f_1 and f_2 of physical order 2:

$$\begin{aligned} Y_{1k} = (i\omega I_{1k} + \alpha_{1k})^{-1}[f_1 L_{1k} + {}^1R_{Nk}(Y_{1k}, Y_{2k}, f_1, f_2) + \\ \ldots + {}^1R_{(N+1)k}(Y_{1k}, Y_{2k}, f_1, f_2)] \end{aligned} \quad (96)$$

${}^1R_{(N+1)k}(Y_{1k}, Y_{2k}, f_1, f_2)$ returns terms of physical order N if the divisors introduced because of the inverse matrix of $(i\omega I_{1k} + \alpha_{1k})$ are of physical order 1. From this we write:

$$Z = P \begin{pmatrix} (i\omega I_{1k} + \alpha_{1k})^{-1}[f_1 L_{1k} + \mathcal{R}_{1,\le(N+1)}] + \mathcal{T}_{1N} \\ f_2 L_{2k} + \mathcal{R}_{2,\le(N+1)}] + \mathcal{T}_{2N} \end{pmatrix} \quad (97)$$

with:

$$\mathcal{R}_{1,\le(N+1)} = {}^1R_{Nk}(Y_{1k}, Y_{2k}, f_1, f_2) + {}^1R_{(N+1)k}(Y_{1k}, Y_{2k}, f_1, f_2) \quad (98)$$

$$\mathcal{R}_{2,\le(N+1)} = {}^2R_{Nk}(Y_{1k}, Y_{2k}, f_1, f_2) + {}^2R_{(N+1)k}(Y_{1k}, Y_{2k}, f_1, f_2) \quad (99)$$

Then we replace all the terms possessing some terms with Y_{1k} or Y_{2k} in order to generate terms of the form $Y_{1k}^M Y_{2k}^N f_1$ or $Y_{1k}^M Y_{2k}^N f_2$. The block bringing back the nonlinear mode in the free vibration case will be given by the matrix:

$$\mathcal{P}_n(\mathcal{S}P \begin{pmatrix} Y_{1k0} \\ 0 \\ Y_{2k0} \\ 0 \end{pmatrix} + \eta(Y_{1k0}, Y_{2k0}) \begin{pmatrix} Y_{1k0} \\ 0 \\ Y_{2k0} \\ 0 \end{pmatrix} \quad (100)$$

where $\eta(Y_{1k0}, Y_{2k0})$ is build from:

$$\mathcal{S} \begin{pmatrix} \mathcal{T}_{1N} \\ \mathcal{T}_{2N} \end{pmatrix} \quad (101)$$

In the forced case, the computations lead to the form:

$$\begin{aligned} &\mathcal{P}_n(\mathcal{S}[PV + \eta(Y_{1k0}, Y_{2k0})V \\ &+PW + DT_N(Y_{1k0}, Y_{2k0})V]) \end{aligned} \quad (102)$$

with:

$$V = \begin{pmatrix} f_1(i\omega I_{1k} + \alpha_{1k} I_{1k})^{-1} L_{1k} \\ 0 \\ f_2(-i\omega I_{2k} + \alpha_{2k} I_{2k})^{-1} L_{2k} \\ 0 \end{pmatrix} \tag{103}$$

$$W = \begin{pmatrix} (i\omega I_{1k} + \alpha_{1k} I_{1k})^{-1} \mathcal{R}_{1,\leq(N+1)} \\ 0 \\ (-i\omega I_{2k} + \alpha_{2k} I_{2k})^{-1} \mathcal{R}_{2,\leq(N+1)} \\ 0 \end{pmatrix} \tag{104}$$

wherein $DT_N(Y_{1k0}, Y_{2k0})$ is the part of the normal transforms containing only the monomials $Y_{1k0}^M Y_{2k0}^N$ without f_1 nor f_2 . We introduce a matrix of nonlinear mode from free vibrations $\mathcal{L}(Y_{1k0}, Y_{2k0})$ of size $n \times n$ such as:

$$\begin{aligned} &\mathcal{SL}(Y_{1k0}, Y_{2k0}) \mathcal{P}_n(\mathcal{S}P \begin{pmatrix} Y_{1k0} \\ 0 \\ Y_{2k0} \\ 0 \end{pmatrix}) = \\ &\mathcal{P}_n(\mathcal{S}P \begin{pmatrix} Y_{1k0} \\ 0 \\ Y_{2k0} \\ 0 \end{pmatrix} + \eta(Y_{1k0}, Y_{2k0}) \begin{pmatrix} Y_{1k0} \\ 0 \\ Y_{2k0} \\ 0 \end{pmatrix}) \end{aligned} \tag{105}$$

9 Free Vibrations and Nonlinear Frequencies

Then we give a generalized nonlinear frequency matrix from free vibrations:

$$-i\Omega = J_{1k}(-i\omega) + \alpha_{1k} I_{1k} + {}^1R_{LNk}(Y_{1k0}, Y_{2k0}) \tag{106}$$

wherein ${}^1R_{LNk}(Y_{1k0}, Y_{2k0})$ denotes the truncation of the nonlinearity obtained by switching off f_1 and f_2 inside the above expressions. Free modes are introduced with a common frequency $\Omega = \Omega_r I_{1k}$ for a positive real Ω_r and compatibility equations:

$$-i\Omega_r Y_{1k0} = J_{1k}(-i\omega) Y_{1k0} + \alpha_{1k} Y_{1k0} + {}^1R_{LNk}(Y_{1k0}, Y_{2k0}) \tag{107}$$

From this we obtain the divisor similar to what happens in the linear case. We consider that the components of Y_{1k0} have the same order than the forcing or the resonant terms. So we introduce:

$$\Delta\Omega = \Omega^2 - \omega^2 \tag{108}$$

where Ω is defined by $\Omega_r I_{1k}$ and we look only at the truncation at order N.

10 Generalized Masses

Afterward, we define the generalized mass matrix:

$$\mathcal{M}(Y_{1k0}, Y_{2k0})^{-1}\Delta\Omega^{-1}\mathcal{L}(Y_{1k0}, Y_{2k0})\mathcal{P}_n(\mathcal{S}P\begin{pmatrix} Y_{1k0} \\ 0 \\ Y_{2k0} \\ 0 \end{pmatrix}) = \mathcal{P}_n(\mathcal{S}[PV + \eta(Y_{1k0}, Y_{2k0})V + DT_N(Y_{1k0}, Y_{2k0})W]) \tag{109}$$

and we replace the vector ${}^t(Y_{1k0}, Y_{2k0})$ of the first part with:

$${}^t(f_1(i\omega I_{1k} + \alpha_{1k})^{-1}L_{1k}, f_2(-i\omega I_{2k} + \alpha_{2k})^{-1}L_{2k}) \tag{110}$$

and we get rid of physical orders higher than N, in order to verify (110) up to order N.

11 Remark For Numerical Computations

We compute only approximate periodic solutions of normal form of the nonlinear systems. We use truncated normalized systems associated to complete series which may diverge. So, the obtained coordinates, called normal coordinates obtained after the change of variable must have not too large amplitudes (for the details see [13], [8]).

12 Conclusions

The initial goal of our work on the nonlinear systems was to reformulate the procedure proposed by Szemplinska-Stupnicka by utilizing the normal form theory in order to introduce more directly non-linear modes [21]. Such an analysis allows us to master the non-linear couplings and hence justify under certain conditions the extension of the modal synthesis to nonlinear cases [21]. Similar studies have been developed by several authors [22], [23] but the advantage of the method of normal form is to use computing algebra generating nonlinear modes and the associated equilibrium equations. In a more general context, this method is a powerful local method. It possesses certain advantages in relation to classical methods. For example, in opposition to the multiple scales method, it requires no a priori hypothesis on the size scale of different terms which appear in the equation system.

References

[1] H. Poincare, 1889, Les Methodes Nouvelles de la Mecanique Celeste. Paris, Gauthier-Villars.

[2] G. D. Birkhoff, 1927, Dynamical Systems. AMS Collection Publications, vol.9, reprinted 1966 .

[3] V.I. Arnold, 1978, Mathematical Methods of Classical Mechanics. New-York, Heidelberg, Berlin , Springer.

[4] J. Moser, 1973, Stable and Random Motion in Dynamical Systems. Princeton, New-Jersey - Hermann Weyl Lectures.

[5] G. Iooss and D.D. Joseph, 1980, Elementary Bifurcation and Stability Theory. New-York Heidelberg Berlin, Springer Verlag.

[6] A.D. Brjuno, 1971, Analytical forms of Differential Equations I 25, 132-198. Transactions of the Moscow Mathematical Society.

[7] A.D. Brjuno, 1972, Analytical forms of Differential Equations II 25,199-299. Transactions of the Moscow Mathematical Society.

[8] S. Wiggins, 1990, Introduction to Applied Nonlinear Dynamical Systems and Chaos, Texts in Applied Mathematics 2, Springer Verlag.

[9] L. Hsu, 1983, Analysis of Critical and Post-critical Behaviour of Nonlinear Dynamical Systems by the Normal form Method. Part I: Normalisation Formulae.89, 169-181. Journal of Sound and Vibration.

[10] L. Hsu, 1983, Analysis of Critical and Post-critical Behaviour of Nonlinear Dynamical Systems by the Normal form Method. Part II: Divergence and Flutter,89, 183-194. Journal of Sound and Vibration.

[11] L. Jezequel, C.-H. Lamarque, 1989, Analysis of Nonlinear Structural Vibrations by Normal Form Theory, Proceedings of the Fifth International Symposium on Numerical Methods in Engineering, Lausanne, vol. 2, 323-328, Springer Verlag.

[12] L. Jezequel, C.-H. Lamarque, 1990, Analysis of Nonlinear Structural Vibrations by Normal Form Theory, Proceedings of IUTAM Symposium, Stuttgart, 133-139, Springer Verlag.

[13] L. Jezequel, C.-H. Lamarque, 1991, Analysis of structural nonlinear mechanical systems by the normal form theory, Journal of Sound and Vibration, 149 (3), 429-459.

[14] C.-H. Lamarque, A. Stoffel, Parametric resonance with a nonlinear term: comparison of Averaging and the normal form method using a simple example, Mechanics Research Communications, vol. 19, 6, 495-504.

[15] P.H. Coullet, E.A. Spiegel, 1983, Amplitude equations for systems with competing instabilities, vol.43, n 4, 776-821, SIAM J. APPL. MATH., august 1983.

[16] R.M. Rosenberg, 1964, On the Existence of Normal Mode Vibrations of Nonlinear Systems with Two Degrees of Freedom, 22, 217-234, Quarterly Applied Mathematics.

[17] R.M. Rosenberg and J.K. Kuo, 1964, Nonsimilar Normal Mode Vibrations of Non-linear Systems Having Two Degrees of Freedom, 283-290.Journal of Applied Mechanics 31E.

[18] W. Szemplinska-Stupnicka, 1962, Resonance modes of a Nonlinear Two-Degree-of-Freedom System,9,123-144.Arch.Budowy Maszyn.

[19] W. Szemplinska-Stupnicka, 1969, On the Normal Coordinates in an Analysis of Steady-State Forced Vibrations of Nonlinear Multiple Degrees-of-Freedom System, 5, 21, 603-621. Archiwum Mechaniki Stosowanej.

[20] W. Szemplinska-Stupnicka, 1979, A Study of Main and Secondary Resonances in Non-linear Multi-Degree-Of-Freedom Vibrating Systems, 10,289-304. International Journal of Nonlinear Mechanics.

[21] L. Jezequel, 1985, Synthese Modale Theorie et Extensions. Ecole Centrale de Lyon: Student Press (in French).

[22] F.G. Gustavson, 1966, On Constructing Formal Integrals of a Hamiltonian System near an Equilibrium Point.71, 670-686.The Astronomical Journal.

[23] H.D. Schrapel, 1979, Entkopplung Nichtlinearer Schwingungen. 48, 289-300. Ingenieur Archiv.

[24] W. Szemplinska-Stupnicka, 1983, Nonlinear Normal Modes and the Generalized Ritz Method in the Problems of Vibrations of Nonlinear Elastic Continuous Systems.18,149-165. International Journal of Nonlinear Mechanics.

[25] R.M. Rosenberg, 1966, Nonlinear Vibrations of Systems with Many Degrees of Freedom, New-York: Academic Press. Advances in Applied Mechanics.

A Damage Localization Algorithm for Nonlinear Structures

N. Stubbs and K.G. Topole
Texas A&M University
Department of Civil Engineering
College Station, TX USA

1 Introduction

This paper addresses some aspects of the problem of detecting and localizing damage in nonlinear structures using changes in modal parameters. The subject is important because the majority of structures exhibit some degree of material or geometric nonlinearity, depending upon the magnitude of the excitation and loading. Previous works have detected and localized damage in structures either via solutions of sensitivity equations in the time or frequency domain [1],[2],[3],[4],[5],[6],[7], or via the use of some form of indicator function [8],[9],[10]. Not only have the methods been applied almost exclusively to linear structures, the extent to which the methods are applicable to nonlinear or time-variant structures has not been rigorously investigated.

The objective of this paper is to demonstrate the extent to which an existing damage detection algorithm, developed on the basis of linear time invariant behavior, can be used to detect and locate damage in nonlinear structures. To meet this objective, the paper is organized into four parts: a derivation of the damage detection algorithm for the linear case, a demonstration of the algorithm for a linear structure, the application of the algorithm to a nonlinear structure, and finally, an evaluation of the localization accuracy as a function of the degree of nonlinearity present, the location of the damage, and how the damage was modeled.

2 Localization Algorithm for Linear Structures

Suppose that we are given an arbitrary skeletal structure with NB elements and N nodes. Assume that the structure behaves linearly. The ith modal stiffness, K_i, of the structure is given by

$$K_i = \Phi_i^T C \Phi_i \tag{1}$$

where ϕ_i is the ith modal vector and C is the system stiffness matrix. The contribution of the jth member to the ith modal stiffness, K_{ij}, is given by

$$K_{ij} = \Phi_i^T C_j \Phi_i \tag{2}$$

where C_j is the contribution of the jth member to the system stiffness matrix. The fraction of energy for the ith mode that is concentrated in member j is given by

$$F_{ij} = K_{ij} / K_i \tag{3}$$

Let Equation (3) represent the undamaged structure. Then the damaged structure can be characterized by the following quantities with asterisks

$$F^*_{ij} = K^*_{ij} / K^*_i;\ [K^*_{ij} = \Phi_i^{*T} C_j^* \Phi_i^*,\ K_i^* = \Phi_i^{*T} C^* \Phi_i^*] \tag{4}$$

On dividing Equation (4) by Equation (3), we obtain

$$\frac{F^*_{ij}}{F_{ij}} = \frac{K^*_{ij}}{K_{ij}} \frac{K_i}{K^*_i} \tag{5}$$

The matrices C_j and C_j^* in Equation (3) may be written in the form

$$C_j = E_j C_{j0};\ C_j^* = E_j^* C_{j0} \tag{6}$$

where the scalar E_j is a parameter representing the material stiffness properties and the matrix C_{j0} involves only geometric quantities (and possibly terms containing Poisson's ratio).

It has been numerically demonstrated and experimentally observed that the modal strain energy for an undamaged element of a structure does not change when the structure is damaged elsewhere. It has also been numerically demonstrated and experimentally observed that relative modal deformations at a given location are larger after damage has occurred. Both of these observations are consistent with the hypothesis that the modal strain energy in an element remains the same before and after damage.

If we assume that the fraction of modal energy in a given member is the same for both damaged and undamaged structures (subjected to the same excitation), on substituting Equation (6) into Equation (5) and rearranging, we obtain

$$\frac{E_j}{E_j^*} = \frac{[\Phi_i^{*T} C_{j0} \Phi_i^*]}{[\Phi_i^T C_{j0} \Phi_i]} \frac{K_i}{K_i^*} = \frac{f^*_{ij}}{f_{ij}} \tag{7}$$

The parameter $\beta_j = E_j/E_j^*$ is taken to be the damage index. All quantities on the right hand side can be obtained from experimental measurement and the geometry of the structure.

The localization scheme used here is essentially a detector which accepts a value of the damage index as input and provides as output a decision regarding the likelihood that the structure is damaged at that location. We hypothesize first that the structure is not damaged at Location j (i.e., hypothesis H_0) and that the values of the damage indices β_i (i=1,NB) represent a sample population of a random variable $\boldsymbol{\beta}$. If H_0 is true, we assume the distribution of the damage indices to be given by $f_\beta(\beta/H_0)$. For a given damage index β_j, the probability of a Type I error (i.e., accept that the structure is damaged at Location j when H_0 is true [11] is given by

$$\alpha_j = 1 - \int_0^{\beta_j} f_\beta(\beta/H)d\beta \qquad (8)$$

or the confidence that j is a damage location is $1 - \alpha_j$. Thus damage localization is accomplished in four steps: 1) compute β_i, 2) compute normalized β_i, 3) select a confidence level, and 4) test for the existence of damage.

As will be demonstrated in the ensuing examples, this method has several appealing features which include the following items: only mode shapes are needed in the analysis, damage may be localized using only one mode, frequency information is not needed, damage is located without solving a system of equations, and, damage may be localized in structures containing many elements.

3 Demonstration of the Algorithm for a Linear Structure

The following procedure was used to demonstrate the usefulness of the algorithm: 1) experiments were designed to demonstrate the efficiency and robustness of the algorithm, 2) a numerical model of the structure was developed, 3) the dynamic response of the structure was generated before and after damage was numerically simulated, and 4) using the modal response of the structure, Equation (7) and Equation (8) were used to localize the damage.

The example structure, a shear beam model of a four story building [12], is shown in Figure 1. Each floor of the building is treated as a single damage location.

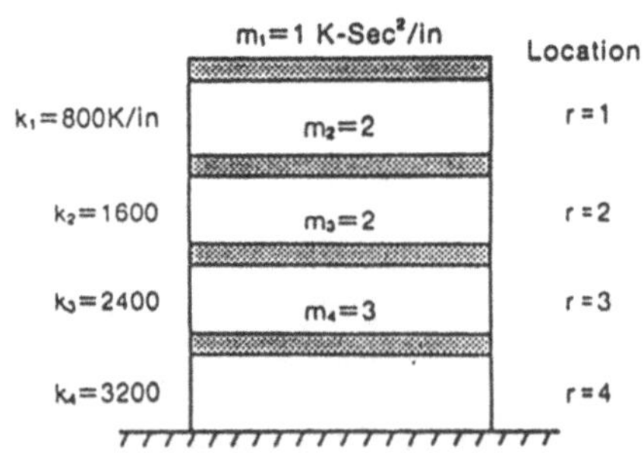

Figure 1 Shear beam model of a four-story building

Relative masses and stiffnesses of each floor are shown in the figure. The modal response of the linear undamaged structure was obtained via standard solution techniques for the eigenvalue problem. In the first scenario, damage was simulated in Location 3 by reducing the stiffness by one-tenth of its original value and the modal response obtained as before. In the second scenario, the same damage magnitude was simulated at Location 4 and the response computed as above. Typical mode shapes are shown in Figure 2.

For the given structure, Equation (7) reduces to the expression,

$$\beta_j = \frac{k_j}{k_j^*} = \frac{({}^j\Phi_i^* - {}^{j+1}\Phi_i^*)^2 K_i}{({}^j\Phi_i - {}^{j+1}\Phi_i)^2 K_i^*} \tag{9}$$

where k_j and k_j^* represent the stiffness of the jth location (note Location 1 is the top floor and Location 4 is the bottom floor) for the pre-damaged and post-damaged structure, respectively; K_i and K_i^* are the respective modal stiffnesses for mode i; and ${}^j\phi_i$ is the jth component of mode i. Note that ${}^1\phi_i$ is the modal component of mass m_1, and ${}^5\phi_i = 0$. The post-damaged modal stiffness K_i^* is estimated by the equation

$$K_i^* = \sum_{j=1}^{NB} ({}^j\Phi_i^* - {}^{j+1}\Phi_i)^2 \, k_j \tag{10}$$

where NB is the number of floors.

Normalized damage indices given by Equation (9) are given in Table 1. The normalized damage indices are obtained by first computing, for each mode i, the mean index ($\mu_{\beta i}$) and standard deviation ($\sigma_{\beta i}$) of the indices for all locations, then using the transformation:

$$z_j = \frac{\beta_j - \mu_{\beta_i}}{\sigma_{\beta_i}} \tag{11}$$

The results for the structure damaged at Locations 3 and 4 are provided in Table 1 for damage indices based on one mode and four modes, respectively. The latter case is warranted since in some problems the f_{ij} term in Equation (7) may be small (e.g., if element j is in the neighborhood of a node of mode i) and unrealistically large values of β_j may result. This problem is circumvented by redefining the damage index as

$$\beta_j = \frac{\sum_{j=1}^{D} f_{ij}^*}{\sum_{j=1}^{D} f_{ij}} \tag{12}$$

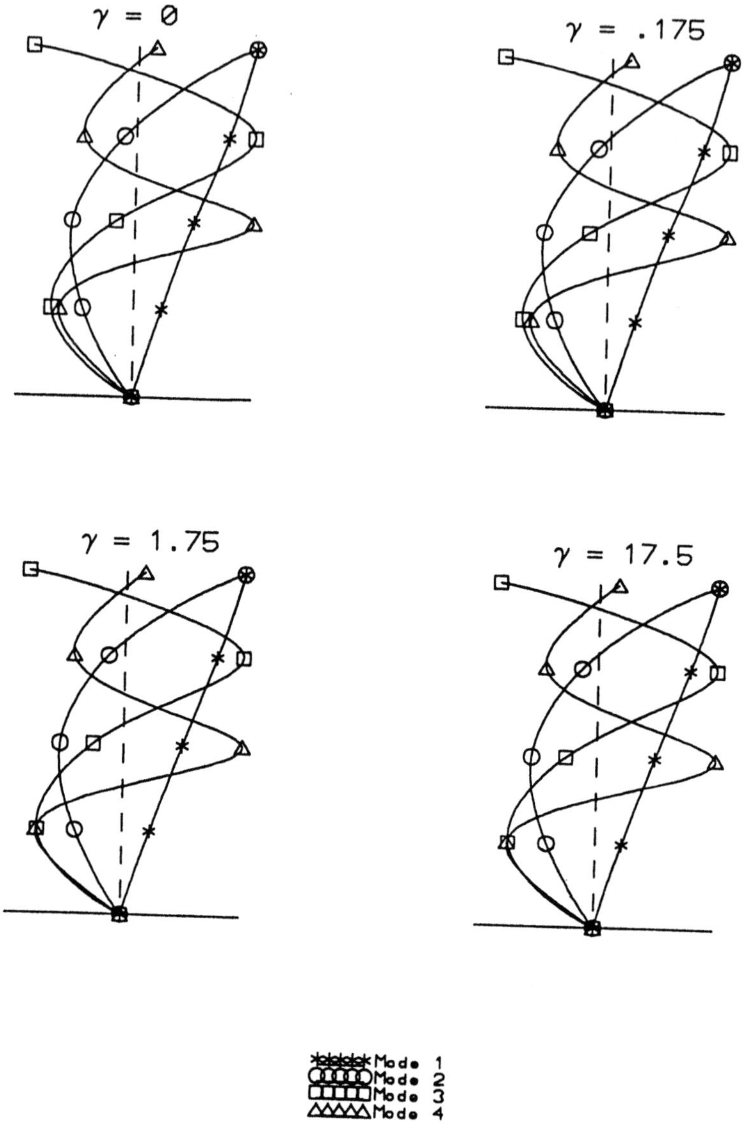

Figure 2 Typical mode shapes

Table 1. Results of Damage Localization, Normalized Damage Index (Z_i)

	Structure Damaged at Location 3		Structure Damaged at Location 4	
Location	Mode 1 Only	All Modes	Mode 1 Only	All Modes
1	-0.56	-0.24	-0.59	-0.58
2	-0.47	-0.72	-0.50	-0.50
3	1.50	1.47	-0.40	-0.42
4	-0.47	-0.51	1.50	1.50

where, D is the number of available modes. From Table 1 we conclude that Locations 3 and 4 are damaged, both at a confidence level of 93% (assuming a normal distribution).

4 Demonstration of the Algorithm for a Nonlinear Structure

Essentially the same procedure used for the linear structure was used to demonstrate the usefulness of the algorithm when applied to a nonlinear structure. The structure given in Figure 1 is selected again, but with the exception that the force-displacement behavior of the fourth location is nonlinear. The nonlinearity is represented by a spring with its restoring force F(x) given by

$$F_4(x) = \alpha_4(x + \gamma x^3) \tag{13}$$

where α_4 is the linear stiffness parameter, γ simultaneously models the type of spring (i.e., hard, $\gamma > 0$, or soft, $\gamma < 0$) and the degree of nonlinearity, and x is the differential displacement of the spring. The stiffness properties of the nonlinear structure are $k_1 = 140$ MN/m, $k_2 = 280$ MN/m, $k_3 = 420$ MN/m, $\alpha_4 = 560$ MN/m, and γ varies from -17.5 to 17.5.

The equations of motion for the structure are given by the following system of equations

$$\begin{aligned} m_1\ddot{x}_1 + k_1(x_1 - x_2) &= f_1(t) \\ m_2\ddot{x}_2 - k_1x_1 + (k_1 + k_2)x_2 - k_2x_3 &= f_2(t) \\ m_3\ddot{x}_3 - k_2x_2 + (k_2 + k_3)x_3 - k_3x_4 &= f_3(t) \\ m_4\ddot{x}_4 - k_3x_3 + (k_3 + \alpha_4)x_4 + \alpha_4\gamma x_4^3 &= f_4(t) \end{aligned} \tag{14}$$

To simulate a modal test we excite the structure with an impulse at Node 1 and measure the response at Nodes 1-4. Accordingly, $f_2(t) = f_3(t) = f_4(t) = 0$ and we select

$$f_1(t) = \begin{cases} 445\ KN & 0 \le t \le .1 \text{ sec} \\ 0 & \textit{otherwise} \end{cases} \tag{15}$$

Note that the excitation force of 445 KN was selected to yield maximum displacements in the range of 2 to 10 mm.

The effective modal response (effective is used here since the system is nonlinear and interpretation of classical modes may change) for the nonlinear structure was obtained by a) solving the system of nonlinear differential equations to give time-domain outputs $x_i(t)$ ($i=1,2,3,4$), b) generating frequency response functions $X_{ij}(f)$ ($j=1,2,3,4$) to give frequency domain representations of the output, and c) generating effective mode shapes by curve fitting the frequency response function and extracting the residues. The differential equations were solved using a Runge Kutta Algorithm with a time step of 0.001 seconds [12]. The frequency response functions were obtained by performing a FFT on the time response [12]. The mode shapes (See Figure 2) were obtained from the frequency response curves via a commercially available modal analysis package [13].

The nonlinear structure was subjected to a total of 32 damage cases. The cases were selected to evaluate the capability of the algorithm to locate damage a) in an environment of a soft spring versus a hard spring, b) in situations in which a linear element or a nonlinear element was damaged, c) as a function of the relative magnitude of the nonlinearity, and d) as a function of the way in which damage was modeled. The 32 damage cases are listed in Table 2.

Typical time histories for degree of freedom $x_4(t)$ are shown for the structure in Figure 3. Note that discernable differences in the response are apparent at the nonlinear measure $|\gamma| = 17.5$; i.e., the coefficient of the cubic term in the spring is 17.5 times greater than the linear term. Typical frequency response curves $X_{14}(f)$ (log(magnitude) vs frequency) are shown in Figure 4. The impact of the nonlinearity on the noise-like behavior is noticeable at $|\gamma| = 1.75$ and is dominant at $|\gamma| = 17.5$. Although modal coefficients vary slightly from case to case, the general shapes extracted for the linear structure maintain their linear interpretation.

The damage index was computed using Equation (9) and the location of damage determined as in the linear case. The results are also summarized in Table 2.

5 Discussion of Results

In all, 32 damage scenarios were analyzed. From Table 2, the damage was incorrectly localized in 7 cases. The overall accuracy of the method for this problem is 78%. Fifty-seven percent (4/7) of the localization errors were associated with the soft spring scenario in which damage was modeled as a reduction in stiffness of the cubic coefficient. The remaining 43% (3/7) of the localization errors were associated with the extreme nonlinearity $|\gamma| = 17.5$. From these observations we conclude that the linear-based algorithm can be used to locate damage in shear beam models if the nonlinearity is less than that indicated by $|\gamma| = 1.75$ and if damage in the nonlinear element is controlled by the linear coefficient.

There are two aspects of originality to this work. First, the damage localization results for a nonlinear structure are presented for the first time. Second, the methodology used to develop the results is applied to a nonlinear structure for the first time. To the authors' knowledge no other modal based study has investigated damage localization in nonlinear structures.

At least two noteworthy implications follow from the results. First, if the nonlinearity in a structure is small, analysis based on the assumption underlying experimental modal analysis (i.e., linear behavior) may be justified. Second, the proposed method of damage localization may be extended to a class of nonlinear structures.

Despite these strong points, the study still has several shortcomings. First, the results have been based on the analysis of a rather simple structure. How the model will fare when applied to more complicated structures has yet to be determined. Secondly, the nonlinearity was limited to one member and to one mode of nonlinear behavior. Thirdly, if the nonlinearity is significant, we have seen that the frequency response functions become useless from an experimental modal analysis point of view. Finally, this study utilized only one set of initial conditions to obtain the frequency response function. Since the structure is nonlinear, we know that the frequency response curves are strongly dependent on the level of excitation.

Finally, this limited study has produced several interesting findings. It is interesting to note that while the method failed to predict the damage location when the damage was modeled as a decrease in the cubic stiffness for the soft spring, the method successfully located the damage for the same kind of damage in the hard spring. We have no explanation for these results at this time. Furthermore, it is surprising that the localization accuracy was so good for the hard spring in general. The only localization errors in the structure with the hard spring occurred in connection with the extreme nonlinear cases. In addition, it is interesting to observe the impact of nonlinearity on the frequency response functions. First, for moderate nonlinearity, the linear characteristics of the frequency response curve dominate. Second, frequency response curves loose their usefulness, in the experimental modal analysis sense, at high nonlinearity. Finally, the noise-like behavior of the frequency response curve is more dominant in the case of the soft spring.

This discussion points to the need for future work in at least three directions: the application of the localization methodology to more complicated structures, the inclusion of more nonlinear elements including nonlinear damping, and an analysis of the impact of

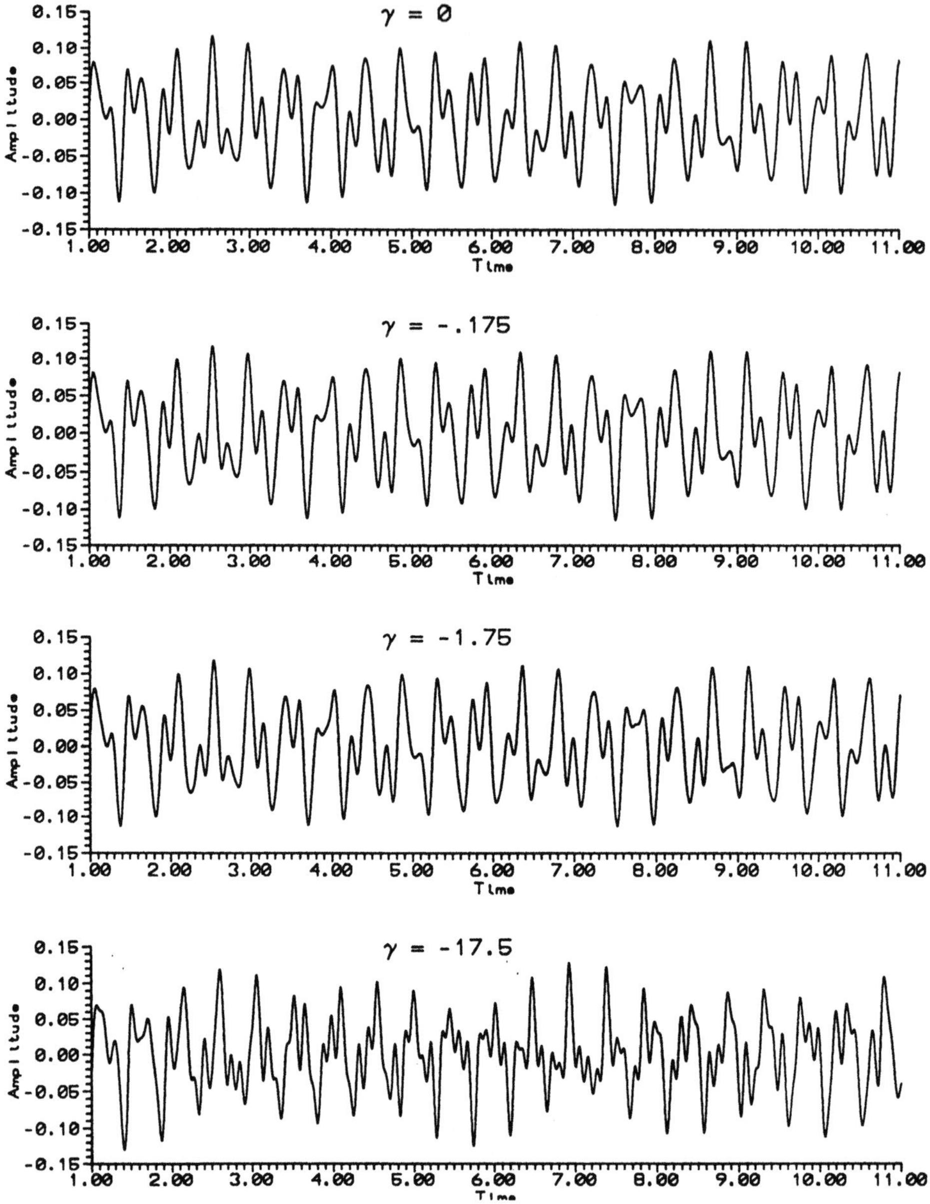

Figure 3 Time history $x_4(t)$ for linear structure, $\gamma = 0$, and nonlinear structure with soft spring, $\gamma < 0$.

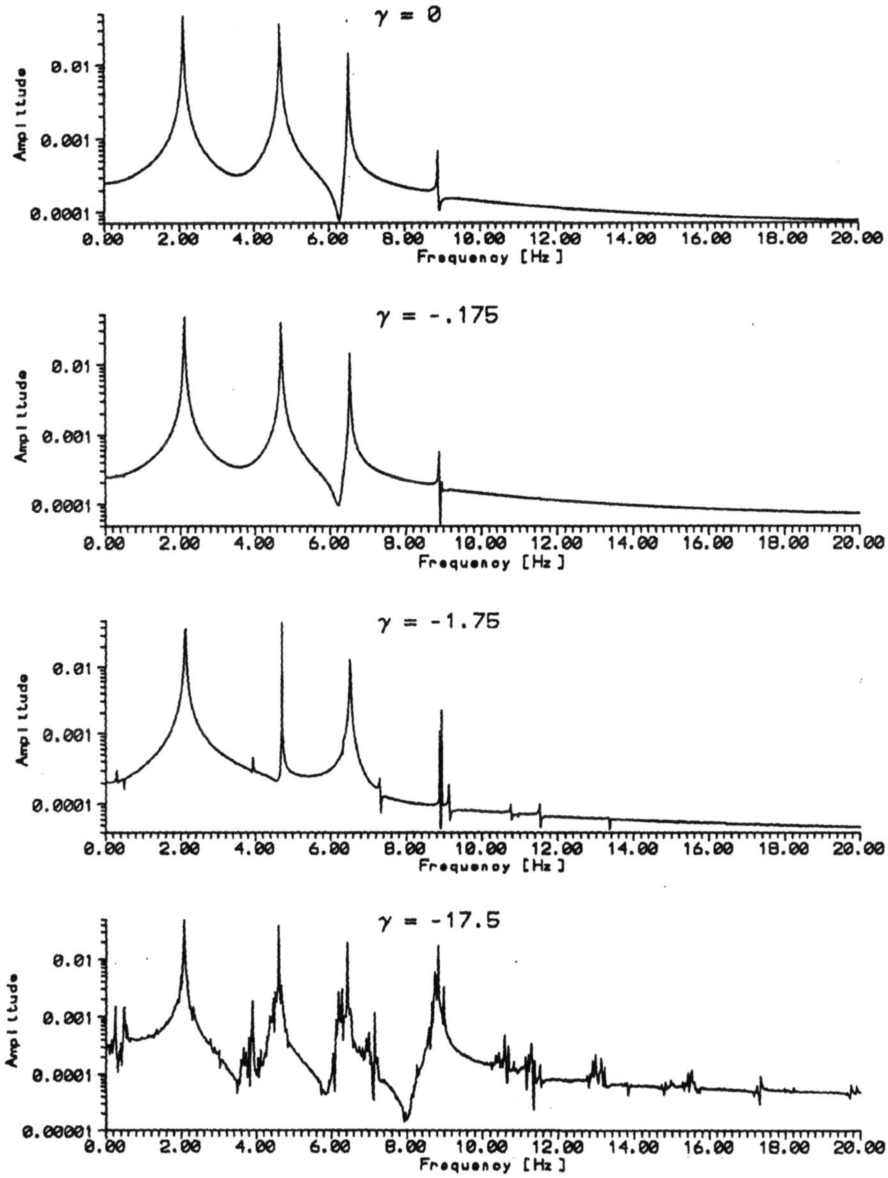

Figure 4 Frequency response function $X_{14}(f)$ for linear structure, $\gamma = 0$, and nonlinear structure with soft spring.

Table 2. Summary of Simulated Damage Cases and Results of Damage Localization for the Nonlinear Structure

Nonlinearity Measure (γ)	Predicted (Simulated) Damage Location			
	I[a]	II[b]	III[c]	IV[d]
-17.5	1 (3)	4 (4)	3 (4)	4 (4)
-1.75	3 (3)	4 (4)	1 (4)	4 (4)
-0.175	3 (3)	4 (4)	3 (4)	4 (4)
-0.0175	3 (3)	4 (4)	3 (4)	4 (4)
0.0175	3 (3)	4 (4)	3 (4)	4 (4)
0.175	3 (3)	4 (4)	4 (4)	4 (4)
1.75	3 (3)	4 (4)	4 (4)	4 (4)
17.5	3 (3)	* (4)	4 (4)	4 (4)

[a]Stiffness at Location 3 reduced by 10%.
[b]Stiffness α_4 at Location 4 reduced by 10%.
[c]Stiffness $\alpha_4\gamma$ at Location 4 reduced by 10%.
[d]Stiffnesses α_4 and $\alpha_4\gamma$ at Location 4 reduced by 10%.
*No prediction.

varying the magnitude of the impact or the initial conditions on the response of the structure and the localization accuracy.

6 Conclusions

The objective of this paper was to demonstrate the extent to which an existing damage localization algorithm, developed assuming linear time invariant structural behavior, can be used to localize damage in nonlinear structures. The theory of the method was presented and applied to a linear and nonlinear structure. Within the limitations of this study, an analysis of the results indicates that the linear-based algorithm can be used to locate damage in shear beam models of structures with hard and soft nonlinear springs so long as the ratio of the coefficient of the cubic term in the force-displacement relation to the coefficient of the linear term is less than approximately 2.

References

[1] M.S. Agbabian, S.F. Masri, R.K. Miller and T.K. Caughey, 1988, A System Identification Approach to the Detection of Changes in Structural Parameters, in Natke, H.G. and Yao, J.T.P. (eds.) *Structural Safety Evaluation Based on System Identification Approaches*, Proceedings of the Workshop at Lambrecht/Pfalz, Braunschweig: Friedr. Vieweg & Sohn, pp. 460-473.

[2] A.D. Dimarogonas and T.G. Chondros, 1980, Identification of Cracks in Welded Joints of Complex Structures, *Journal of Sound and Vibration*, Vol. 69, No. 4.

[3] P. Gudmundson, 1982, Eigenfrequency Changes of Structures Due to Cracks, Notches or Other Geometrical Changes, *Journal of Mechanics and Physics of Solids*, Vol. 30, pp. 339-353.

[4] F.D. Ju, E.T. Wong and T.L. Paez, 1982, Modal Method in Diagnosis of Fracture of Damage in Simple Structures, *Productive Applications of Mechanical Vibrations*, ASME Publications.

[5] H.G. Natke and J.T.P. Yao, 1988, System Identification Approaches in Structural Safety Evaluation, in Natke, H.G. and Yao, J.T.P. (eds.) *Structural Safety Evaluation Based on System Identification Approaches*, Proceedings of the Workshop at Lambrecht/Pfalz, Braunschweig: Friedr. Vieweg & Sohn, pp. 460-473.

[6] N. Stubbs, 1987, A General Theory of Non-Destructive Damage Detection in Structures in Leipholz, H.H.E., ed., *Structural Control*, Proceedings of the Second International Symposium on Structural Control, Waterloo, Canada, 1985, Dordrecht, Netherlands: Martinus Nijhoff, pp. 694-713.

[7] N. Stubbs and R. Osegueda, 1990, Global Non-Destructive Damage Evaluation in Solids, *The International Journal of Analytical and Experimental Modal Analysis*, Vol. 5, No. 2, pp. 67-79.

[8] P. Cawley and R.D. Adams, 1979, The Location of Defects in Structures from Measurement of Natural Frequencies, *Journal of Strain Analysis in Engineering Design*, Vol. 14, pp. 49-57.

[9] N. Stubbs, and D. Rials, 1985, Non-Destructive Evaluation of Damage in Multi-Story Buildings, *4th International Conference on Structural Safety and Reliability*, Kobe, Japan, May 27-29, 1985, II-517-522.

[10] N. Stubbs, J.T. Kim and K.G. Topole, 1992, An Efficient and Robust Algorithm for Damage Localization in Offshore Platforms, *Structures Congress '92*, Compact Papers, Tenth Structures Congress, San Antonio, Texas, pp. 543-546.

[11] J.D. Gibson and J.L. Melsa, 1975, *Introduction to Nonparametric Detection with Applications*, New York: Academic Press.

[12] R.R. Craig Jr., 1981, *Structural Dynamics An Introduction to Computer Methods*, New York: John Wiley & Sons.

[13] W.H. Press, B.P. Flannery, S.A. Teukolsky, and W.T. Vetterling, 1989, *Numerical Recipes, The Art of Scientific Computing*, Cambridge: Cambridge University Press.

[14] *Star Reference Manual, The Star System*, Milpitas, CA: Structural Measurement Systems, January 1990.

Dynamic Damage Indicators for Composite based Structures

Use of stationary and progressive waves
at various frequencies

T. Vinh
Institut Supérieur des Matériaux
et de la Construction Mécanique
Laboratories of Rheology and Structures
Saint-Ouen, France

(Contribution submitted in the published style)

Abstract : For composite based structures, stationary waves and progressive waves are used. Damage theories are reviewed. Damage indicators using ultrasonic waves are presented.

Debonding at the interface between two half spaces gives rise to slip wave which is theoretically analysed for its potentiality in applications of non destructive evaluation.

I - INTRODUCTION To detect, localise and evaluate defects in composite based structures, by dynamic damage indicators, two kinds of waves are utilised :

i) **Stationary waves at low frequencies** (f < 5 000 Hz) give rise to dynamic charaterisation of damaged structure : Rigidity matrix, damping matrix, force and displacement components near or at the boundaries are systematically analysed by Natke [1]. Cracked rotors are examined by Dimarogonas [2]. Eigenvalues and eigenvectors of damaged structure are compared to those of undamaged structure [1].

ii) **Progressive waves at higher frequencies** (f > 100 000 Hz) are extensively used in ultrasonic non destructive testing. The two classes of methods are : The first one uses **wave propagation parameters** i.e **group (or phase) velocity and attenuation** which are evaluated and serve to correlate with the damage structure, Gross and Zhang [3].

The second one concerns the **inverse problem** and tries to localise and evaluate more or **less complex geometry of defects.** It is then closely related to fracture mechanics. Scattering and diffraction of waves allow evaluation of dimensional characteristics of single inhomogeneities and some simple distribution characteristics of inhomegeneities (in a plane for example). The results are related to the residual strength of structure, Achenbach [4] and serve as scientific foundation of failure prevention.

Diffraction and scattering of ultrasonic waves are extensively investigated, Pao [5]. Unfortunately the range of practical problems for which solutions are available remains limited. In many cases, the problems are analytically intractable. Approximate solutions are developed, Bond [6] and numerical modeling using finite difference of finite elements is very active, Temple [7].

The majority of non-destructive techniques (NDT) originally developed for metallic structures are extended to composite based structures for number of years with more or less success. Measured non destructive parameters are more difficult to correlate with damages in composite materials at microscopic and/or macroscopic levels. Globally, deffects in structures are divided into two classes [8].

- Manufactured induced effects
- Fracture and fatigue damage due to generalised external forces and environmental agents

Figure (1) collects some manufacturing defects among others and related micrograhies.

Figure (2) gives some fracture mechanisms on one dimensional or two dimensional carbon-carbon composite.

If we analyse three dimensional composites, fracture mechanisms become extremely complex and are dominated by strong interactions between elementary defects, figure (3). The three figures give a quick overview of qualitative physical interpretation of defects by destructive methods.

In the domain of non destructive evaluation, we try to focuss attention, in this paper, on the following three aspects :

Resin rich area (initiation of cracks) Kevlar - Epoxy composite	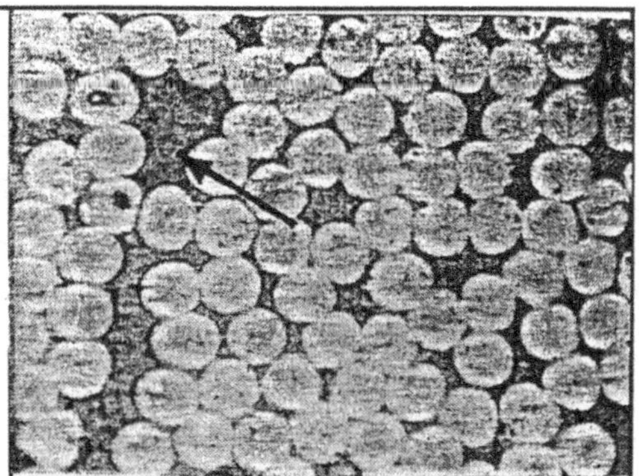
Voids (strength reduction) (Kevlar - epoxy composite)	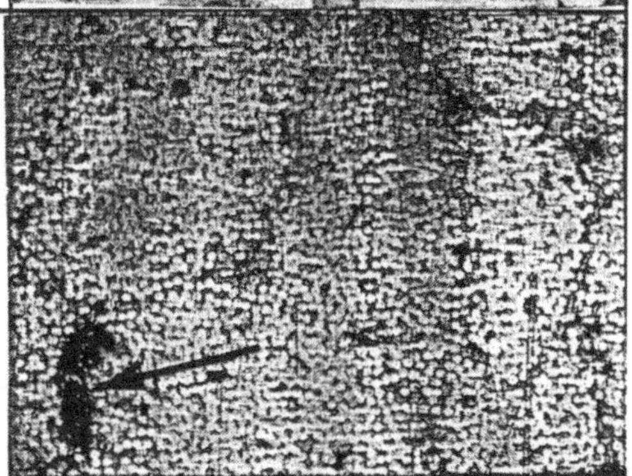
Fibre kink or ply waviness (Reduced residual strength , in compression (carbon - carbon 2 D composite)	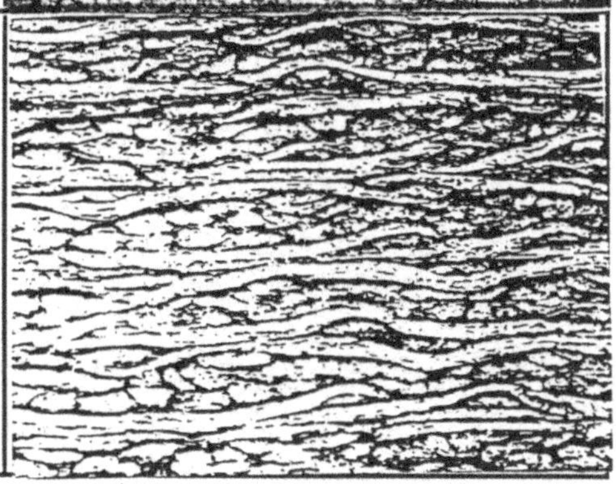

Figure (1) Some manufacturing induced defects and their related micrographies

Delamination . Reduction of compression strength . Propagation at ply interface . Reduction of the stability of load bearing fibres (carbon woven composite)	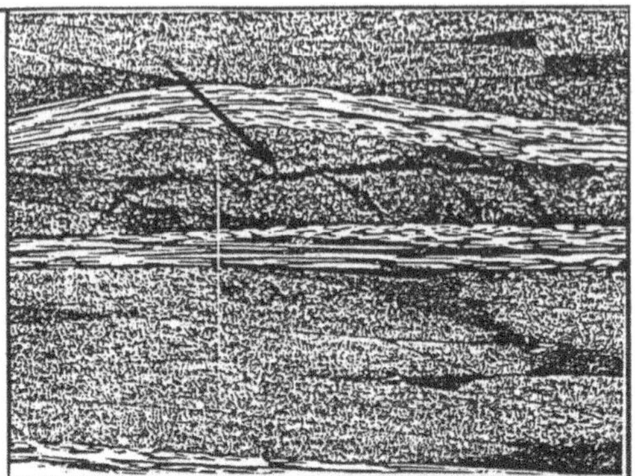
Fibre fracture combined with delamination (carbon - carbon 2 D composite)	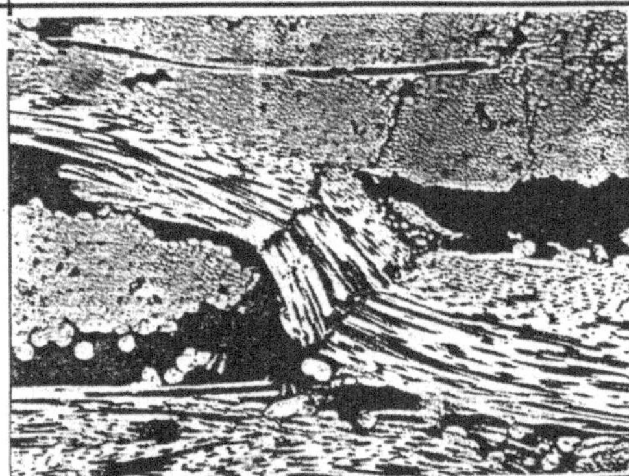
Fibre fracture and pull - out . (adhesion between matrix and fibers) (carbon - carbon 2 D composite)	

Figure (2) **Some fracture mecanisms on unidirectional and 2 dimensional composites .**

Compressive fatigue test .

Fracture happens for the whole areas called " tribe - resistant populations "

Compressive fatigue test .

Fracture changes in direction or plane .

Compressive nearly static test .

Fracture facets may be different compared to those in fatigue test .

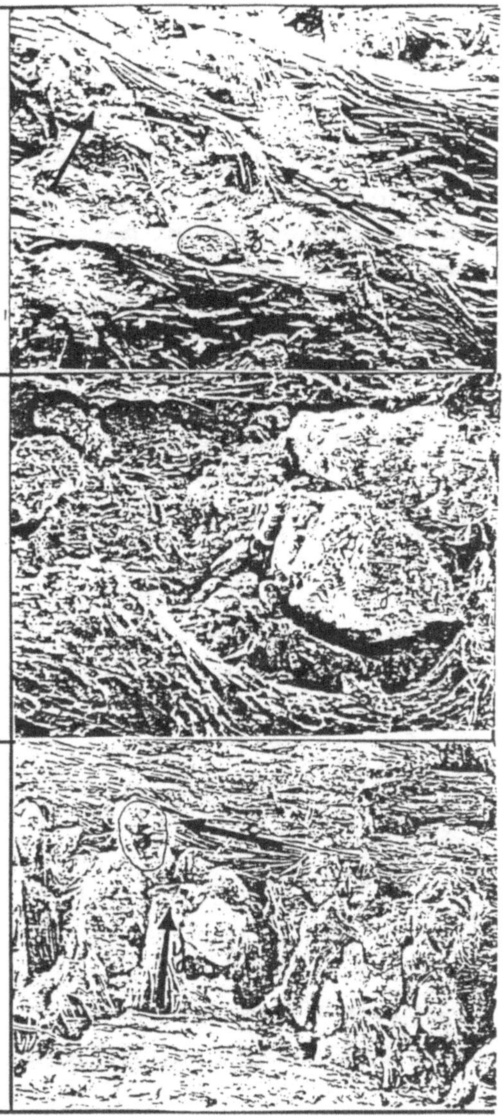

Figure (3) Three dimensional carbon - carbon composite .
Fracture mechanisms are more complex due to interaction between elementary defects .

- Existing damage theories in continuum mechanics
- Wave propagation un cimposite based structure
- Signal processing and some related non-linear phenomena

II - DAMAGE THEORY It concerns the first stage of evolution of material. In presence of external forces are created microvoids and microcracks whose dimensions are not too large compared to the dimension of elementary representative volume of matter. Homogenisation theory permits characterisation of damaged material by equivalent one.

Figure (4) shows schematically formulation of constitutive equation by using invariant theory to describe fracture criteria. Invariance concerns coordinate group transformation by defining the irreducible integrity base. From state variables (temperature T, deformation gradient F, temperature gradient g) and damage parameters, entropy and specific Helmholtz free energy φ are defined.

Figure (5) represents two approaches : empirical and quantitative ones. Kachanov [9] was the first to formulate empirical damage parameter D for metal which is written as

(1)

$$D = 1 - \frac{\hat{A}}{A}$$

A effective area, A damage section area

and the effective stress for homogenised material is defined as

(2)

$$\tilde{\sigma} = \frac{\sigma}{1 - D}$$

σ real stress.

And Young modulus is :

(3)

$$E = E_0 \ (1 - D)$$

E_0 Young modulus of undamaged material.

Evolution of damage is expressed under differential form which can be integrated

(4)

$$dD = \left[\frac{\sigma - \sigma_D}{S\ (1 - D)}\right]^s \frac{d\sigma}{S}$$

σ_D is the lower bound of stress under which damage does not occur. S and s are damaged parameters evaluated from experiments [10]

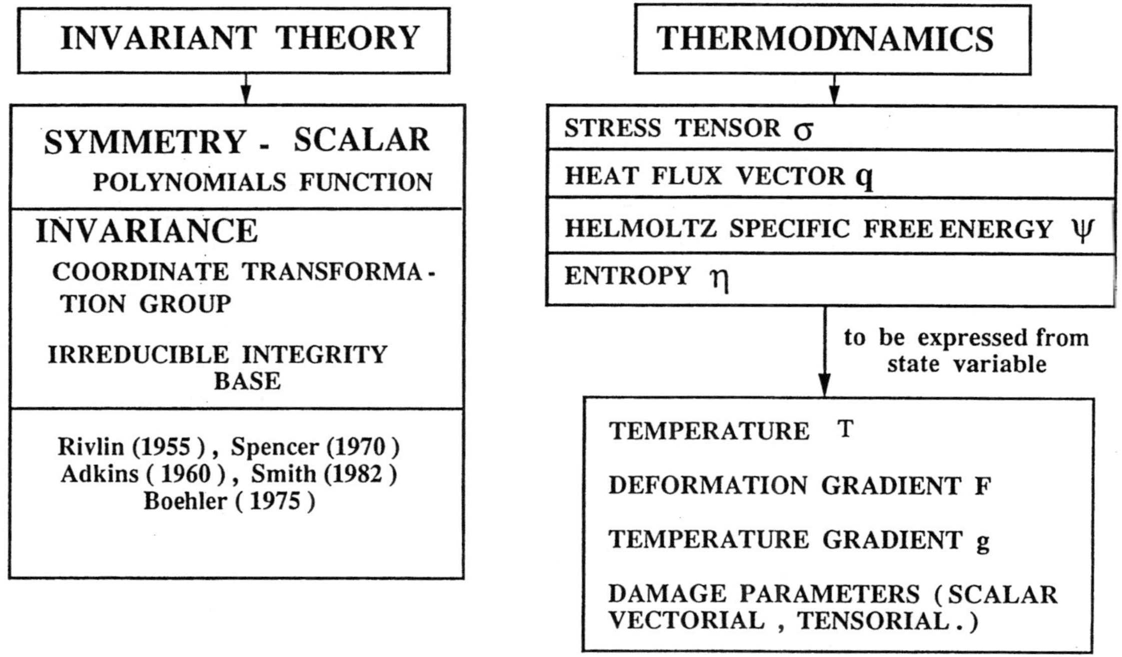

Figure (4) Formulation of damaged material constitutive law.

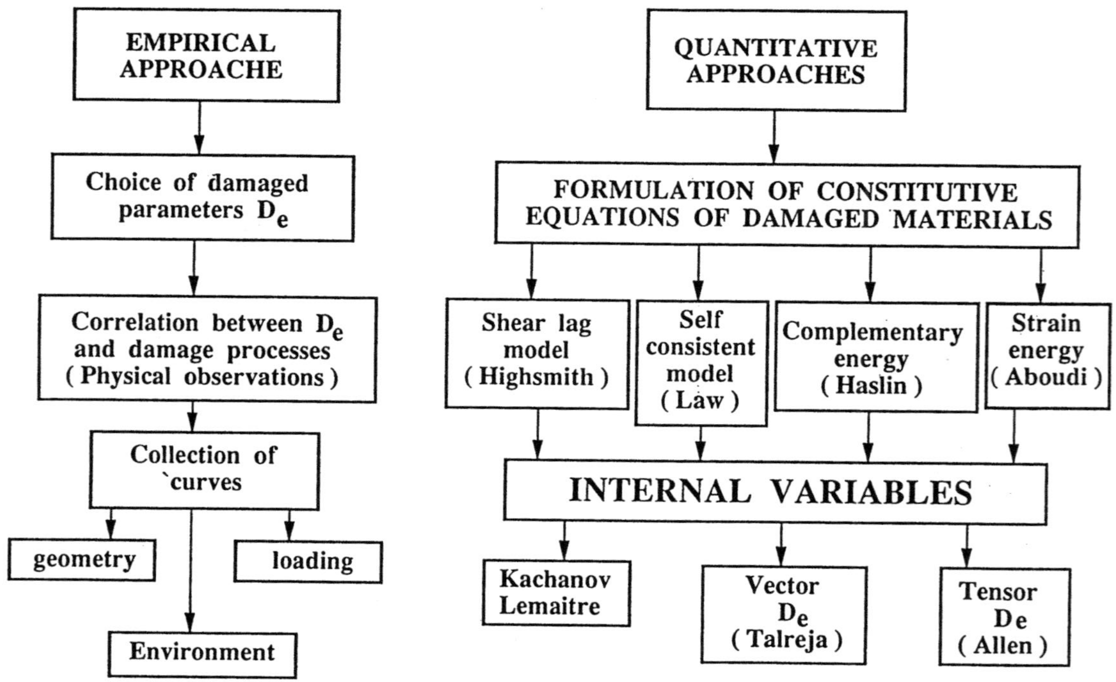

Figure (5) Damage theories of composite materials.

(5)

$$D = 1 \left[1 - \left(\frac{\sigma - \sigma_D}{S} \right)^{s+1} \right]^{\frac{1}{s+1}}$$

To apply D parameter to composite material (which is essentially anisotropic) tensorial formulation is referred to. A variety of models are proposed (shear lag model, selfconsistent model, complementary energy model etc ...). Using appropriate internal variables, Talreja [11] has presented a vectorial theory and proposed application to unidirectional fiber composite. He shows that, for the type of material symmetry (transverse isotropic) at initial state, cracks can be produced normally to fibres, material remains transverse isotropic. If cracks are produced parallel to fibres damaged material then becomes orthotropic, figure (6).

Allen and Harris [14] proposed tensorial characterisation of damage I^{η}_{ijkl} as a fourth order tensor in the constitutive equation :

(6)

$$\sigma_{ij} = C_{ijkl} \; \varepsilon_{kl} + I^{\eta}_{ijkl} \; \alpha^{\eta}_{kl}$$

σ_{ij} , ε_{kl} stress and strain components.

C_{ijkl} Hook tensor component, η indicates damage mode.

α_{kl} is defined as a tensor as follows :

(7)

$$\alpha^{\eta}_{ij} = \frac{1}{V} \int_{s^{\eta}} u_i \; n_j \; d\,S$$

u_i displacement vector, n_j unit normal vector, S^n area of the cracked surface, V local volume.

Allen evaluated reduced rigidity of multilayered composite vs. crack density.

The previous damage theories of damage are to be compared to the recent study of Natke and Cempel [15] concerning mechanical structure using D's parameters to characterise stiffness k, inertia m and damping C :

(8)

$$k(D) = k_0 \, (1 - a_k \, D(\sigma))^{\alpha_k} \qquad 0 < \alpha_k < 1$$

$$m(D) = m_0 \, (1 - a_m \, D(\sigma))^{\alpha_m} \qquad 0 < \alpha_m < 1$$

$$c(D) = C_0 \, (1 + a_c \, D(\sigma))^{\alpha_c} \qquad 0 < \alpha_c < 1$$

a_k , a_m , a_c are coefficients to be evaluated.

σ is the life time, D is defined as the ratio

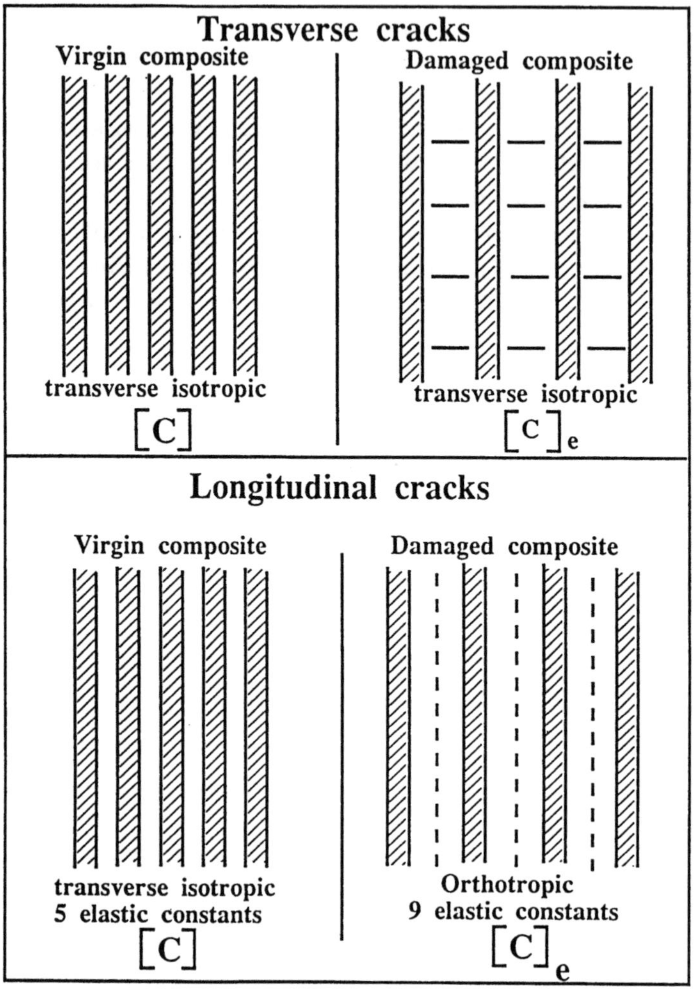

Figure (6) Talreja ' s theory of damage applied to unidirectional fibre composite .

(9)

$$D = \frac{e_t\ (\sigma)}{E_{tl}}$$

of the dissipated energy E_t to its limit value E_{tl} .

Through the various works on damage, we draw the following remarks for composite structures.

i) As those structures are essentially anisotropic, a **tensorial formulation is more appropriate** than a scalar formulation.

ii) **D** may be a fourth order tensor which can be decomposed into isotropic part and anisotropic part. The last one corresponds to alignment of cracks along a privileged direction.

iii) **D** may be a second order tensor (symmetrical or not).

i IV) To go from initial stress tensor σ to $\hat{\sigma}$ effective stress tensor, we have to use an operator which is symmetric. Many possibilities exist

(10)

$$\hat{\sigma} = (1 - \mathbf{D})^{-1/2}\ \sigma (1 - \mathbf{D})^{-1/2}$$

$$\hat{\sigma} = \left\{\sigma\ (1\ \mathbf{D})^{-1}\right\}^{s}$$

$$\hat{\sigma} = 2\ \chi\ (1 - \mathbf{D})\ (\sigma)$$

χ being a linear operator, function of **(1 - D)**.

i V) **D** operator is chosen so as it is **compatible with experiments**. The main difficulty resides in the evaluation of non-diagonal components of **D** and also in the number of components in (10) to be evaluated experimentally.

i VI) To extend existing damage theories to dynamics structural problems, formulation proposed by Natke [16] merits further studies.

III - WAVE PROPAGATION IN DAMAGED COMPOSITE STRUCTURE

III-1 **Low frequency stationary waves** are used by many researchers to tests structures, Williams [17], Natke [15] among others. Natke focussed attention on the interest of higher modes of vibration which are more sensitive to damages and consequently easier to detect and localise. Lack of room does not permit us to present this section in details.

We will emphasize on the interest of **elastodynamic higher mode of vibrations in plates and rods.** In dynamics of structure, in the majority of cases, by vibration techniques, one uses the **first elastodynamic mode** which has not the same significance as higher structural modes (eigenvalues). There is two classes of elastodynamic modes : symmetric and antisymmetric, figure (7).

Structure dynamics uses generally elastodynamic modes of order O which are characterised by zero modal lines in the section (for symmetric vibrations) and one modal line (for flexural antisymmetric vibrations), figure (7).

If we have to deal with vibrations, of **multilayered plate or shell,** we realise that higher elastodynamic modes are very interesting to use. Those kinds of waves are very sensitive to defects particularly to the lack of adhesion between layers.

Figure (8) represents special piezoelectric transducers mounted on composite rod as emitter and receiver with metallic electrodes to be connected to oscilloscopes. Shear waves and dilatational waves are created with polarisation unit vector normal (or parallel) to propagation unit vector.

Figure (9) shows recording of signals from piezoelectric receivers for the three kinds of waves (flexural, torsional, extensional).

Figure (10) represents elastodynamic spectrum (i.e. circular frequency ω versus the Real value of wave number k. Higher elastodynamic modes (or branches) are presented.
At our knowledge, those kinds of waves are not yet utilised in mechanical structures to study defects. It opens a new field of experimental research.

III - ULTRASONIC PROGRESSIVE WAVES when propagating in a damaged solid is accompanied by amplitude **attenuation** and **velocity dispersion**. Those two parameters are related, in a complex manner, to the obstacles the wave finds along its path. Let us enumerate some of them :

i) In composite material, presence of many constituents (fibre, matrix) creates **deviation of the direction of propagation. Conversion of modes** is frequent. Wave polarisation changes in direction and consequently energy of transmitted waves is distributed in many types of waves (quasi longitudinal, quasi shear etc ...).

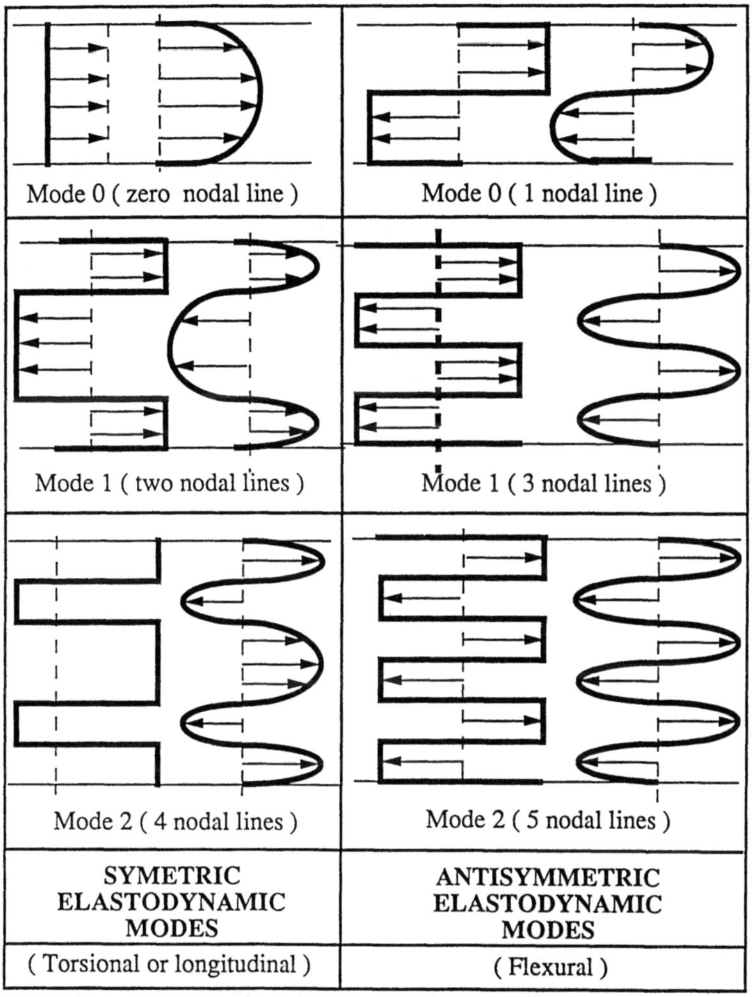

Figure (7) Elastodynamic modes are characterised by the number of nodal lines of vibration in a section .

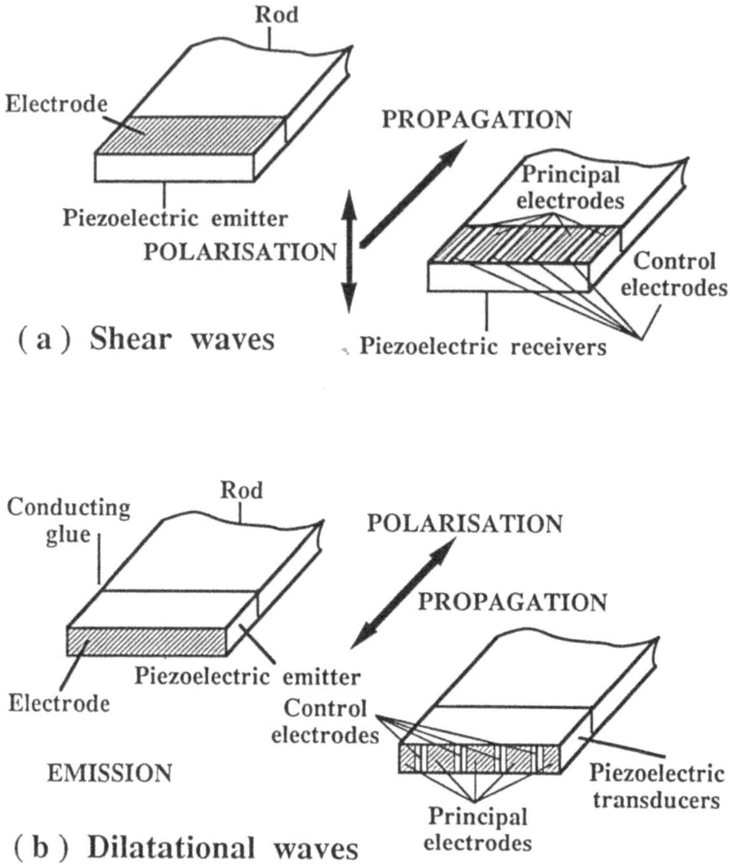

Figure (8) **Special piezoelectric transducers are used to create in rod higher elastodynamic modes.**

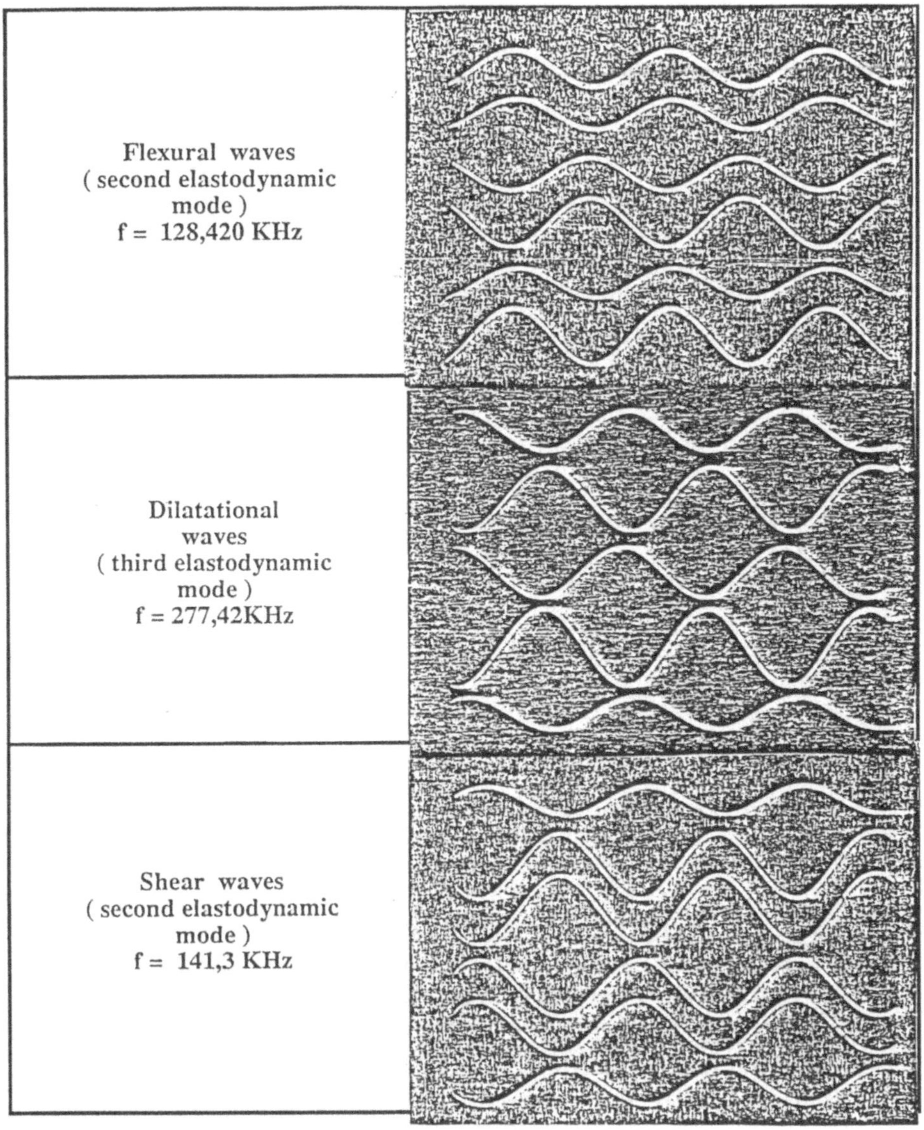

Figure (9) **Recorded signals from piezoelectric receivers, nodal lines are created in the section, Touratier [18]**

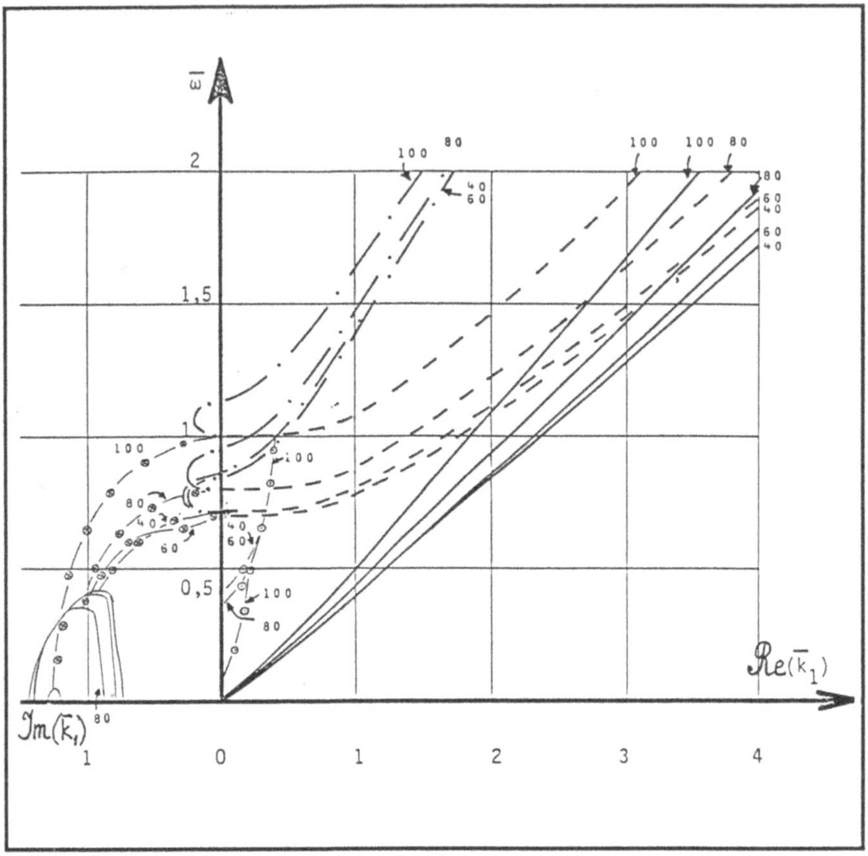

Figure (10) Curve $\bar{\omega}$ ($\bar{k}_1$) of the circular frequency vs wave number of a transverse isotropic composite rod. Higher elastodynamic modes correspond to dotted lines (second branch and third branch) with various volumic percentage of boron fibres.
Touratier [18]

ii) **Diffraction of wave** creating the spreading of wave beam, **scattering of wave** which can be simple (in the case of an isolated defect) and often multiple (due to the presence of many defects). This last case is more diffiuclt to study. However some attempt is made to exploit the phase coherent nature of wave propagation through random media, Feng and Sornette [22].

Literature concerning this problem is very rich [4], [6]. Elastic wave inverse scattering remains an open field of active research, Devaney [23], Varadan [24].

The main difficulty in the inversion problem is the adaptation to complex geometry of defects.

iii) **Attenuation of waves** is successfully correlated to micromechanisms (which resort to dislocation theory) in metals. It constitutes a good non-destructive tool to analyse many microscopic physical phenomena in metals.

For composite materials, the situation is not so optimistic. Most works were restricted to Rayleigh regime where the wavelength is much larger than the characteristic dimension of the defects. This is the case where low frequency approximation is applicable. In this region dispersion behaviour of wave is weak or negligible.

Many non-linear phenomena are the causes of wave attenuation :

Solid friction between fibres and matrix.

Debonding concerning fibres or layers.

Difference in the behaviour of cracked material in tension and in compression.

i IV) **The viscoelastic property** of some composite matrix (expoxy resin for example) can be described by **linear convolution** (if the material is linear) or by **non-linear multiple convolution** (if the material is non-linear). It remains difficult to delineate the two domains.

Figure (11) presents the two categories of ultrasonic inspection techniques.

III-1 Dispersion of waves in composite periodic material

Dependency of wave velocities with respect to frequency is well known in structural dynamics when we have to deal with **bounded medium**. It is inherent to guided waves. Dispersion phenomena however is more difficult to analyse in composite based structures. The following mechanical parameters are to be taken into account :

a) Geometric dispersion

i) The anisotropic nature of the material creates strong interaction between many kinds of waves.

ii) The material itself constitutes a real mechanical structure and gives rise to dispersion phenomena even in unbounded medium. The composite material is constituted of a piling of layers and a series of interfaces.

In the plane of a layer, the composite is often doubly periodic. Consequently the triple periodicity of composite material creates dispersion.

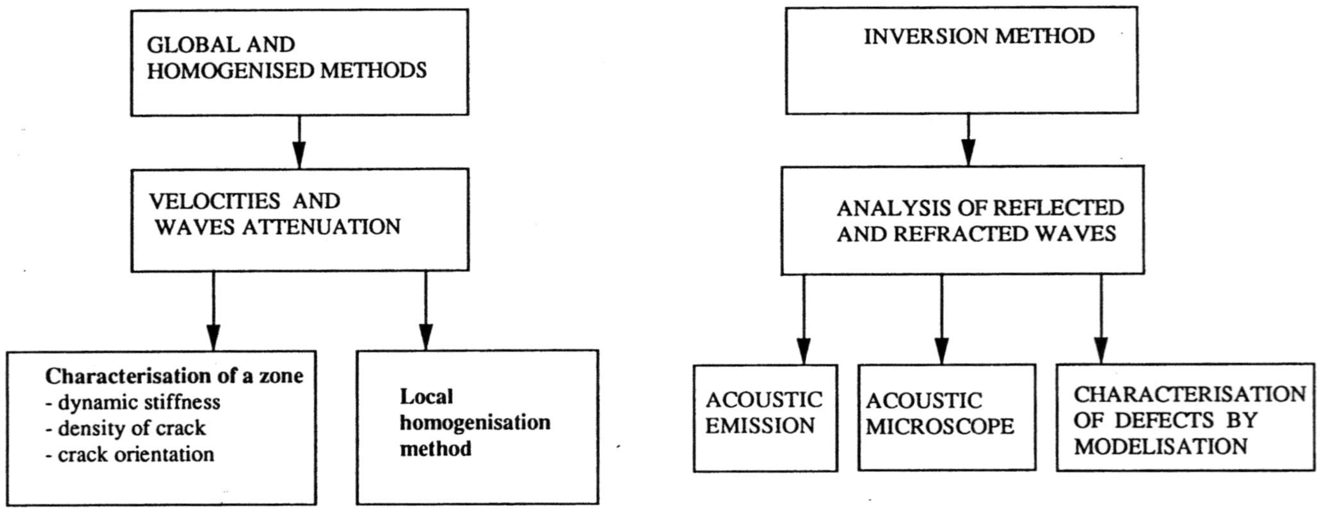

Figure (11) : The two classes of ultrasonic inspection techniques

b) Viscoelastic dispersion By viscoelastic dispersion we mean viscoelastic behaviour of composite constituents (fibers and/or matrix) whose velocities increase with frequency. Viscoelastic dispersion may vary in the same vay as geometric dispersion. It also may vary in the opposite way depending on the nature of the wave itself.

Geometric and viscoelastic dispersions in composite material and structure indeed constitute an active field of theoretical research since 1960. A large variety of theories are used :

- micropolar theory (Eringen [26])
- mixture theory (Bedford and Stern [27])
- microstructure theory (Mindlin [25])
- variational approach (Hashin Strikman [28], Willis [29]).

It is out of the scope of this paper to review all the existing theories. We want to emphasize on the practical interest of the quantitative formulation of dispersion phenomena in multilayered composites in the framework of **previsional calculations**. The interest resides in the fact that experimentalists who use ultrasonic non destructive methods, must have informations concerning wave dispersion of **damage and defect free structure, taken as a reference** for further comparison.

Figure (12) shows longitudinal phase velocity in special composite material evaluated by warious methods. An exact elastic solution proposed by Rytov [30] exists.. We present Chevalier's solution [31] which used mixed fields and stress polarisation tensor. Globally the majority of existing theories predicted quite well dispersion curve at low frequency (or reduced wave number) but not at medium and high frequency.

A formulation of wave dispersion by Datta [32] merits our attention particularly on the **influence of bond layer** figure (13). Phase velocity dispersion is presented with 13 branches (or elastodynamic modes) ant with various hypothesis (springmodel, density model) for bonds.

III-2 Damage indicators applied to advanced composites

The method presented herein does not resort to either acoustic imaging or to acoustic emission.

We will present some results of research carried out in our laboratories on advanced composites for aeronautical and astronautical applications.

a) Ultrasonic equipment and signal processing

Figure (14) shows schematic diagrams of the type of ultrasonic waves which can be used to characterise damage. Volume waves (i.e. dilatational and shear waves) are commonly used in a through-transmission bench with two transducers and in a pulse echo bench with one transducer.

To create guided waves (Lamb waves) in plates or shells the ultrasonic beam angle has to be adjusted by a special transducer in which the position of the piezoelactric element is variable with respect to a perspex half cylinder, figure (14) [33]. To create Stoneley waves in a interface, the same kind of transducer is used. The ultrasonic incident beam angle in the perspex cylinder is adjusted to a critical value so as to obtain a grazing ultrasonic beam for the refracted wave parallel to the interface, figure (14 c). Chevalier, Louzar and Maugin [34] have presented a theoretical study using Stoneley waves which is potentially interesting for the quantitative study of interfacial adhesion.

Figure (15) presents two types of transducer. The first one (a) is characterised by a narrow frequency bandwidth. The second one (b) produces an isolated impulse whose Fourier spectrum is larger. It is used to

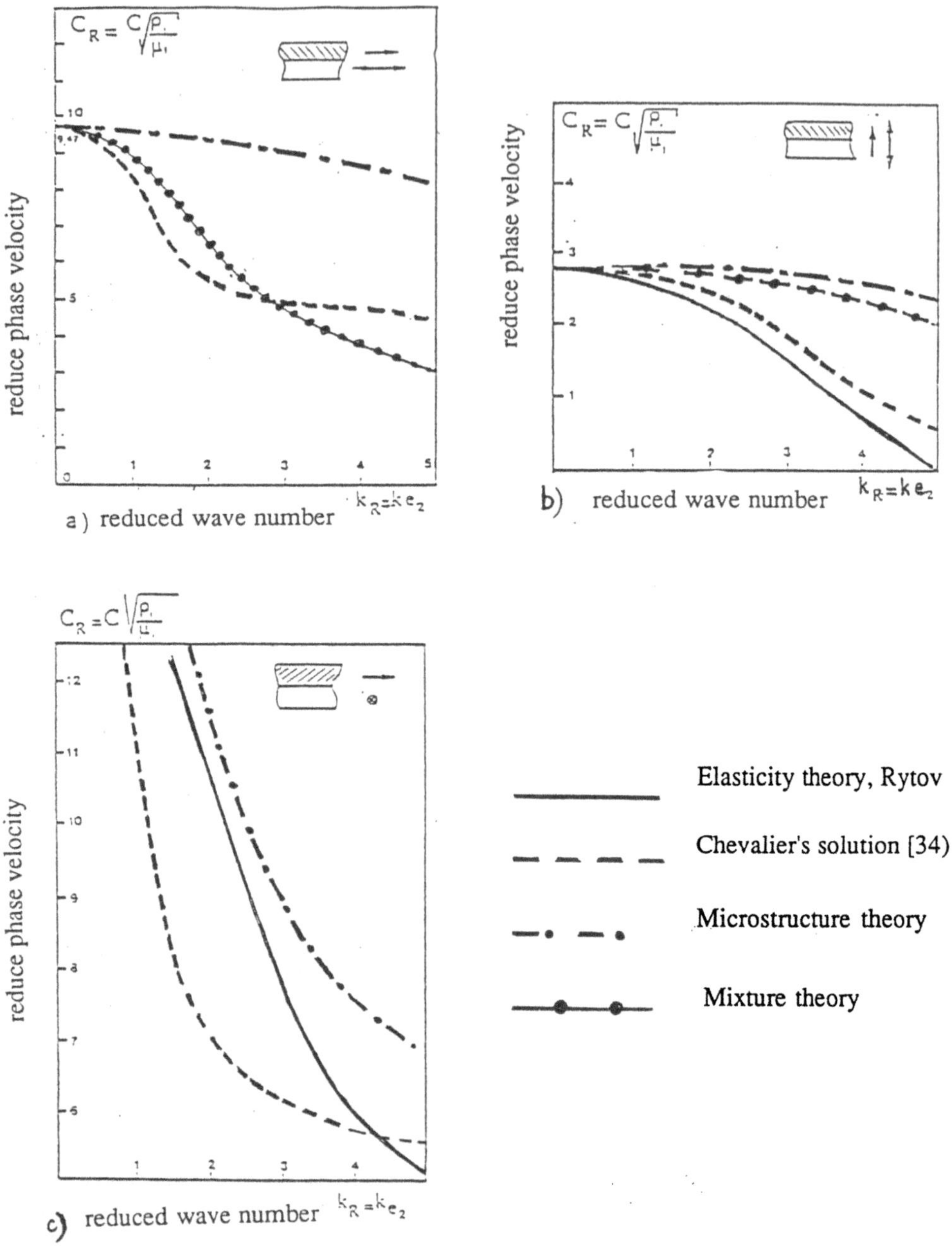

Figure (12) : Reduced phase velocity vs, reduced wave number $k_R = k\ell_2$ in a multilayered medium (Sun's material) - First mode : a) Dilatational wave propagating in the direction of layers - b) Dilatational wave perpendicular to layers - c) Transverse wave parallel to layers. After Chevalier [34]

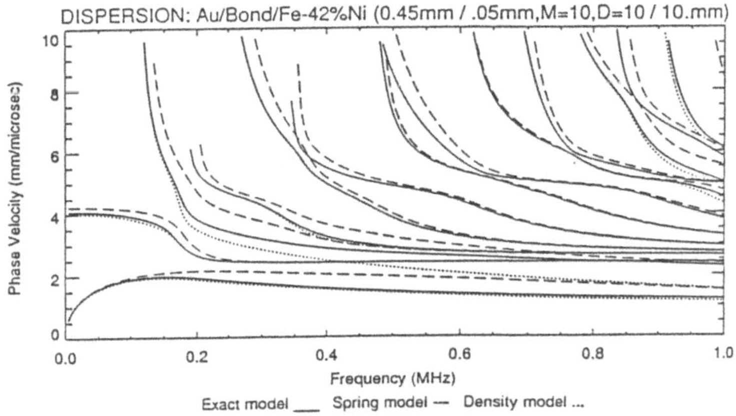

Figure (13 a) : Effect of a heavy stiff bond layer on guided waves, Datta [32]

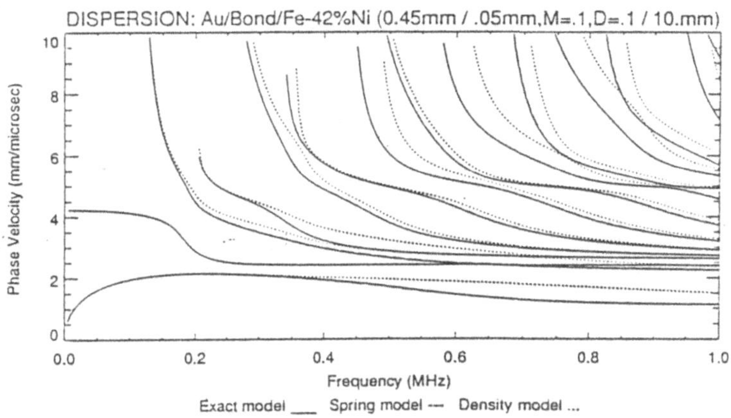

Figure (13 b) : Effect of a light soft bond layer on guided waves, Datta [32]

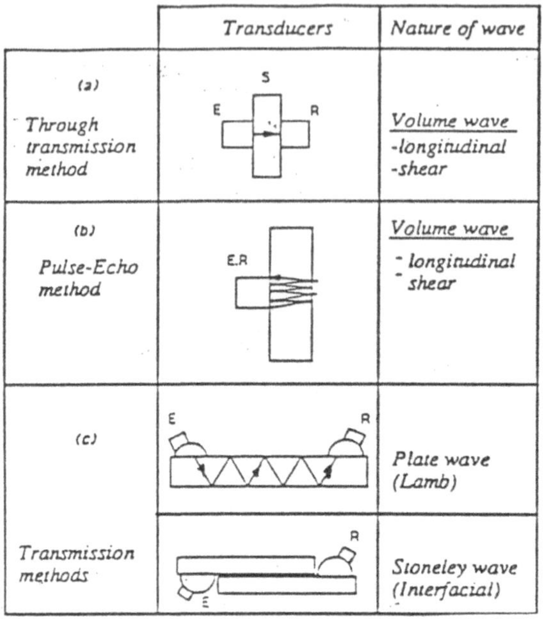

Figure (14) : Three types of ultrasonic waves obtained by four experimental set-up

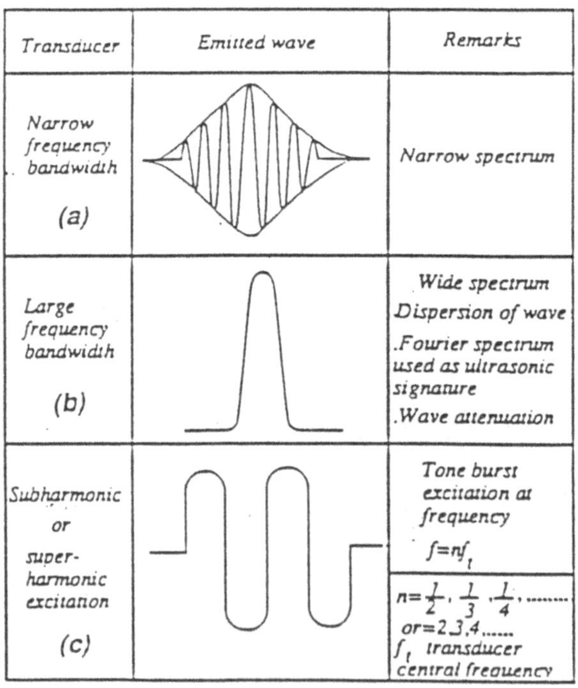

Figure (15) : Shapes of ultrasonic wave

study the dispersion of a wave i.e. the dependence of wave velocity on frequency. To extend the frequency range, many transducers with different central frequencies have to be used. Another possibility is to use the second class of transducer with special excitation consisting of sinusoïdal packets (figure 15 c) whose frequency can be adjusted as a harmonic or subharmonic of the central frequency of the piezoelectric element [35]. Relevant features of some various composite types are presented in figure (16). Waves emitted from a transducer (the emitter) and received at another transducer (the receiver) after propagation in a composite material, are recorded on a digital oscilloscope and analysed. As we focus our attention on measurements of wave velocities and wave attenuations, the "unrectified" waves are used. We do not use the envelope detection of waves.

If the wave shape is conservative and undistorted, velocity (v) measurements are effected via transit time (t) evaluations, d being the distance.

(11)

$$v = \frac{d}{t}$$

The simplest formula for wave attenuation is :

(12)

$$\alpha_{(naper/m)} = \frac{1}{d} \; Log_N \left(\frac{A_i}{A_r} \right)$$

where A_i and A_r designate respectively the incident wave and trasmitted wave amplitudes evaluated on an oscilloscope reconrding and Log_N is the naperian logarithm. In practice, equation (12) is to be replaced by a more complicated formula taking into account the difference of acoustic impedances of various layers, diffraction loss and scattering loss.

Diffraction phenomena are pronounced when the diameter D of the transducer is small compared to the wavelength. The scattering effect due to the variation of the elastic constants of the material along the ultrasonic trajectory, is frequency dependent.

If the trasmitted signal y(t) is distorted, direct measurement in the time domain as presented above is not possible.

Figure (17) represents incident x(t) and transmitted y(t) signals in high temperature resistant composite (carbon tissue in silicon carbide SiC matrix).

Clearly, the wave shapes are not similar. To evaluate the time of flight of the wave, it is necessary to evaluate an intercorrelation function defined as follows

(13)

$$C_{yx}(\tau) = \frac{1}{T} \int_{-T}^{+T} y(t)\; x(t+\tau)\; dt$$

where τ is the time delay. We evaluate the cross-spectrum and, by inverse Fourier transform $F^{-1}[\,.\,]$ obtain $C_{yx}(\tau)$

COMPOSITE	DEFECTS AND DAMAGES	ULTRASONIC SIGNATURE AND PHYSICAL PHENOMENA
Unidirectional fibres+matrix	- Misalignment - Porosities - Voids	• *Velocities and attenuation of waves*
Tissues woven composite	- Cracks { translaminar / interlaminar	• *Diffraction*
	- Lack of adhesion (fibre-matrix adhesion interlaminae adhesion)	• *Scattering*
Three dimensional composite	- Incorrect volume fraction - Discontinuities at ply end - Moisture propagation - Fiber breaking - Warp or wift breaking	• *Conversion of wave*
	Attachment failure	

Figure (16) : Various types of composites of defects and damages and principal ultrasonic phenomena

SIGNAL PROCESSING

SHEAR WAVE PROPAGATING IN FIRE PROOF COMPOSITE

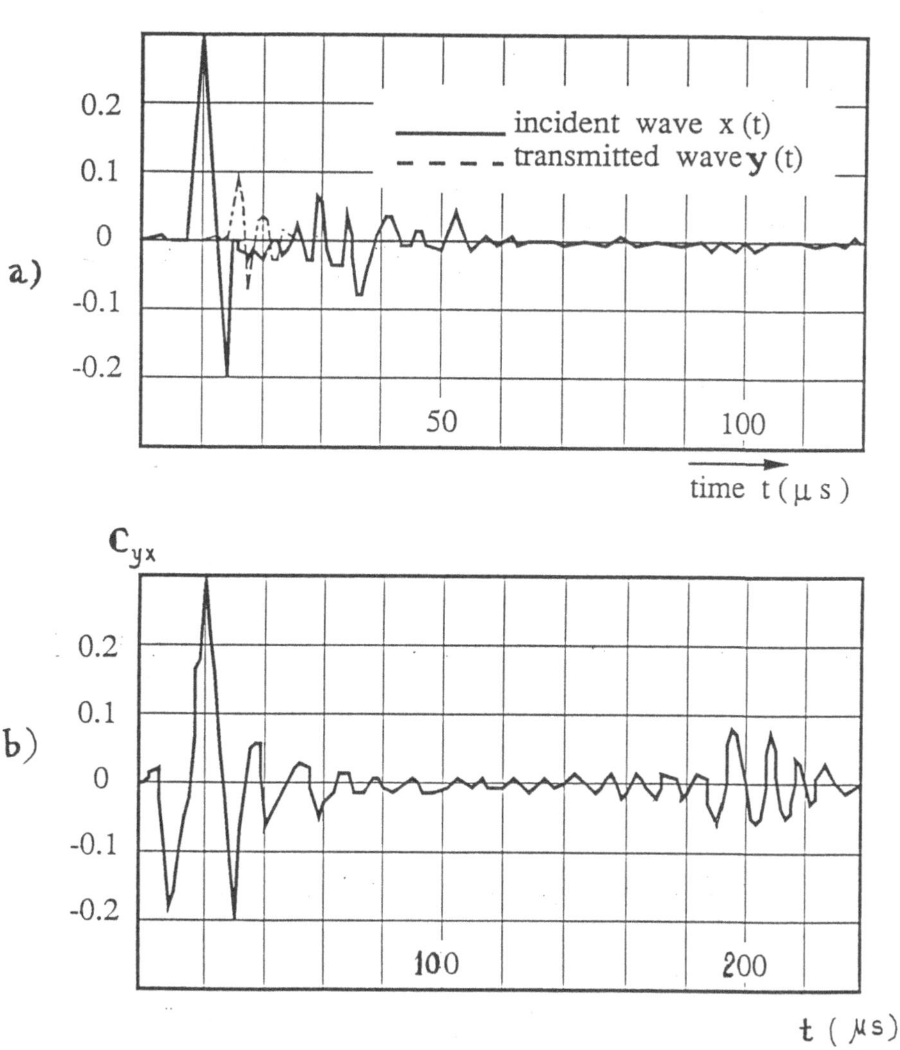

Figure (17) : (a) Incident longitudinal wave and transmitted waves - (b) Intercorrelation function $C_{yx}(\tau)$ permits culculation of time delay τ between incident and transmitted waves

(14)

$$S_{yx}(f) = F\left[C_{yx}(\tau)\right] = Y(f) \,.\, X^*(f)$$

$$C_{yx}(\tau) = F^{-1}\left[S_{yx}(f)\right]$$

Capital letters represent Fourier transforms and a star designates the conjugate quantity.
Figure (17 c) shows the intercorrelation of the two signals.
The maximum peak permits evaluation of transit time from which the group velocity is determined.

Attenuation of the distorted wave also requires signal processing. Diffraction and/or scattering of waves in composites (due to transducer geometry, bounded medium represented by the structure, porosity and defects which act as ultrasonic wave absorbers and wave scatterers) are the two phenomena which influence the ultrasonic signature of the material. Dispersion of the wave is another phenomenon to be taken into account. It is due to the periodicity of the medium itself.

Figure (17c)shows the Fourier spectra of input and output signals for the same composite. Using these two curves, we can evaluate the wave attenuation versus frequency from equation (12). The difference between the spectra can be used as the ultrasonic signature of the composite to charactrise defects and porosity.

b) **Influence of moisture and porosity on ultrasonic volume wave**

Moisture might be introduced during the fabrication of a composite or might penetrate the structure during service. It propagates along the fibres and creates local debonding between the fibres and the matrix.

The choice of the type of wave and the central frequency of the transducers depend on the mechanical parameters to be evaluated and the physical phenomena encountered. Table (1) gives data for a unidirectional long fibre kevlar composite with epoxy resin as the matrix. We used volume waves with a long wavelength λ (compared with the dimension of an elementary representative volume of the array d_a , $\lambda/d_a > 1$). Two classes of samples including dry and wet composites were tested.

SIGNAL PROCESSING

SHEAR WAVE PROPAGATING IN FIRE PROOF - COMPOSITE

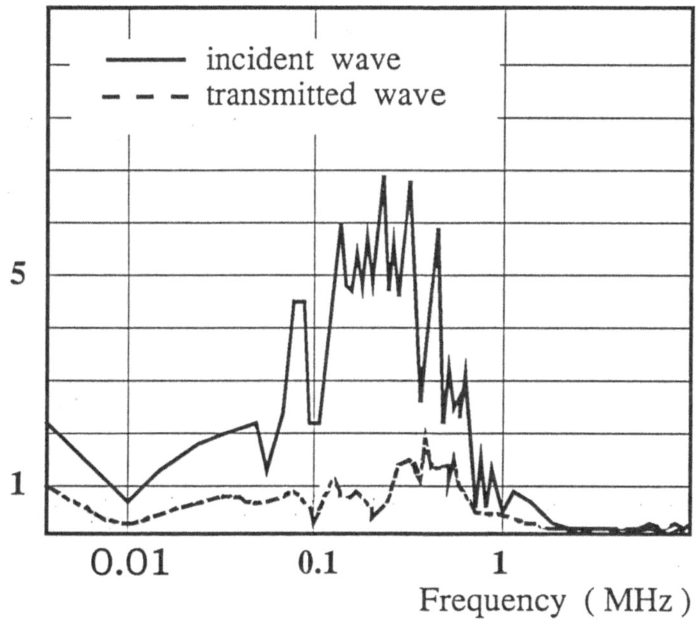

Figure (17 c) : Fourier spectra of incident wave and trasmitted wave

Sample	K_{11}	K_{12}	K_{13}	K_{14}
V_1 (m/s) longitudinal velocity	2410 (2480)	2363 (2557)	2379 (2511)	2414 (2442)
α_1 (naper/m) attenuation	64 (81)	53 (93)	63 (55)	64 (68)
$V_{T(m/s)}$ transverse velocity	1148 (1172)	1141 (1190)	1146 (1193)	1156 (1165)
α_T (naper/m)	105 (142)	103 (95)	104 (155)	101 (110)
C_{11} (MPa) longitudinal	7530 (8210)	7250 (8740)	7290 (8360)	7520 (7950)
C_{55} (MPa) shear	1710 (1830)	1690 (1960)	1690 (1890)	1720 (1810)

Table (1) Velocities, wave attenuations and elastic constants of dried Kevlar composite at frequency of 500 kHZ : Values without brackets. Between brackets, values corespond to a Kevlar composite with moisture.

c) **Debonding characterisation by volume waves**

The velocity of a dilatational wave perpendicular to the fibres and the velocity of a shear wave were measured and the elastic constants C_{11} $(=C_{1111})$ and C_{55} $(=C_{3131})$ were derived. For the dry specimen, wave attenuations are less than for the wet specimen. This remark applies for longitudinal wave attenuation α_L and transverse wave attenuation α_T. Cracks and porosity in the matrix, along the ultrasonic wave trajectory, influence α_L and α_T. The utilisation of shear (transverse) waves is particularly interesting in the study of damage due to debonding or lack of adhesion between fibre and matrix. The reduction of transmitted energy consequently is due to partial wave reflection in this zone.

d) **Porosity distributed in the volume**

The porosity volume percentage contributes to reduce the strength of the matrix. To evaluate the weakening of the composite, we suggest using previsional elastic calculations, Chevalier [36]. We then obtain for an ideal composite (with a regular array of fibres) a set of elastic constants C_{ijkl} which serve as reference to quantify the modification of material properties.

III-3 **Damage indicators during tests on Carbon-SiC composites**

This composite is used, for high temperature applications (T > 800°C), as an external coating of the space shuttle. The sample is divided into rectangular meshes whose maximum dimension must exceed the transducer diameters. In order to follow the evolution of damage, the elastic constants and wave attenuations have to be evaluated before and during the fatigue test.

Volume waves presented in figure (14) can be used.

Figure (18 b) shows, schematically, the sample clamped on the right, with additional mass at the free end.The rod is divided into three zones. Ultrasonic measurements were carried out in each zone, before and after fatigue tests. The dilatational rigidity constants C_{3333} (3 indicates the direction of the warfs of the tissue) are presented. Weakening is not systematic. In zones (1) and (3), C_{3333} increases after fatigue tests, a consequence of a rather surprising strengthening effect which is attributed to the variation of local porosity volume percentage. Flexural dynamic loading might reduce this percentage and con sequently incresea C_{3333} Similar remarks can be made for damping capacities $tg\ \delta_c$ defined as the ratio of the imaginary part to the real part of C_{3333} .

In the accelerated fatigue tests, flexural vibrations of samples were maintained at resonance frequencies. We indicate in figure (19) the variations of Young's modulus [E*] and the related damping capacity $tg\ \delta_E$ versus the number of cycles.

These two last curves show the complementarity of the low frequency stationary flexural waves with respect to the ultrasonic damage indicator method using high frequency progressive waves.

To evaluate the efficiency of the two methods, we have to use the following ratio as a guide.

$$\frac{\lambda}{D_d} = \frac{\text{wave length}}{\text{damage dimension}}$$

If $\lambda/D_d >> 1$ corresponding to long wave and small damage, the stationary long wave method is not efficient.

On the other hand, for small size damage, the ultrasonic method is more sensitive. The wavelength in this case can be chosen in the order of the damage dimension $D_d << 1$ mm .

III-4 **Remarks**

Lack of room does not permit us to present the theoretical aspects of our research. However through four programs of tests on different advanced composites we present the following remarks.

1 - Quantitative ultrasonic wave measurements (velocities and attenuations) can be used as local damage indicators.

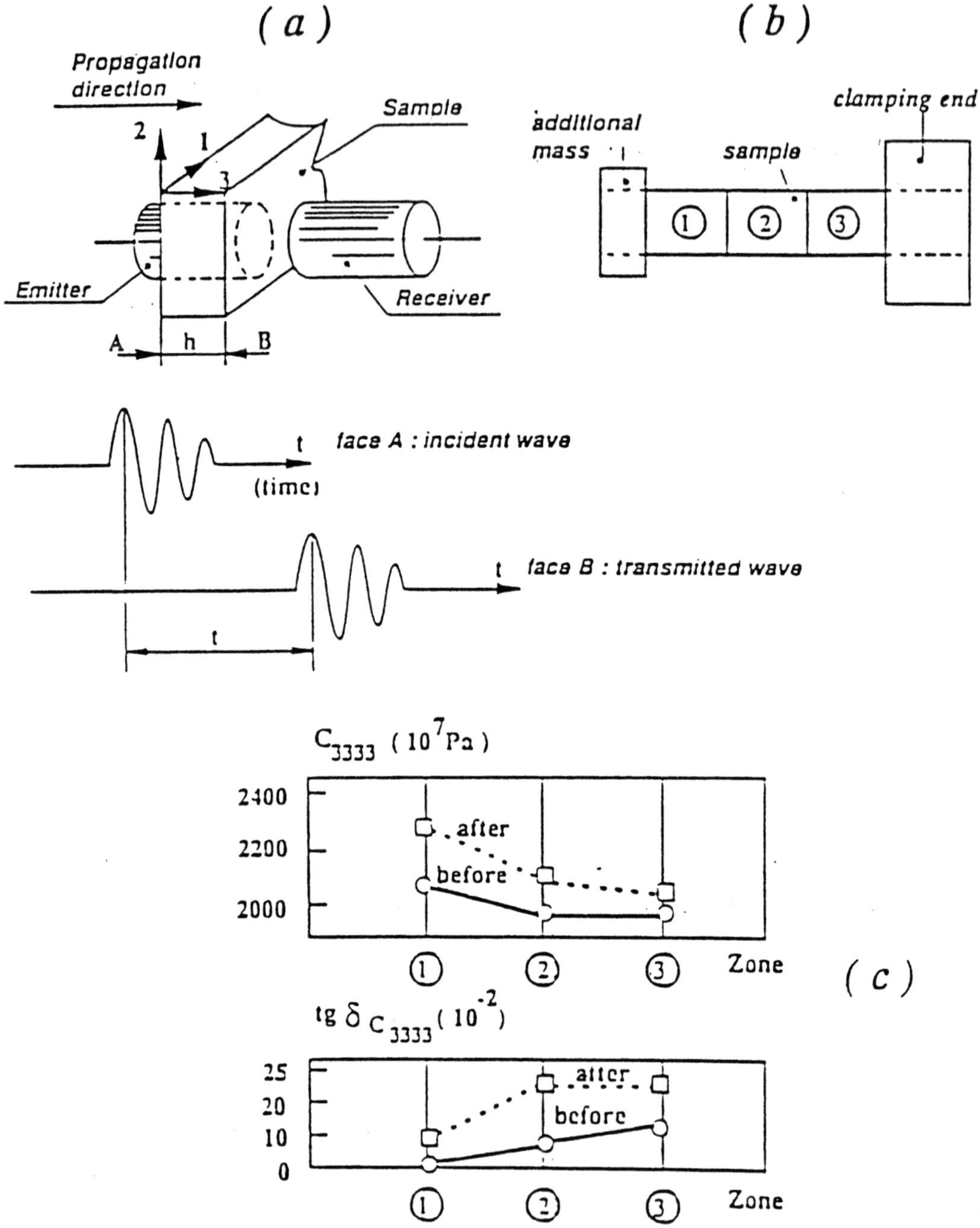

Figure (18) : (a) Through transmission ultrasonic bench,
(b) Fatigue sample of (Carbon-SiC, composite) is divided in three zones
(c) Dilatational constant C_{3333} and damping in each zone

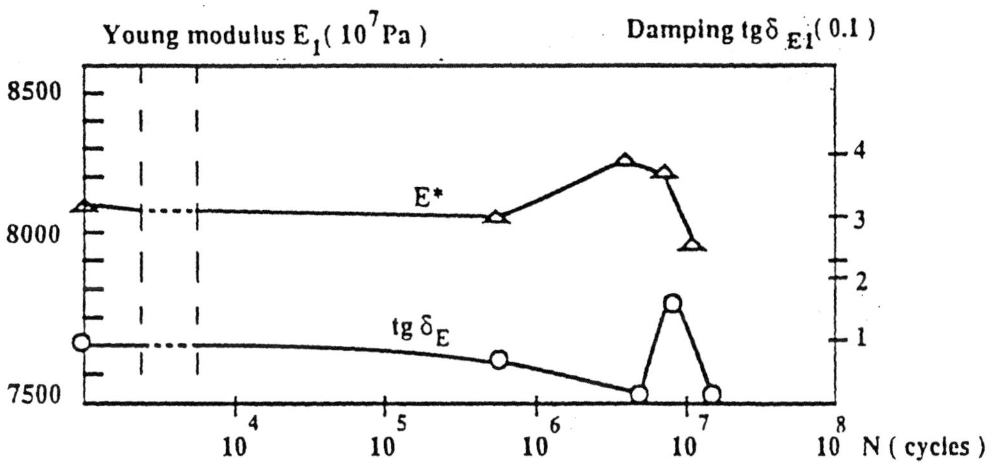

Figure (19) : Evolution of Young modulus and damping during fatigue test on sample Carbon-SiC indicated in Figure (17 b)

2 - The experimental set-up has to be carefully chosen taking into account :

a) The nature of the defects, and their spacial position in the structures.

b) The transducer central frequency has to be chosen with careful regard of the size of the defect.

c) Debonding and lack of adhesion have a strong influence on a shear wave with appropriate polarisation direction (particle motion).

d) For absorbing materials containing many defects, transmitted waves are distorted, and evaluation of velocity and attenuation requires digital signal processing with Fourier transform.

e) A catalogue of defects and ultrasonic signatures has to be established to permit, by inverse methods, the practical quantification of defects.

III-5 Debonding characterisation by special ultrasonic wave

Debonding or delamination in multiply composite can be characterised by Stoneley wave and slip wave, Chevalier [34], Murty [37], Barnett at al. [38].

We present herein Chevalier's approach concerning **slip waves** .

At the interface, continuity of normal components of displacements and stress is written for Stoneley wave :

(15)

$$[u_N] = 0 \quad , \quad [\sigma_N] = 0$$

Brackets designate the jump.

If tangential components is discontinuous, we have to deal with **slip wave**

(16)

$$[u_T] = b_1 < u_T >$$

(17)

$$[\sigma_T] = b_2 < \sigma_T >$$

<.> designate mean value.

The displacement is expressed as a linear combination of plane surface harmonic waves.

(18)

$$U^{(m)} = \sum_{k=1}^{3} \xi_k^{(m)} u_k^{(m)}$$

Elastic slip waves then give rise to the following equations.

$$[\sigma_N] = 0 \qquad [\sigma_N] = 0$$

(19)

$$\sigma_T^{(1)} = \sigma_T^{(2)} = b[u_T] \quad \text{at} \quad x_3 = 0$$

Exponents (1) (2) designate the two media in contact, figure (20), b is the elastic slip stiffness.

Existence of slip and Stoneley waves

▨ existence □ noexistence

Materials	Generalized Stoneley waves		Elastic slip waves	
	Sag . pol .	SH	Sag . pol .	SH
Al - W	existence		existence	existence
Glass / epoxy Glass / epoxy			existence	existence

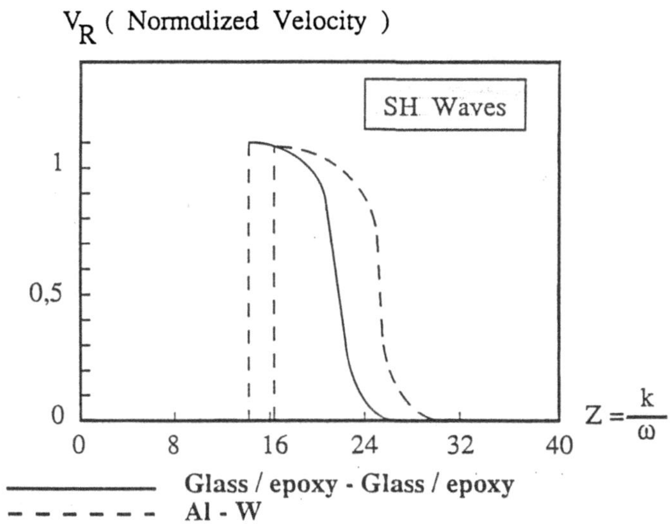

Figure (20) : (a) Existence of slip wave and Stoneley waves, Chevalier [34]
(b) Normalized shear wave against
$Z = \frac{k}{\omega}$ for two interfaces, Chevalier [34]

The existence of SH (shear horizontal) vaves is expressed by, v being velocity :

(20)

$$\frac{b}{\varpi} F_{ST1}(v) + \frac{1}{v} F_{GL1}(v) = 0$$

and the existence of wave polarised in the sagittal plane is described by the following equation

$$i \frac{k}{\omega} F_{ST2}(v) + \frac{1}{v} F_{GL2}(v) = 0$$

In (20) and (21) F_{ST1} , F_{ST2} , F_{GL1} , F_{GL2} are functions deduced from Chirstoffel equations, taking into account boundary conditions [34].

Remarks

a) The quality of the bonding is characterised by SH waves. There exists a cut-off frequency f_c which increases with coefficient of bonding k in (19).

b means perfectly bonded media.

b) f_c could be used as an indicator of the bonding between two media.

c) The study could be extended to more complex interface involving cracks.

d) The study could be extended by accounting the viscoelastic nature of the materials or the misalignement of orthotropic material fibres.

e) The E.S.W. (SH or sagitally polarised) wave exists for all couple of orthotropic materials. An example is given in the figure (20). This wave always exists and is dispersive.
This shear wave is very interesting to characterise bonding whereas the Stoneley wave is guaranteed only for **peculiar assemblies of materials and perfect bonding.**

IV - NON PARAMETRIC METHODS AND SIGNAL PROCESSING

In spite of recent success in the analysis of non-linear structures by time series methods, Billings [39] among others, non parametric methods of analysis constitute the majority of techniques employed in structural dynamics. It remains the obligatory gateway to deal with black box system before using identification techniques.

In this section, we try to present potentialities and limitations of existing methods using linear analyser. We present some of our ideas and researches on special non-linear analyser.

IV-1 Linear analyser

Figure (21) summarises somes time and frequency functions used in characterisation of damages.

Correlation function presented in section III,2 is used to evaluate the time of flight for progressive ultrasonic waves.

Cepstrum is interesting. By using the deconvolution technique, we try to analyse groups of signals with different times of propagation. This programme may be interesting in the analysis of ultrasonic signal to decompose signals into refracted and reflected signals. That is one possible way to localize defects.

Linear transfer function is used to characterize weakly non linear structures by evaluating its distorsion of global response.

Hilbert transform gives rise to extensive studies and applications Vinh [40], Tomlinson [41]. Special algorithms exist to be implemented in existing Fast Fourier transform analyser. They permit non-linearity detection and characterisation in low frequency structural dynamics.

In the domain of non destructive ultrasonic methods, there exists a similar version of Hilbert transform (in frequency domain).

Causality and passivity of medium permit relate real part and imaginary parts of the complex wave number $K(\omega)$.

So attenuation of wave $\alpha(\omega)$ is related to phase velocity c_p, Beltzer [42], as follows, ω being circular frequency :

(22)

$$(\omega) = c(o)\left[1 + \frac{2c(o)\omega^2}{\pi}\int_0^{\infty}\frac{\alpha(\Omega)d\Omega}{\Omega^2(\Omega^2 - \omega^2)}\right]^{-1}$$

(23)

$$\alpha^H(\omega) = -\frac{2\omega}{\pi^2}\int_0^{\infty}\left[\frac{1}{C(\Omega)} - \frac{1}{C(0)}\right]\frac{d\Omega}{\Omega^2 - \omega^2}$$

superscript H designates Hilbert transform.

In (22) and (23) integrals represents Cauchy principal value integrals.

In (22) C(0) is directly related to static effective stiffness of damage solids.

If we have at our disposal both experimental infrmations concerning $c_p(\omega)$ and $\alpha(\omega)$ by numerical Hilbert transform, we can evaluate $c_p^H(\omega)$ and $\alpha^H(\omega)$ and compare to exprimental results. Discrepancy between real responses and Hilbert responses means existence of non-causality which is essentially of computational nature. It serves as non-linear behaviour indicator.

IV-2 **Non-linear analyser** is not commercially available yet.

However elements of this analyser exist and are used in some domains (electrical and electronic engineering, image processing). In mechanical domain some attempts have bean made recently, Vinh [43], Powers [44] among others.

Here we try to indicate the guidelines.

There are three necessary ingredients to be taken into account to construct this analyser : mathematical framework, special signal processing, and testing methods .

a) **Mathematical framework** If we want to extend modal analysis to non-linear system, the only tractable way is to use functionals as for linear system. The difference in this case is to use non-linear

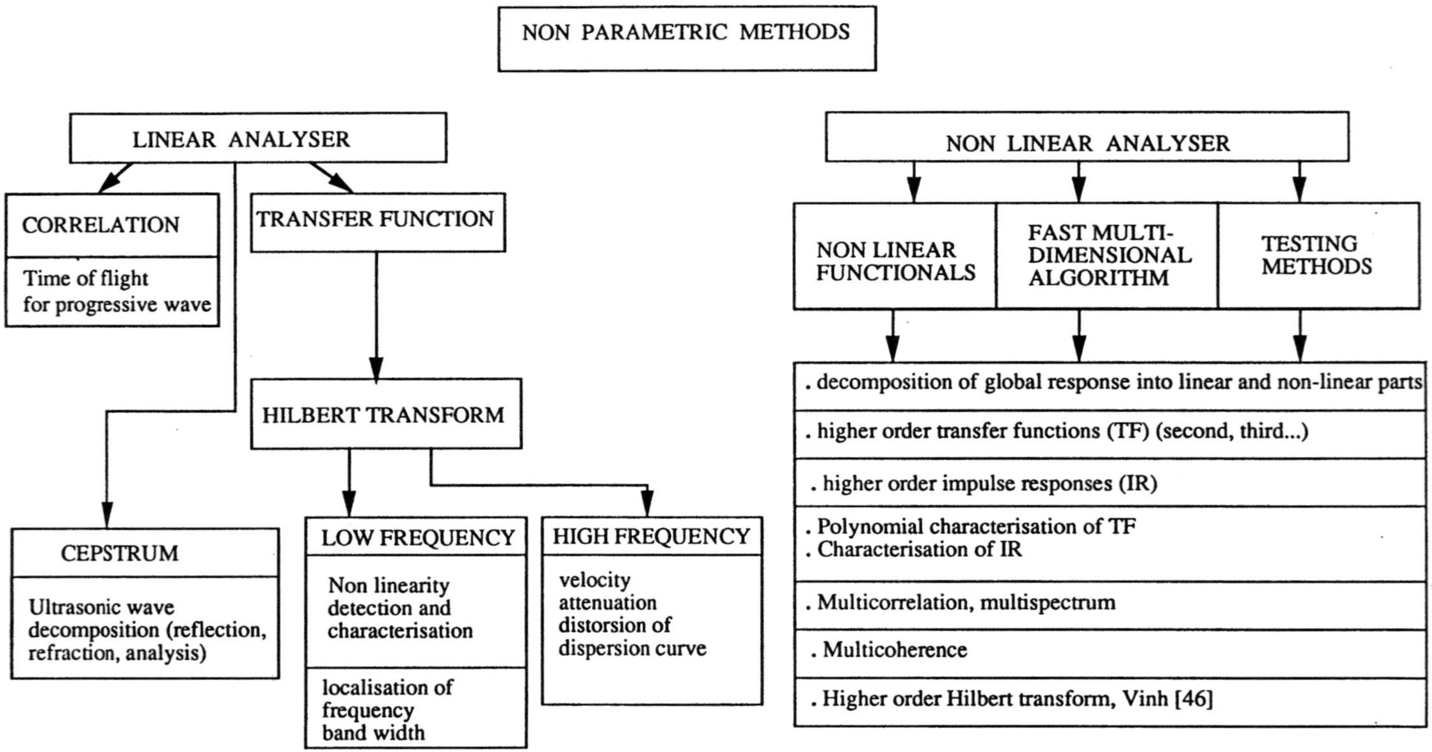

Figure (21) : Linear and non-linear analysers for non parametric study of structures

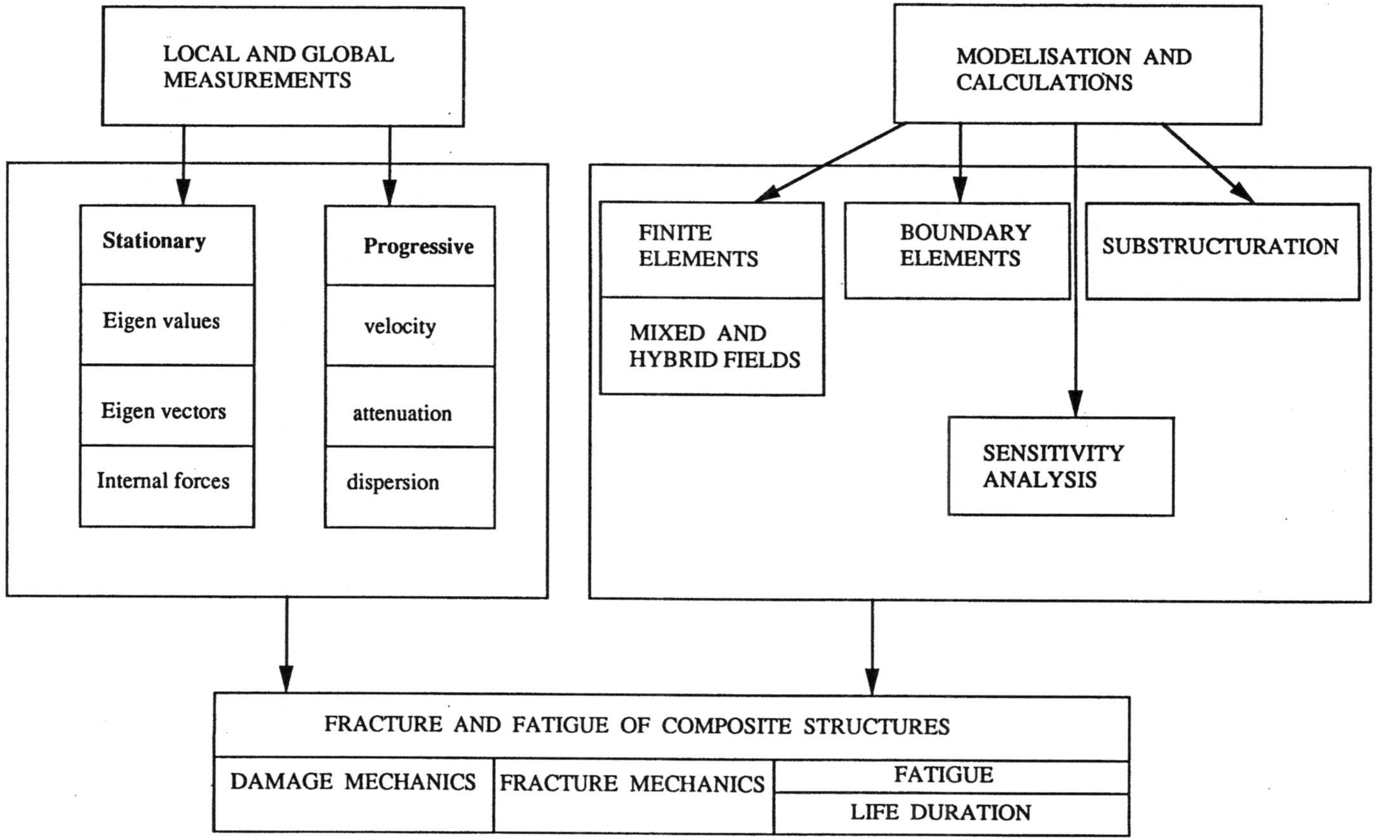

Figure 22 : Three aspects of the study of damage in composite based structure

functionals or better a series of non-linear functionals. Volterra's series is at our knowledge the best mathematical framework to describe non-linear structural behaviour for applications, Liu [45].

The practical interests of this series are the following :

- **Global response is decomposed into linear component and non-linear components** of various orders.
- Damages in structure create more or less non-linear components in the response. We have to extract them from global response.
- The multiple manifestations of non-linear structural behaviour are known in non-linear mechanics. Harmonic, subharmonics, combinational (integer or fractional) resonance frequencies, transfer of energy from one frequency to another these phenomena can be described thoroughly using Volterra's series.
- Modal analysis, initially tailored for linear systems, can be extended to non-linear systems by using this series and mathematical formulations of higher order transfer functions and impulse responses [45].

b) **Multidimensional (time and frequency) Fourier transforms and fast algorithms**

Popular Cooley Tuckey FFT algorithm is not the fastest.

It is possible to use faster algorithm [43] to reduce time of computation in a drastic manner. Multidimensional FFT permits evaluation of transfer functions and/or impulse responses of order higher than one.

Figure (21) collects some principal functions in non-linear analyser.

V - CONCLUDING REMARKS

We have presented some of our researches on composite based structures.

We have not covered all the existing methods using dynamic damage indicators. in this seminar, many lectures are also devoted to this problem (i.e. special integral transforms : Wigner-Ville, wavelets, time Hilbert transform, etc...).

To be complete analysis of damage composite structure requires three group of studies.

a) **Global and local measurements of damage using stationary waves and progressive ultrasonic waves** respectively.

b) **Fracture and fatigue of composite structure** which remains an open subject. **Life duration has to be related to static fracture of equal** rank. Damage evolution in composite material has been analysed but not completely.

c) **Modelisation and calculations** constitute the third aspect of the problem.

Finite elements although used with success to analyse structure require additional works (i.e. mixed or hybrid fields) to take into account characteristics inherent to presence of a number of plies.

REFERENCES

[1] **N.G. Natke, C. Cempel, A. Ziolkowski** : (1992) "Applications of transformed normal modes for damage location in structures" in Structural Integrity assessment - Edited by P. Stanley - Elsevier applied Science edit London - pp 246-255

[2] **A.D. Dimarogonas, C.A. Papadopoulos** : (1983) "Vibration of cracked shafts in bending" Jour. Sound and Vibrations - vol 91 - pp 583-593

[3] **D. Gross, Ch. Zhang** : (1992) "Wave propagation in damaged solids" - Int. J. Solids Structures vol 29 N° 14/15 - pp 1763-1779

[4] **J.D. Achenbach** : (1990) "From ultrasonics to failure prevention" in Elastic waves and Ultrasonic Non desctructive Evaluation S.K. Datta, J.D. Achenbach, Y.S. Rajapakse editors North-Holland Amsterdam

[5] **Y.H. Pao** : (1983) "Elastic waves in solids" J. Appl. Mechanics Trans. ASME vol 50 - pp. 1152-1164

[6] **L.J. Bond** : (1990) "Numerical techniques and their use to study wave propagation and scattering. A review" in Elastic waves and Ultrasonic Nondestructive Evaluation SK Datta, J.D. Achenbach, Y.S. Rajapakse Editors North-Holland Amsterdam

[7] **J.A.G. Temple** : (1987) "European developments in theoretical modelling of NDE for pipework" Int J. Pres. Ves. and Piping - vol 28 - pp.227-267

[8] **W.J.Cantwell, J. Morton** : (1992) "The significance of damage and defects and their detection in composite materials : a review" - Jour. of Strain Analysis - Vol 27 n° 1 - pp 29-42

[9] **L.M. Kachanov** : (1958) Izv. AN. SSR. Otd. Tekhn. Nauk. 8 p. 26

[10] **J. Lemaitre and J.L. Chaboche** : (1978) "Aspects phénoménologiques de la rupture par endommagement" - Journal de Mécanique Appliquée Vol 2, n° 3 - pp 317-365

[11] **R. Talreja** : (1987) "Fatigue of composite materials" Technomic Publishing Compagny Lancaster USA

[12] **K. Kamimura** : (1985) "Modélisation théorique de la croissance de l'endommagement appliquée à la théorie des plaques stratifiées" Journal de Mécanique Appliquée vol. 4 n°4 - pp 537-553 - pp 83-112

[13] **A. Highismith , K.L. Reifsnider** : (1982) "Damage in composite materials" ASTM STP 775 Edi : KL Reifsnider Philadelphia - pp 103-117

[14] **D.H. Allen, C.E. Harris, S.E. Grives, R.G. Norvell** : (1988) "Characterisation of stiffness loss in crossply laminates with curved matrix cracks" Journ. Composite Materials Technomic, vol 22 n°1 - pp 71-80

[15] **H.G. Natke, C. Cempel** : (1990) "Damage measures and evolution report" CRI-B-1/90 Curt Risch Institut, Univ. Hannover

[16] **C. Cempel, H. Natke, A. Ziolkowski** : (1992) "Application of transformed normal modes for damage location in structures" in Structural Integrity Assessment Edit. P.Stanley, Elsevier Applied Science, London

[17] **C. Williams** : (1992) "Testing of large structures using vibration techniques" in Structural integrity assessment Edit. P.Stanley, Elsevier Applied Science London, pp 290-299

[18] **M. Touratier** : (1979) "On the propagation of waves on transverse isotropic rectangular rods" Docteur-Ingenieur thesis (in French) University of Paris VI

[19] **A. Mal, L. Knopoff** : (1967) "Elastic wave velocities in two component systems" J. Inst. Math applicns vol 3 - pp 376-387

[20] **J. Aboudi** : (1988) "Wave propagation in damaged composite materials" Int. J. Solids Structures vol 24 - pp 117-138

[21] **S.K. Datta, H.M. Ledbetter, et al** : (1988) "Phase velocity and attenuation of plane elastic waves in a particle reinforced composite medium, Wave motion vol 10 - pp 171-182

[22] **S.C. Feng, D. Sornette** : (1991) "Acoustical non destructive evaluation fo heterogenous materials in the multiple scattering regime" J. Acoust. Soc. America vol 90 n° 4 - pp 1742-1748

[23] **A.J. Devaney** : (1990) "Elastic wave scattering" in Elastic waves and non destructive Evaluation Edit S.K. Datta, J.D. Achenbach, Elsevier Science Publ. - pp 29-38

[24] **V.K. Varadan, V.V. Varadan** : (1985) "A multipliscattering theory for elastic wave propagation in discrete random medium" J. Acoust. Soc. Ameri. vol 63 - pp 1310-1319

[25] **R.D. Mindlin** : (1964) "Microstructure in linear elasticiy" Arch. for Rat Mechanics and Analysis vol 16 - pp 51-78

[26] **A.C. Eringen** : (1968) " Theory of micropolar elasticity in fracture" Edited by H. Liebowitz vol 2 Academic Press New York

[27] **A. Bedford, M. Stern** : (1971) "Toward a diffusing continuum theory of composite materials" Journ. Appl. Mechanics vol 38n° 1 - pp 8-14

[28] **Z. Hashin, S. Shtrikman** : (1963) "A variational approach to the theory of elastic behaviour of multiphase materials" J. Mech. Phys. of Solids vol 11 n° 2 - pp 127-140

[29] **J.R. Willis** : (1980) "A polarisation approach to the scattering of elastic waves"
I - Scattering of an simple inclusion - pp 287-305
II - Multiple scattering from inclusions - pp 307-327
J. Mech. Phys. Solids vol 28

[30] **S.M. Rytov** : (1956) "Acoustical properties of thinly laminated medium" Soviet Physical Acoustics vol 3 n° 2 - pp 68-80

[31] **Y. Chevalier** : (1986) "Wave dispersion in elastic periodic composite by use of polarisation tensors" Ph. D. thesis University of Paris VI, prepared at ISMCM Laboratory of Rheology (in French)

[32] **S.K. Datta, A.H. Shah, W. Karunasena** : (1990) "Wave propagation in composite media and material characterisation" in Elastic Waves and Ultrasonic Non destructive Evaluation Edited by Datta and Achenbach, Elsevier Publ. North Holland

[33] **D. Noiret, J. Roget** : (1989) "Calculation of wave propagation in composite materials using wave concept" Journ. Composite Materials vol 23 - pp 195-206

[34] **Y. Chevalier, M. Louzar, G.A. Maugin** : (1991) "Surface wave characterisation of interface between two anisotropic media" Journ. Acoust. Soc. Americ. vol 90 n°6 - pp 3218-3227

[35] **A. Satour** : (1981) "Ultrasonic wave propagation in multilayered media" Ph. D. thesis - University of Paris VI (in French)

[36] **Y. Chevalier** : (1975) "Various methods of previsional calculation on composite materials" Sciences et technique de l'armement 493ème fasc. - pp 417-466 (in French)

[37] **G.S. Murty** : (1975) "Wave propagation at an unbonded interface between two elastic half spaces" J. Acoust. Soc. Am. vol 58 - pp 1094-1095

[38] **D.M. Barnett, J. Lothe, S.D. Gavazza, M.J.P. Musgrave** : (1985) "Consideration of the existence of interfacial (Stoneley) waves in bonded anisotropic half spaces" Proc. R. Soc. London Ser.A vol 402 - pp 153-166

[39] **S.A. Billing** : (1987) Short Course Note, Heriot Watt University - Parametric methods of analyzing linear and non-linear system

[40] **T. Vinh** : (1985) "Theory and practice of Hilbert transform" Course note Simon Engineering Laboratory University of Manchester

[41] **G. Tomlinson** : (1987) "Developments in the use of Hilbert transforms for detecting and quantifying non -linearity associated with frequency response" Mechanical Systems and Signal Processing vol 1 - pp 151-171

[42] **A.I. Beltzer** : (1989) "The effective dynamics response of random composite and polycrystals a survey of the causal approach" Wave motion vol 11 - pp 211-229

[43] **T. Vinh, H. Liu** : (1991) "Multidimensional signal processing for non-linear structural dynamics" Mechanical Systems and Signal Processing vol 5 n°1 - pp 61-80

[44] **E.J. Powers, C.H. Choi, R.W. Miksad** : (1985) "Application of digital cross-bispectral analysis techniques of model the non-linear responses of a moored vessel system in random sea" Journ. of sound and Vibrations vol 99 - pp 309-326

[45] **H. Liu** : (1989) "Multidimensional signal processing and Volterra series" Ph. D. thesis CNAM Paris (in French)

[46] **T. Vinh, T. Chouychai** : (1991) "Multidimensional frequency Hilbert transform to validate higher order transfer function" Proceedings of the Florence Modal Analysis Conference - Italy Septembre 10-11-12 - Palazzo degli Affari - Florence - pp 81-88

Identification of Parameters, Ill-Posedness and Adaptive Systems

J. Baumeister
Johann Wolfgang Goethe University
Department of Mathematics
Frankfurt/Main, Germany

Abstract: In this survey some issues in solving ill-posed nonlinear problems and some aspects of parameter identification are discussed. Moreover, we present a collection of inverse problems which may serve as examples of (nonlinear) ill-posed problems.

1 Inverse Problems

Classical mathematical physics had traditionally been devoted to the mathematical description of natural processes. While direct problems are usually associated with the mathematical description of processes or phenomena in nature, inverse problems may be described as problems of determining the internal structure of systems or causes of phenomena from indirect measurements as well as possible:

Everybody wants something for nothing !

(see [Fi]). Of course one cannot get something for absolutely nothing but one can get something (a solution) for something (some sort of a priori information).

1.1 Ill-posedness

It turns out that usually inverse problems are ill-posed in the sense of Hadamard. A problem is defined to be **well-posed** if the following conditions are satisfied:

Existence: The problem has a solution.

Uniqueness: The solution is uniquely determined.

Stability: The solution depends continously on the data.

Otherwise the problem is called **ill-posed.**

Notice that this definition depends on the notion of a solution (classical solution, weak solution, quasi-solution, least-squares solution, ...) and on the topologies chosen in the solution and data space.

The question of existence and uniqueness is of great importance in testing the assumption behind any mathematical model. Stability is necessary if one wants to be sure that a variation of the given data in a small range leads to an arbitrary small change in the solution. Clearly, stability is an essential sufficient condition for getting good results in a computational method.

Ill-posed problems which are governed by linear equations may be investigated by using linear functional analysis and numerical linear algebra. The state of the theory for such problems may be considered as almost complete; see for instance [B] and [L]. The theory for nonlinear problems, however, is by far not so well-devoloped due to the large number of varieties of phenomena in nonlinear problems. A preliminary discussion of ill-posedness in nonlinear problems can be found in [Na] and [SV].

1.2 Examples

We survey some examples of inverse problems (which lead to ill-posed problems). The overview is not intended to be complete. Further information on inverse problems can be found in [A] and [Ki].

Can one hear the shape of a drum?

This is the title of a famous article by M. Kac [K] in which he showed that geometric properties of a domain (area, length and curvature of its boundary,...) may be elicited from the asymptotic expansion of a spectral function of the Laplacian operator for that region with appropriate boundary conditions; see [P], [Z] and [Pi].

The inverse boundary value problem of geophysical geodesy.

This problem consists in the determination of the figure of the earth and its gravity field from gravimetric data. The analysis ot this problem is related to hard implicit function theorems; see [H].

The inverse problem of geomagnetic induction.

The problem of revealing the electrical conductivity profile of a layered earth from geomagnetic induction data is solved using a modified version of the method of Gel'fand and Levitan, devised for the solution of Sturm-Liouville problems; see [W].

The inverse problem of potential theory.

Find the mass distribution of matter within the earth, given the gravitational potential on the surface of the earth. The problem is important in locating high or low density regions within the earth; see [Br].

Reflection seismology.

In exploration seismology an impulsive or vibrating load is applied at the ground surface and the particle velocity which is induced at this surface by the reflection of waves at inhomogeneties is measured. The inverse problem here is to calculate from the measurements the mechanical properties of the material in the underground; see [BB].

The inverse problem of acoustic scattering.

Given a domain in $\mathbb{R}^n$ (obstacle), n = 2 or 3, and an incident field. Then a scattered field (far away from the obstacle) may be observed. The so-called far-field amplitude of this field contains important geometrical information about the scattering object; see [CK]. An application is the reconstruction of aircrafts, missiles and submarins.

Locating of cracks by electrical prospection.

This problem is concerned with the determination of an unknown internal conductivity profile from boundary measurements of voltage potentials and corresponding current fluxes. When some structural information is already known then one can determine characteristic features like cracks by making only a finite number of boundary measurements; see [BV].

The inverse problem of electrocardiography and magnetocardiography.

The problem consists in finding out the cardiac generators of the electric activity from knowledge of external potential and electromagnetic fields, respectively; see [CGM] ,[CJ]. This techniques is playing an increasing role in medicine.

The final value problem for the heat equation.

One of the classical ill-posed problems in the sense of Hadamard is the solution of a parabolic equation backwards in time. The image reconstruction problem with Gaussian-like point spread function may be reformulated in terms of solving a heat equation backwards in time; see [CSH].

The inverse heat conduction problem.

This problem deals with extracting useful information about unknown boundary conditions in diffusive systems from limited measurements within the interior of the region: see [Di] and [BBC].

Compartment analysis.

The use of compartmental models has become a standard tool in fields of biomedical engineering. The problem consists in finding out which parameters in the model can be reconstructed from input-output data; see [An].

Parameter identification for fluid flow in porous media.

Mathematical modelling is playing an increasing role in understanding prediction and control of groundwater hydrology and contaminat transport in porous media. The inverse problem consists in the determination of parameters which appear in equations which govern the flow of the fluid; see [EL] and [B2].

X-ray tomography.

Computarized tomography is a technique to reconstruct the density of a medium from projections. The rate of absorption of a parallel beam of X-rays passing through materials is closely related to the density of the materials. The reconstruction is done by inverting the so called Radon transform; see [N1].

NMR-tomography.

In nuclear magnetic resonance tomography plane integrals of a spin density of hydrogen nuclei are measured. The reconstruction of the density from measurements can be interpreted as the inversion of the so-called Radon transform in three dimension; see [N1].

Ultrasound tomography.

Ultrasound tomography is an example of a technique currently used in medical practice which is not electromagnetic. Properties of the tissue under investigation are reconstructed by inverting a wave equation; see [G] and [N1].

Shape from probing.

Identifying and understanding objects from sensory data is a fundamental problem in robotics and computer vision. An X-ray probe through a polygon measures the length of intersection between a line and the polygon. How many probes are necessary to determine a convex polytop? See [ES].

Appearance potential spectra.

Inversion of the autoconvolution is an important problem in appearance potential spectroscopy. This is a technique to investigate electronic properties of solids in their surface region which are related to the density of unoccupied states; see [SS].

Can one construct a digital sundial?

This picturesque question has a positive answer at least in (geometric measure) theory: One can construct a set in $\mathbb{R}^3$ such that at almost all times of a day the shadow of the set gives the digits of the time. This is a consequence of the fact that one can construct subsets of $\mathbb{R}^n$ which prescribed projections onto almost all k-dimensional subspaces; see [F].

1.3 Ill-posedness in nonlinear problems

Let X and Y be Banach spaces and let $N : X \to Y$ be a (nonlinear) mapping with domain of definition $D \subset X$ and range $R \subset Y$. Consider the equation

(1) $$Nx = y$$

Following the definition in Section 1.3, the problem to solve this equation is said to be **well-posed** in the sense of Hadamard, relative to the spaces X and Y, if the following conditions are satisfied:

i) $R = Y$;

ii) $Nx = Ny$ implies $x = y$;

iii) the inverse operator N^{-1} is continous.

Otherwise the problem is said to be **ill-posed**.

Assume that for $y^0 \in Y$ there exists a solution $x^0 \in D$ of equation (1):

$$Nx^0 = y^0$$

Since nonlinear mappings satisfying the conditions above in a global sense are very rare it is reasonable to consider N only in a neighborhood of x^0, i.e. D is now considered as a neighborhood of x^0. If the problem (1) is ill-posed due to the fact that ii) and/or iii) is not satisfied some **a-priori restrictions** for the solution may restore these conditions:

Suppose that

$$x^0 \in Q$$

where $Q \subset D$ describes the a priori restrictions and consider now N as a mapping from Q into N(Q). Obviously, injectivity of N and continuity of N^{-1} is restored if the following condition is satisfied:

(2) There exists a continuous and increasing function $\Gamma : [0,\infty) \to [0,\infty)$ with $\Gamma(0) = 0$, $\Gamma(r) > 0$ for $r > 0$, such that

$$\|x - x'\| \le \Gamma(\|Nx - Nx'\|) \quad , x, x' \in Q.$$

Γ is called the **rate of regularity of N relative Q**. Rates of regularity which occur frequently in application are given by the following functions:

$$\begin{array}{lll} \Gamma(r) = \varkappa r^{\gamma} & : \text{Hölder-continuity} & \\ & & ;\ \varkappa,\gamma > 0. \\ \Gamma(r) = \varkappa|\ln(r)|^{-\gamma} & : \text{Logarithmic continuity} & \end{array}$$

Now, if $y^{\varepsilon} \in Y$ is a measured approximation of y^0 satisfying

$$\|y^{\varepsilon} - y\| \le \varepsilon$$

then the worst case in approximation of x^0 by an element in

$$M_{\varepsilon}(Q) := \{x \in Q \,|\, \|Nx - y^{\varepsilon}\| \le \varepsilon\}$$

can be estimated by

$$\omega(Q,\varepsilon) := \sup\{\|x - x^0\| \,|\, x \in M_{\varepsilon}(Q)\}.$$

From (2) then follows

$$\omega(Q,\varepsilon) \le \Gamma(2\varepsilon).$$

In order to apply the estimate above, we need a specific method to construct x^{ε} in $M_{\varepsilon}(Q)$. The least-squares method

$$\text{minimize } J(x) := \|Nx - y^{\varepsilon}\|^2 \quad \text{subject to } x \in Q$$

is such a method since $x^0 \in Q$. Tikhonov's method may be used too; we come back to this method in Chapter 3.

Now, it is clear that the main analytic problem consists in finding a subset Q of D and a function Γ such that (2) is satisfied. One might expect that the method of linearization may be used to derive such estimates. However, examples show that the connection between ill-posednes of a nonlinear problem and its linearization is not as strong as one might expect; see [EKN]. But some justification of the method of linearization is provided in Chapter 3.

2 Parameter identification in distributed systems

Identification of parameters is an important part of system analysis and is therefore a well-studied problem if the system is governed by ordinary differential equations.

The identification of coefficients in distributed parameter systems arises in a variety of applications such as heat conduction [KT], seismic exploration [BB], reservoir simulation [E], determination of cracks [BV], transport theory [N2], electrical impedance tomography [D], and models for flexible structures [BGRW]. The term "distributed systems" means that these applications are discribed by partial differential equations; a source of references is [BK].

2.1 Ill-posedness

Let us consider a differential equation of evolutionary type

(3) $$x' = A(q)x + f(t)\ ,\ t \geq 0,$$

with an equation of observation

$$y = Cx.$$

The interpretation is clear: f is the input, x is the state variable, y is the output and q is a parameter which characterizes the system operator $A : H \to H$.
We assume:

$$x \in H,\ y \in Y,\ p \in P$$

where H and Y are normed spaces and P is a subset of a normed space $\mathbb{H}$. Of course, one has to develop a conception what should be a solution of (3), especially if the operator A(q) is an unbounded operator as it is the case if A(q) is a partial differential operator.
The **identification problem** consists in finding out the true parameter from observed data y on a time interval [0,T]. In the following, we restrict ourselves to the case that it is possible to observe the state itself; see the remark at the end of this chapter. In terms of the parameter-to-solution mapping

$$N : q \mapsto x(q;.)\ ,\ x(q;.) \text{ solution of (3)},$$

we want to consider the injectivity of N as well as aspects of stability, which refers to the continous invertibility of N. Notice that, in general, the range of N has no nice geometric properties like linearity or convexity.

To get some insight into the ill-posed nature of parameter identification it is enough to consider the most simple example

(4) $$x' = qx + f(t),\ t \geq 0,$$

where $H = Y = \mathbb{H} = \mathbb{R}$, $A(q)x = qx$; we assume that $f : [0,\infty) \to \mathbb{R}$ is continous. As it is well-known, the solution x(q;.) of (4) is uniquely determined and given by the formula

(5) $$x(q;t) = e^{qt}x(q;0) + \int_0^t e^{q(t-s)}f(s)\,ds\ ,\ t \geq 0.$$

Let $T > 0$. Then we have the following well-defined parameter-to-solution mapping which we denote by L_T instead of N:

$$L_T : P \ni q \mapsto x(q;.) \in L_2(0,T;H)$$

Let $p \in P$ and let $z := L_T p$. We have the following observations:

a) If $q \in P$ with $L_T q = z$ then $(q-p)z(t) = 0$ for all $t \in [0,T]$.
   From this we may conclude:
   If there exists $t^* \in [0,T]$ with $z(t^*) \neq 0$ then $q = p$.
   If $z(t) = 0$ for all $t \in [0,T]$ we cannot decide whether $q = p$ holds. The reason is that the experiment represented by z is not rich enough.

b) Suppose that there exists $t^* \in [0,T]$ with $z(t^*) \neq 0$.
   Then the equation $L_T q = z$ is drastically overdetermined.

c) Suppose that $z(t) = 0$ for all $t \in [0,T]$.
   Then the equation $L_T q = z$ is underdetermined.

d) Let $f(t) = 0$ for all $t \geq 0$ and let us define for each $n \in \mathbb{N}$:

$$q_n := -n, \; z_n(t) := z(q_n;t) := e^{-nt}, \; t \geq 0.$$

   We have:

$$\lim_n z_n = 0 \text{ in } L_2(0,T;H), \quad |q_n - q_{n+1}| = 1 \text{ for all } n \in \mathbb{N}.$$

   This shows that the mapping L_T is not continously invertible. The reason is that we allow a noncompact intervall of parameters.

There is a narrow ridge between overdetermination and undetermination. In order to avoid underdetermination one has to design the experiment represented by f (and the initial value z(0)) in an appropriate way. Sensitivity analysis may help to find those experiments which give the "best" identification results.

2.2 Identification in systems with a bilinear structure

Suppose that H and $\mathbb{H}$ are Hilbert spaces with inner products $(.,.)_H$ and $(.,.)_{\mathbb{H}}$, respectively, and let V be the domain of definition of the operator A(q); as a consequence, $A(q)v \in H$ for all $v \in V$.

We make the assumption that the following **decoupling**

$$(A(q)x,v)_H = (q,b(x,v))_{\mathbb{H}} \quad \text{for all } q \in \mathbb{H} \text{ and all } v \in V,$$

is possible for each x in V.

Let us sketch some examples which demonstrate that sufficiently many problems belong to this class of identification problems.

1) **Identification of a matrix.**

$\mathbb{H} = \mathbb{R}^{n,n}$, $H = \mathbb{R}^n$; $\quad A(q)z := qz.$

Then: $\quad b(z,v) = vz^t$

2) **Identification of the diffusion coefficient in a parabolic equation.** *)

$\mathbb{H} = L_2(0,1)$, $H = L_2(0,1)$; $A(q)z := \partial_\xi(q\partial_\xi z)$ for $z \in V := H^2(0,1) \cap H_0^1(0,1)$.

Then: $\quad b(z,v) = \partial_\xi z \, \partial_\xi v.$

3) **Identification of a "nonlinear diffusion coefficient" in a parabolic equation.**

$\mathbb{H} = \mathbb{R}^n$, $H = L_2(0,1)$; $\quad A(q)z := \partial_\xi(k(q,z)\partial_\xi z)$ for $z \in V := H^2(0,1) \cap H_0^1(0,1)$ where $k(q,z) = \sum_{i=1}^{n} q_i z^{i-1}$ describes a polynomial dependence of the diffusion coefficient on the solution.

Then: $\quad b(z,v) = (z^{i-1} \partial_\xi z \partial_\xi v)_{1 \le i \le n}.$

Let $T > 0$ and let N be again the coefficient-to-solution mapping in the time interval $[0,T]$. Let $p \in D$ and let $z = Np$. As we know from the preceding section, injectivity in a neighborhood of p holds only under additional assumptions. Such an assumption may be formulated with the family of mappings

$$L_{t_1,t_2} : V \ni v \mapsto -\int_{t_1}^{t_2} b(z(s),v)ds \in \mathbb{H} \quad , 0 \le t_1 \le t_2 \le T,$$

which is well-defined if z is a sufficiently regular solution of the differential equation (1).

Let $u = Nq$ be a solution of the differential equation governed by the parameter $q \in D$. Then by formal calculations we obtain for each $v \in V$

$$\text{(6)} \qquad \int_{t_1}^{t_2} (u'(s) - z'(s) - A(q)(u(s) - z(s)),v)_H ds = (q - p, L_{t_1,t_2}v)_{\mathbb{H}}$$

and we may conclude that $q = p$ if $u = z$ provided

(7) $\quad$ range (L_{t_1,t_2}) is dense in $\mathbb{H}$ for some pair (t_1,t_2).

This shows that (7) is a sufficient condition for identifiability of p.

Suppose that the condition

(8) $\quad L_{t_1,t_2}$ has a continous inverse on $\mathbb{H}$ for some pair (t_1,t_2)

is satisfied. (Obviously, (8) is stronger than (7)). Now, the equation

$$L_{t_1,t_2} v = q - p$$

*) $H^2(0,1)$, $H_0^1(0,1)$ are the usual Sobolev spaces.

can be solved in a stable way. Using (6), an estimate of the following type may be derived:

$$\|q - p\|_{\mathbb{H}} \le \|u - z\|_{H}^{\nu} \quad (\nu \in (0,1)).$$

This is a stability estimate of Hölder type.

In the finite dimensional case - see example 1) above - the condition (7) is satisfied if one of the following condition holds:

a) $\operatorname{span}\{z(s) \mid t_1 \le s \le t_2\} = \mathbb{R}^n$;

b) $\int_{t_1}^{t_2} z(s) z(s)^t ds$ is positive definite;

c) There exists $\varepsilon > 0$ such that for all $h \in \mathbb{H}$ $\left\| \int_{t_1}^{t_2} h z(s)\, ds \right\| \ge \varepsilon \|h\|$.

These conditions play an important role in the convergence-analysis of adaptive identification methods; see for instance [Kr], [ID] and [J]. In the next section we sketch such an adaptive methods in a not necessarily finite-dimensional setting.

2.3 Model reference adaptive systems

Adaptive methods (on-line methods) have been developed by engeneers for problems which are governed by ordinary differential equations. The idea of such methods is the following:

Adjust dynamically a parameter $q(t)$ on the basis of information obtained during the observation procedure up to time t so that $\lim_{t\to\infty} q(t) = p$.

Necessarily, we have to know the output z for all times $t \ge 0$. This is not a serious restriction since we may continue data on an interval $[0,T]$ to a periodic function on $[0,\infty)$.

We sketch three methods. Each such method consists of model equation for a reference quantity x and an adaption rule for the parameter q:

1) $x' = A(q^*)x + A(q)z(t) - A(q^*)z(t) + f(t)$

$q' = - b(z(t), x - z(t))$

Here q^* is a fixed parameter in P.

2) $x' = A(q)x + f(t)$

$q' = - b(x, x - z(t))$

3) $x' = A(q)z(t) + f(t)$

$q' = - b(z'(t), x' - z'(t))$

Each method may be considered as a certain variant of an gradient method where the objective of minimization is an appropriate functional of the defect $x - z$. The first method has been introduced in [NK], the second one is a well-known procedure in adaptive control, the third one is the so called equation error method.

If we introduce the error quantities

$$w := x - z, \; r := q - p$$

we may derive in each case a system of differential equations for the variables w,r. Usually, method 1) leads to a simpler error system than 2). Moreover, even in the finite-dimensional case, method 2) and 3) are much more difficult to analyze than method 1). The main difficulty in the proof of the assertion $\lim_{t\to\infty} q(t) = p$ is related to the fact that each method leads to a non-autonomouos system for the error quantities w,r. As it is well-known, the long-time behaviour of such systems is much harder to analyze. Convergence results for the method 1) in an infinite-dimensional setting can be found in [S] and [B2]. Method 2) is used in [AHS] but no complete analysis has been given. It should be not surprising that in each method a condition of the type (8) which says that the output z has to be sufficiently **exciting** in order to adjust the parameter during the adaption plays an important role in order to obtain convergence results.

Two problems arise when the data z are erroneous. Firstly, we have to differentiate the data z ($\partial_\xi z, z'$, ..). This is an operation which is ill-posed. Secondly, the adaptive method has to be endowed with an appropriate stopping criterion as it usually the case in the iterative solution of ill-posed problems. Up to now, no rigorous analysis of the question how to choose such a criterion is available in the infinite-dimensional setting. In the finite-dimensional case considerations concerning this problem can be found in [ID].

A very strong disadvantage of the adaptive methods as formulated above lies in the fact that we need a distributed observation of the state. This is for problems which are governed by partial differential equations a serious requirement. Up to now, adaptive observer which are used in ordinary differential equations to overcome these difficulties are not fully developed.

Numerical results are reported in the case that (1) is a partial differential equation with state variable in $\mathbb{R}$; see [S]. They are in good aggreement with the theoretical results.

3 Regularization in nonlinear ill-posed problems

In this chapter we sketch some aspects in the regularization of a nonlinear problem, especially of Tikhonov's regularization method. The problem of parameter identification is an interesting field of application; see [KS] and [EKN].

3.1 Least-squares

Let us consider the equation

(9) $\quad Nx = y$

where X,Y are Hilbert spaces, $N : X \to Y$ is a (nonlinear) operator with domain of definition D.
Suppose that we have a solution $x^0 \in D$ of the equation for the right-hand side $y^0 \in Y$. Moreover, assume that $y^\varepsilon \in Y$ is an available approximation of y^0 with

$$\|y^\varepsilon - y^0\| \le \varepsilon.$$

The customary way to solve the problem has been least-squares:

(10) $\quad$ Minimize $J(x) := \|Nx - y^\varepsilon\|^2$ subject to $x \in D$.

But if the problem (9) is ill-posed the optimization problem has several serious difficulties even if D describes the unknown x^0 in a resonable way:

- i) The objective functional J is not convex.
- ii) There are usually many local minima.
- iii) The global minimum may be very flat.
- iv) There may be low sensitivity of interesting parts of the unknown x^0.

Contributions from the literature which are influenced by these symptoms are [C] and [CK]. Chavent defines the concept of **least-squares stability**: The problem (10) is **stable** if there exists $\varepsilon^0 > 0$ such that for each $\varepsilon \in [0, \varepsilon^0)$ the problem (10) has a unique solution which depends continuously on y^ε. Sufficient conditions for this type of stability involve the diameter of D and conditions on the first and second Frechet-derivatives of N. Observe that uniqueness of a solution requires uniqueness of the projection πy^ε of y^ε on N(D) as well as uniqueness of the inverse of N at πy^ε.
Instabilities and symptoms of ill-posedness suggest to regularize the problem. **Regularization** of a problem refers in general to solving a related problem, called the regularized problem. In this regularized problem stability should be restored.

3.2 The method of Tikhonov

Instead of (10) we consider the following optimization problem:

(11) Minimize $J_\beta(x) := J(x) + \beta\|x - x^*\|^2$ subject to $x \in D$.

Here $\beta > 0$ is the regularization parameter and x^* is an estimate of x^0.
It is interesting that the problem (11) has a unique solution in a generic sense: There is an everywhere dense subset G of X such that (11) has a unique solution for every $x^* \in G$ provided the functional J is lower semicontinuous and D is a closed bounded subset of X; see [R].
The role of Tikhonov regularization to stabilize an ill-posed problem has been studied in [KS], [SV] and [EKN]. Here we present a result which is similar to the main result in [EKN]; the difference consists in the assumptions we use for the result.

Consider the following conditions

A1 There exist Hilbert spaces V_+, V_- satisfying $V_+ \subset X \subset V_-$ such that (V_+, V_-) is a dual pair with respect to the inner product in X.

A2 There exist constants $q \geq 0$, $c > 0$, $r > 0$ such that

$$\|x - x^0\|_- \leq q\|x - x^0\|^2 + c\|Nx - Nx^0\|$$

for all $x \in D$ with $\|x - x^0\| < r$ where $rq < 3/2$.

A3 $x^0 - x^* \in V_+$ and $\|x^* - x^0\|_+ < r/3$.

Let $x^{\varepsilon,\beta,\eta}$ be an element in D satisfying

$$J_\beta(x^{\varepsilon,\beta,\eta}) \leq \inf\{J_\beta(x) | x \in D\} + \eta$$

Then the following **convergence result** holds:

Suppose that the conditions A1 - A3 are satisfied and assume that the parameter β and the accuracy η are chosen as follows:

(12) $\beta = O(\varepsilon)$, $\varepsilon^2/\beta(\varepsilon) = o(1)$, $\eta = O(\varepsilon^2)$, $\eta/\beta(\varepsilon) = o(1)$ as ε goes to zero.

Then

$$\| x^{\varepsilon,\beta,\eta} - x^0\| = O(\sqrt{\varepsilon}), \quad \|N x^{\varepsilon,\beta,\eta} - Nx^0\| = O(\varepsilon).$$

The proof of this result will be given in [B3]. The order of convergence is the same as in [EKN]. The main difference in the assumption is that we avoid the use of the space range$(dN(x^0)^*dN(x^0))$ where $dN(x^0)$ is the Frechet derivative of N at x^0. A2 is the crucial assumption. It can be verified by a linearization technique using the Frechet-differentiability of N in the following way:

There exist constants $c_1 > 0$, $c_2 \geq 0$, $r' > 0$ such that

A2' $\|x - x^0\|_- \leq c\|dN(x^0)(x - x^0)\|$, $\|Nx - Nx^0 - dN(x^0)(x - x^0)\| \leq c_2\|x - x^0\|^2$ for all $x \in D$ with $\|x - x^0\| < r'$.

The "gap" between the spaces V_+ and V_- in connection with A3 may be considered as a measure of the degree of ill-posedness of the problem (9).

Let now $x^{\varepsilon,\beta}$ denote a solution of the minimization problem (11). In the proof of the convergence result above the normal equation

$$dN(x^{\varepsilon,\beta})^*(y^\varepsilon - Nx^{\varepsilon,\beta}) = \beta(x^{\varepsilon,\beta} - x^*).$$

is not used; here $dN(x^{\varepsilon,\beta})^*$ is the (Hilbert) space adjoint of $dN(x^{\varepsilon,\beta})$. This equation should be helpful to derive a better rate of convergence; see [Ne].

(12) is an a priori strategy, a posteriori strategies may be formulated by various so-called discrepancy principles; we mention two of them ($\varkappa > 0$):

Choose $\beta > 0$ such that $\|Nx^{\varepsilon,\beta} - y^\varepsilon\| = \varkappa\varepsilon$.

Choose $\beta > 0$ such that $\|dN(x^{\varepsilon,\beta})^*(N(x^{\varepsilon,\beta}) - y^\varepsilon)\| = \varkappa\varepsilon/\beta$.

Some investigations have been done in [EKN] and [KS] for the first strategy, the second principle is well-known in the linear theory where it is optimal in a certain sense, in nonlinear problems no results seem to be known.

References

[A] Anger, G.: Inverse and improperly posed problems in differential equations. Akademie Verlag, Berlin, 1989.

[An] Anderson, D.H.: Compartmental modelling and tracer kinetics. Lecture Notes in Biomathematics, Vol. 50, Springer Verlag, 1983.

[AHS] Alt, H.W., Hoffmann, K.-H., Sprekels, J.: A numerical procedure to solve certain identification problems. In: Optimal control of differential equations, 11-43, eds. Hoffmann, Krabs, ISNM 68, Birkhäuser, Boston, 1984.

[B1] Baumeister, J.: Stable Solution of Inverse Problems. Vieweg Braunschweig, 1987.

[B2] Baumeister, J.: Parameter identification by model reference adaptive systems. Adv. in Water Resources 14 (1991), 98-103.

[B3] Baumeister,J.: A note on Tikhonov regularization for nonlinear problems. In preparation.

[Br] Brodsky, M.A.: On the uniqueness of the inverse potential problem for homogeneous polyhedrons. SIAM J. Appl. Math. 46 (1986), 345 - 350.

[BB] Bube, K.P., Burridge,R.: The one-dimensional inverse problem of reflectior seismology. SIAM Review 25 (1983), 497 - 559.

[BK] Banks, H.T., Kunisch,K.: Estimation Techniques for Distributed Parameter Systems. Birkhäuser Boston, 1989.

[BV] Bryan,K., Vogelius,M.: A uniqueness result concerning the identification of a collection of cracks from finetely many electrostatic boundary measurements. SIAM J. Math. Anal. 23 (1992), 950 - 958.

[BBC] Beck, J.V., Blackwell, B., St. Clair, C.R.Jr.: Inverse heat conduction problems. Wiley, New York, 1985.

[BHZ] Böcker,J., Hartmann,I., Zwanzig,Ch.: Nichtlineare und adaptive Regelungssysteme. Springer, Berlin, 1986.

[BGRW] Banks,H.T., Gates,S.S., Rosen,I.G., Wang,Y.: The identification of a distributed parameter model for a flexible structure. SIAM J. Control and Optimization 26 (1988), 743 - 762.

[C] Chavent,G.: On the theory and practice of non-linear least-squares. Adv. in Water Recources 14 (1991). 55 - 63.

[CJ] Clarke,C.J.S., Janday,B.S.: The solution of the biomagnetic inverse problems by maximum statistical entropy. Inverse Problems 5 (1989), 483 - 500.

[CGM] Colli Franzone,P., Guerri,L., Magenes,E.: Oblique double layer potentials for the direct and inverse problems of electrocardiology. Math. Biosci. 68 (1984), 23 - 55.

[CK] Colonius,F., Kunisch,K.: Output least squares stability in elliptic systems. Appl. Math. Optim. 19 (1989), 33 - 63.

[CKr] Colton,D., Kress,R.: Integral Equation Method in Inverse Scattering Theory. Academic Press New York, 1983.

[CSH] Carasso, A.S., Sanderson, J.G., Hyman; J.M.: Digital removal of random media image degradations by solving the diffusion equation backwards in time. SIAM J. Numer. Anal. 15 (1978), 344 - 367.

[D] Dobson,D.C:: Convergence of a reconstruction method for the inverse conductivity problem. SIAM J. Appl. Math. 52 (1992), 442 - 458.

[Di] Dinh Nho Hao: Inverse heat conduction problems. Dissertation, Freie Universität Berlin, 1991.

[E] Ewing,R.E.: Determination of coefficients in reservoir simulation. In: Numerical treatment of inverse problems in differential and integral equations; eds. Deuflhard, Hairer; Birkhäuser, 1983.

[EL] Ewing, R.E., Lin, T.: A class of parameter estimation techniques for fluid flow in porous media. Adv. Water Resources 14 (1991), 68 - 97.

[ES] Edelsbrunner, H., Skiena, S.S.: Probing convex poygons with X-Rays. SIAM J. Comput. 17 (1988), 870 - 882.

[EKN] Engl,H.W., Kunisch,K., Neubauer,A.: Convergence rates for Tikhonov regularization of non-linear ill-posed problems. Inverse Problems 5 (1989), 523 - 540.

[F] Falconer, K.J.: Sets with prescribed projections and Nikodym sets. Proc. London Math. Soc. 53 (1986), 48 - 64.

[Fi] Fienup, J.R.: Feature issue on signal recovery: Foreword. J. Opt. Soc. Am. 73 (1983), 1412.

[G] Greenleaf, J.F.: Computerized tomography with ultrasound. Proc. of the IEEE 71 (1983), 330 - 337.

[H] Hörmander,l.: The boundary poblems of physical geodesy. Arch. Rat. Mech. Anal. 62 (1976), 1 - 89.

[HS] Hsiao,G.C., Sprekels,J.: A stability result for distributed parameter identification in bilinear systems. Math. Methods Appl. Sci. **10** (1988), 447 - 456.

[ID] Ioannou,P.A., Datta,A.: Robust adaptive control: A unified approach. Proc. of the IEEE 79 (1991), 1736 - 1768.

[J] Janecki,D.: Persistency of excitation for continous - time systems. Time domain approach: Systems & Control Letters 8 (1987), 333-344.

[K] Kac, M.: Can one hear the shape of a drum? Amer. Math. Monthly 73 (1966), 1 - 23.

[Ki] Kirsch, A.: Inverse problems. Institut für Numerische und Angewandte Mathematik, Universität Göttingen. NAM-Bericht Nr. 55, 1986.

[Kr] Kreisselmeier,G.: Adaptive observers with exponential rate of convergence. IEEE Trans. Aut. Contr. Vol. 22 (1977), 509 - 535.

[KK] Kohn,R.,V., Mc Kenney,A.: Numerical implementation of a variational method for electrical impedance tomography. Inverse problems **6** (1990), 389 - 414.

[KS] Kravaris,C., Seinfeld,J.h.: Identification of parameters in distributed parameter systems by regularization. SIAM J. Control Opt. 23 (1985), 217 - 241.

[KT] Kitamura,S., Tanigushi,Y.: Parameter estimation of one and two-dimensional heat conduction processes using a minicomputer. Int. J. Contr. 34 (1977), 785 - 802.

[L] Louis,A.K.: Inverse und schlecht gestellte Probleme. Teubner Studienbücher, 1989.

[M] Morgan,A.P.: On the construction of non-autonomous stable systems with applications to adaptive identification and control. SIAM J. Control and Optimization 17 (1979), 400 - 431.

[N1] Natterer,F.: The Mathematics of Computerized Tomography. Wiley, 1986.

[N2] Natterer,F.:Identifying a coefficient in a stationary Boltzmann equation. Publ. dell'Instituto di Analisi Globale e Applicazioni, n. 38, Firenze, 1988.

[Na] Nashed,M.Z.: On nonlinear ill-posed problems I: Classes of operator equations and minimization of functionals. In: Nonlinear Analysis and Applications, ed. Lakshmikantham, 1987.

[Ne] Neubauer,A.: Tikhonov regularization for non-linear ill-posed problems: optimal convergence rates and finite-dimensional approximation. Inverse Problems 5 (1989), 541 - 557.

[NA] Narendra,K.S., Annaswamy,A.M.: Persistent excitation in adaptive systems. Int.J.Contr. 45 (1987), 147 - 160.

[NK] Narendra,K.S., Kudva,P.: Stable adaptive schemes for system identification and control I,II. IEEE, SMC - 4 (1974), 542 - 560.

[Pi] Pietsch, A.: Can one hear the structure of a Banach space? Integral Equations and Operator Theory 12 (1989), 751-756.

[R] Revalski,J.P.: Generic properties concerning well-posed optimization problems. Compt. rend. l'Acad. bulg. des Sci. 38 (1985), 1431 -1434.

[S] Scondo,W.: Ein Modellabgleichsverfahren zur adaptiven Parameteridentifikation in Evolutionsgleichungen. Dissertation, Universität Frankfurt/M., 1987.

[SS] Schleicher,K.-Th., Schulz,S., Gmeiner,R., Chun,H.-U.: A computational method for the evaluation of highly resolved DOS-functions from APS -measurements. J.Electron Spectrosc. and Relat. Phenom. 32 (1983). 33 - 45.

[SV] Seidman,T.I., Vogel,C.R.: Well-posedness and convergence of some regularization methods for nonlinear ill-posed problems. Inverse Problems 5 (1989), 227 - 238.

[W] The inverse problem of geomagnetic induction. Zeitschrift für Geophysik 38 (1972), 257.

[Z] Zayed, E.M.E.: Hearing the shape of a general convex domain. J. Math. Anal. and Appl. 142 (1989), 170 - 187.

Estimation of Physical Parameters for Dynamic Processes with Application to Fault Detection of an Industrial Robot

R. Isermann
Technical University Darmstadt, Germany
Institute of Control Engineering

Abstract

The unknown parameters of physical laws which govern the dynamics of processes can be gained from measured input and output signals in a two step procedure. First the parameters of differential equations are estimated and then the physical parameters are calculated based on algebraic relations. After considering the model structures a suitable parameter estimation method for continuous time signals is described and the influence of a priori known parameters on the convergence is shown. Then the determination of physical parameters is considered based on nonlinear algebraic relations known from theoretically modeling. Applying the implicit function theorem identifiability conditions for the unique determination of physical parameters are given. Then applications are shown for the estimation of the ratios of inertia, d.c. motor damping coefficients and Coulomb friction coefficients of an 6-axis-industrial robot and the use of parameter changes for the detection of faults.

1. Introduction

Many contributions do exist for the estimation of parameters $\underline{\theta}$ of input-output models from measured input-output signals e.g. Eykhoff (1974), Sorensen (1980), Young (1984), Isermann (1988). However, only some few publications are known for the dermination of *physically defined parameters* $\underline{p}$ of the laws which govern the process dynamics, Walter (1982), Raksanyi et al. (1985), Dasgupta et al.(1988) see Fig. 1.

These physical parameters will be called *process coefficients*. For usual tasks in control engineering, e.g. the design of control systems, the knowledge of the process parameters $\underline{\theta}$ in general is sufficient. However, the knowledge of process coefficients $\underline{p}$ is required for following problems:

- Determination of nonmeasurable coefficients in natural sciences
- Checking of performance data for technical systems

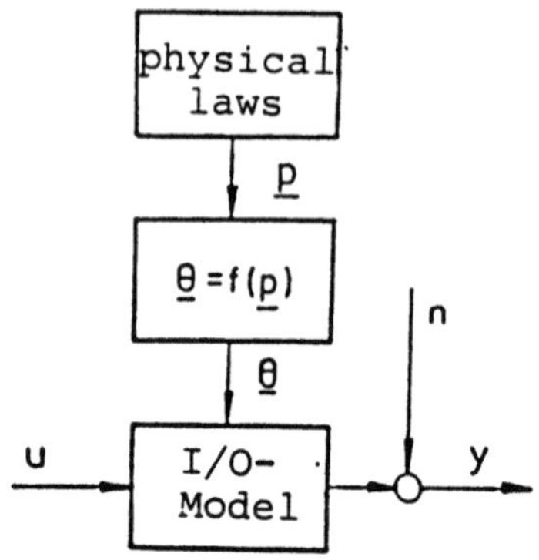

Fig. 1 Dynamic process model with model parameters $\underline{\theta}$ and physical process coefficents $\underline{p}$

- Supervision and fault diagnosis during on-line operation of technical processes
- Quality control in manufacturing.

The determination of process parameters $\underline{\theta}$ from measured input and output signals with parameter estimation methods has obtained a mature status during the last 20 years. For linear and some nonlinear processes with stochastic disturbances several methods are known. They result in a good convergence, if the model structures fits with the process structure, the input sufficiently excites the dynamics and the number of parameters is not more than 6 to 8 for SISO-systems.

The process model parameters now depend on physical process coefficients according to more or less complicated algebraic relations

$$\underline{\theta} = f\ (\underline{p}). \qquad (1)$$

The relations are known from theoretically modeling. The task now consists in determining the physical process coefficients based on measured input and output signals u(t) and y(t). A straightforward possibility is to first estimate the model parameters $\underline{\theta}$ and then to use the inverse relationship

$$\underline{p} = f^{-1}(\underline{\theta}). \qquad (2)$$

Then following questions arise:

a) Can the process coefficients $\underline{p}$ be determined uniquely (process coefficient identifiability)?

b) Which signals must be measured in order to determine the process coefficients $\underline{p}$?

c) What is the influence of a priori known process coefficients p_i on the variance of $\underline{\theta}$ and $\underline{p}$?

In this context Raksanyi, Walter, et al (1985) consider the identifiability of biological models, especially compartmental models, with computer algebra. Dasgupta, Anderson and Kaye (1988) proposed a two step procedure to estimate physical parameters under the assumption that a considerable number of physical parameters is known. The unknown parameters are then estimated by using a gradient method with the known parameters as boundary condition. Nold and Isermann (1986) gave some basic relationships between physically defined process elements and the resulting model structure and show that the model parameters are the sums of products of process coefficients.

2. On the model structure for processes with lumped parameters

The basic equations of dynamic lumped processes in form of balance equations, physical state equations and phenomenological laws can be represented by unified relations Isermann (1988).

It is assumed that the process is linear and posesses M measurable signals $\eta_j(t)$ and N nonmeasureable signals $\xi_j(t)$. For L elements the Laplace transformed *process element equations* become

$$\sum_{j=1}^{M} g_{ij}\eta_i(s) = \sum_{j=1}^{N} h_{ij}\xi_j(s) \qquad i = 1,2,\ldots, L \qquad (3)$$

with

$$g_{ij} \in (0, \pm 1, \alpha_{ij}, s^{\kappa})$$

$$h_{ij} \in \{0, \pm 1, \beta_{ij}, s^{\kappa}\} \qquad (4)$$

$$\kappa \in \{-1, 0, 1\}$$

In matrix notation this is

$$\underline{G}\,\underline{\eta} = \underline{H}\,\underline{\xi} \qquad (5)$$

with

$$\underline{\eta}^T = [\eta_1, \eta_2, \ldots, \eta_M] \qquad (6)$$

$$\underline{\xi}^T = [\xi_1, \xi_2, \ldots, \xi_N] \qquad (7)$$

$\dim(\underline{G}) = L \times M$; $\dim(\underline{H}) = L \times N$.

It is further assumed that the process element equations are linearly independent.

For the parameter estimation a model structure with only measurable signals is required. The nonmeasurable signals are eliminated by transforming the matrix $\underline{H}$ in an upper triangular form or by subsequent insertion of the single equations. One then obtains an input/output differential equation

$$a_n^* y^{(n)}(t) + \ldots + a_1^* y^{(1)}(t) + a_0^* y(t)$$
$$= b_0^* u(t) + b_1^* u^{(1)}(t) + \ldots + b_m^* u^{(m)}(t). \qquad (8)$$

However, for parameter estimation the following form is required

$$a_n y^{(n)}(t) + \ldots + a_1 y^{(1)}(t) + y(t)$$
$$= b_0 u(t) + b_1 u^{(1)}(t) + \ldots + b_m u^{(m)}(t) \qquad (9)$$

such that the regression model

$$y(t) = \underline{\psi}^T(t)\underline{\theta}, \qquad (10)$$

with

$$\underline{\psi}^T(t) = [-y^{(1)}(t) \ldots -y^{(n)}(t) \;\vdots\; u(t) u^{(1)}(t) \ldots u^{(m)}(t)]$$

$$\underline{\theta}^T = [\theta_1, \theta_2, \ldots, \theta_r] = [a_1 \ldots a_n \;\vdots\; b_0 \ldots b_m]$$

can be obtained. Hence, all parameters in eq.(8) have to be multiplied by $1/a_0^*$ and the number of parameters θ_i reduces by one.

In the system of basic equations (5) the process coefficients

$$\underline{p}^T = [p_1, p_2, \ldots, p_l] \tag{11}$$

appear separately in the original form. The elements of $\underline{G}$ and $\underline{H}$ are

$$g_{ij} = p_{gij}s^{\kappa}; \; h_{ij} = p_{hij}s^{\kappa}; \quad \kappa \in (-1, 0, 1). \tag{12}$$

After transformation into upper triangular form the input-output models appears in the last row Nold and Isermann (1986). The model parameters then get the form

$$\theta_i = \sum_{\mu=1}^{q} c_{i\mu} \prod_{\nu=1}^{l} p_{\nu}^{\epsilon_{\mu\nu}}$$

$$= c_{i1}p_1^{\epsilon_{11}} \cdot p_2^{\epsilon_{12}} \ldots p_l^{\epsilon_{1l}} + c_{i2}p_1^{\epsilon_{21}} \cdot p_2^{\epsilon_{22}} \ldots p_l^{\epsilon_{2l}} + \ldots + c_{iq}p_1^{\epsilon_{q1}} \cdot p_2^{\epsilon_{q2}} \ldots p_l^{\epsilon_{ql}}$$

Hereby it is assumed that a_0^* can be represented by

$$a_0^* = \prod_{\nu=1}^{l} p_{\nu}^{\epsilon_{\mu\nu}}.$$

With the abbreviation

$$z_{\mu} = \prod_{\nu=1}^{l} p_{\nu}^{\epsilon_{\mu\nu}} \tag{13}$$

the model parameters can be written as

$$\theta_i = \sum_{\mu=1}^{q} c_{i\mu} z_{\mu} \tag{14}$$

Therefore the parameters are algebraic functions of the process coefficients p_ν.
Hence the z_μ are abbreviations for the existing products and single values of the process coefficients $p_1 \ldots, p_l$.
Eq.(14) can now be turned into matrix form

$$\begin{bmatrix} \theta_1 \\ \theta_2 \\ \vdots \\ \theta_r \end{bmatrix} = \begin{bmatrix} c_{11} & \cdots & c_{1q} \\ c_{21} & \cdots & c_{2q} \\ \vdots & & \vdots \\ c_{r1} & \cdots & c_{rq} \end{bmatrix} \begin{bmatrix} z_1 \\ z_2 \\ \vdots \\ z_q \end{bmatrix}$$

$$\underline{\theta} = \underline{C} \cdot \underline{z} \qquad (15)$$

In the following first the process model parameters θ_i are determined by parameter estimation via the measured signals. Then the unknown process coefficients p_ν are calculated.

3. Parameter estimation for processes with continuous-time signals

In this section a method for parameter estimation of continuous-time models is described which has demonstrated its applicability in several applications to electrical and mechanical processes. First all parameters are assumed to be unknown. Then the influence of a-priori-known parameters is considered.

The measured variables are assumed to be deviations from steady-state values U_{00}, Y_{00}

$$\left.\begin{aligned} u(t) &= U(t) - U_{00} \\ y(t) &= Y(t) - Y_{00} \end{aligned}\right\} \qquad (16)$$

and the output signal is disturbed by noise n(t)

$$y(t) = y_u(t) + n(t) \qquad (17)$$

After introducing the equation error e(t) the regression model eq. (10) becomes

$$y(t) = \underline{\psi}^T(t)\underline{\theta} + e(t) \qquad (18)$$

Now the signals are measured at discrete times $t=kT_0$, $k=0,1,2,\ldots, N$ with the sampling time T_0. Then N+1 equations result

$$y(k) = \underline{\psi}^T(k)\hat{\underline{\theta}} + e(k). \tag{19}$$

or in matrix representation

$$\underline{y} = \underline{\Psi}\,\hat{\underline{\theta}} + \underline{e}. \tag{20}$$

After minimization of the loss function

$$V = \underline{e}^T\underline{e} \tag{21}$$

the well known LS-parameter estimate results Eykhoff (1974), Sorenson (1980), Young (1984):

$$\hat{\underline{\theta}}(N) = [\underline{\Psi}^T\underline{\Psi}]^{-1}\underline{\Psi}^T y. \tag{22}$$

The main problem is the generation of the derivatives of the signals for $\underline{\Psi}$ and $\underline{y}$. Several investigations have shown that good results can be obtained by applying state variable filters (SVF) Young (1984)

$$F(s) = \frac{y_f(s)}{y(s)} = \frac{1}{f_0 + f_1 s + \ldots + f_{n-1}s^{n-1} + s^n} \tag{23}$$

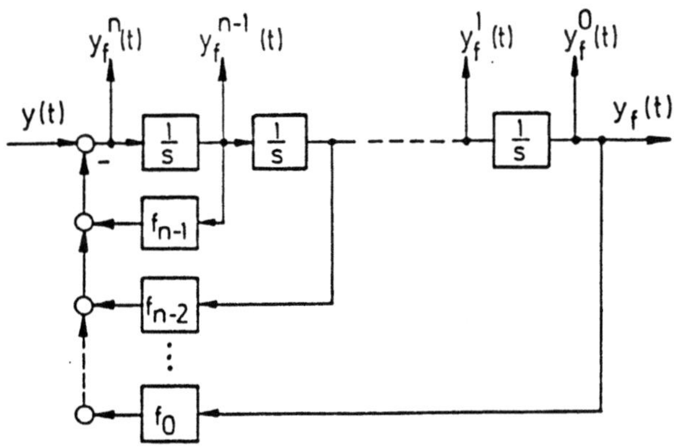

Fig. 2 State variable filter for generation of the signal derivatives

for both, the input and output signal, see Fig. 2. They may be designed as Butterworth-filters in order to lowpass filter the noise and are then z-transformed in order to obtain recursive filter algorithms. The LS-method is programmed in the form of a discrete square root filter in information form (DSFI).

Despite the fact, that the LS-method produces biased estimates for any noise n(t), good results were obtained in many practical cases if the signal-to-noise ratio is comparatively large Specht and Isermann (1988), Peter and Isermann (1989), (1990).

If some parameters $\underline{\theta}''$ are known, the parameter vector is split

$$\underline{\theta} = [\underline{\theta}' \ \underline{\theta}'']^T \tag{24}$$

where $\underline{\theta}'$ are the unknown parameters to be estimated and the signal derivatives with known parameters are separated

$$y''(t) = y(t) - \underline{\psi}''^T(t)\underline{\theta}'' = \underline{\psi}'^T(t)\underline{\theta}' + e(t). \tag{25}$$

The LS-estimate then becomes

$$\hat{\underline{\theta}} = [\underline{\Psi}'^T\underline{\Psi}']^{-1}\underline{\Psi}'^T y'' \tag{26}$$

Simulations with a 2nd order process have shown, that a considerable improvement of the convergence is obtained by:

(i) One parameter is known: a_2 or b_1 is known, i.e. parameters with largest variances are known.

(ii) More parameters are known: a_2 or b_1 must be contained in the set of known parameters. If other parameters are known, they must be known very precisely.

4. Calculation of the physical process coefficients

As the process parameters $\underline{\theta}$ are nonlinear algebraic functions of the process coefficients $\underline{p}$ no general applicable solution for the unknown process coefficients p_ν in form of eq. (2) can be given. For models of 1st or 2nd order in most cases a direct solution can be found. For higher order systems successive resolution for the unknowns p_ν can be tried or the use of computer algebra Walter (1982), Schumann (1990).

However, independent on the solution method an identifiability condition for the process coefficients can be given. The basic relation eq.(15) is written in implicit form

$$\left.\begin{array}{l} \underline{q} = \underline{\theta} - \underline{C}\,\underline{z} = \underline{0} \\ \text{where } \underline{z} = g(\underline{p}) \end{array}\right\} \qquad (27)$$

The implicit function theorem (Ortega and Rheinbolt 1970) now states that a necessary condition for a solution for $\underline{p}$ in the neighborhood of the solution $\underline{p}_0$ is that the functional determinant

$$\det \underline{Q}_p \neq 0 \qquad (28)$$

where $\underline{Q}_p$ is the functional matrix

$$\underline{Q}_p = \frac{\partial \underline{q}^T}{\partial \underline{p}} = \begin{bmatrix} \frac{\partial q_1}{\partial p_1} & \frac{\partial q_2}{\partial p_1} & \cdots & \frac{\partial q_r}{\partial p_1} \\ \frac{\partial q_1}{\partial p_2} & \frac{\partial q_2}{\partial p_2} & \cdots & \frac{\partial q_r}{\partial p_2} \\ \vdots & & & \vdots \\ \frac{\partial q_1}{\partial p_l} & \frac{\partial q_2}{\partial p_l} & & \frac{\partial q_r}{\partial p_l} \end{bmatrix} \qquad (29)$$

This implies r=l, which means that the number l of process coefficients must be equal the number r of model parameters.

5. Model based fault diagnosis

In the following the further development of the model based approach via parameter estimation for the incipient fault diagnosis of industrial robots according to Isermann et al (1990) is described. Some advantages characterize this method:

- No additional sensors are required. Only the signals of motor armature current, motor angular velocity and axis position are necessary, available for control in each industrial robot control system.
- Varying load can be taken into account. Accurate load determination can provide additional information for diagnosis (e.g. permanent overload).

- The overall procedure works in real time during normal point-to-point movements of one axis of the device. The described method can be applied to all robots with electric drives.
- The estimated and statistically classified parameters are finally treated as symptoms in the diagnostic sense. They lead to an early and reliable fault detection and support a knowledge based evaluation and an increased depth of diagnosis.

By the model based parameter identification approach variations of the inner physical coefficients of the robots axes are determined, e.g. friction coefficients, damping, moments of inertia and gravitational torques. Changes of these quantities indicate mechanical defects at an early stage caused e.g. by wear, maintenance deficiencies, or collision. The aim is to prevent possible breakdowns of the robot by a condition based maintenance strategy. The method is easily transferable to robots with different kinematic chains.

Theoretical modelling

The static and dynamic behaviour of a given industrial robot mechanical assembly is described by

$$M_A^* = J^*(\underline{r}, m_L)\ddot{r} + M_D^*(\dot{r}) + M_G^*(\underline{r}) \qquad (30)$$

M_A: Actuating drive torque

J : Pos. + load dependend mom. of inertia

M_D: Dissipative torque (torque of friction)

M_G: Gravitational torque

m_L: Mass of load at end effector

$r, \dot{r}, \ddot{r}$: Arm position, velocity and acceleration

The structure of this equation is generally valid for each of the robot's axes. It is important to include the dominant terms of friction and gravitation. For clarity of presentation the model will be derived - without loss of generality - for one specific position of the device, the so called reference position $\underline{r}_0$.

In the case of electrical DC-motors (as mostly used in industrial robot drives) the signals of armature current and motor angular velocity can be easily measured. With the motor flux linkage, assumed to be known from motor data sheet, a proportional relation holds between armature current I_A and actuating torque $M_A(t) = \Psi_A I_A(t)$.

The angular velocity of the respective robot arm is related to the signal of the tacho generator by $\dot{\varphi}(t)=\nu\,\omega(t)$ (ν denotes the total gear ratio of the drive chain). As a third measurement quantity the arm position $\varphi(t)$ is available. This results in the single axis model equation according to the assumed assembly of Fig. 3.

$$M_A(t)=J(\underline{\varphi}_0,m_L)\dot{\omega}(t)+M_D(\omega(t))+M_G(\varphi(t)) \qquad (31)$$

The structure of the gravitational term depends on the considered axis i, i = 1..6, and consists of a superposition of trigonometric functions Specht, Isermann (1988), Freyermuth (1990). This model equation establishes a correlation between the easy measurable quantities and the unknown process coefficients, which depend on the mechanical conditions of the respective axis, Fig. 4.

In order to describe the friction characteristic as accurate as possible, a third order algebraic equation in $\omega(t)$ was chosen:

$$M_D(\omega(t))=M_{D0}\,\mathrm{sign}(\omega(t))+M_{D1}\omega(t)+M_{D3}\omega^3(t) \qquad (32)$$

This equation represents an approximate description of rolling bearing characteristics. Its validity results from several experimental investigations at the different axes, see Freyermuth (1991 and 1990). M_{D0} is a Coulomb friction coefficient, while M_{D1} and M_{D3} characterize the influence of viscous friction.

Eq. (31) and (32) are linear in the parameters, therefore a least squares (LS) parameter estimation algorithm for the identification of the unknown coefficients can be applied.

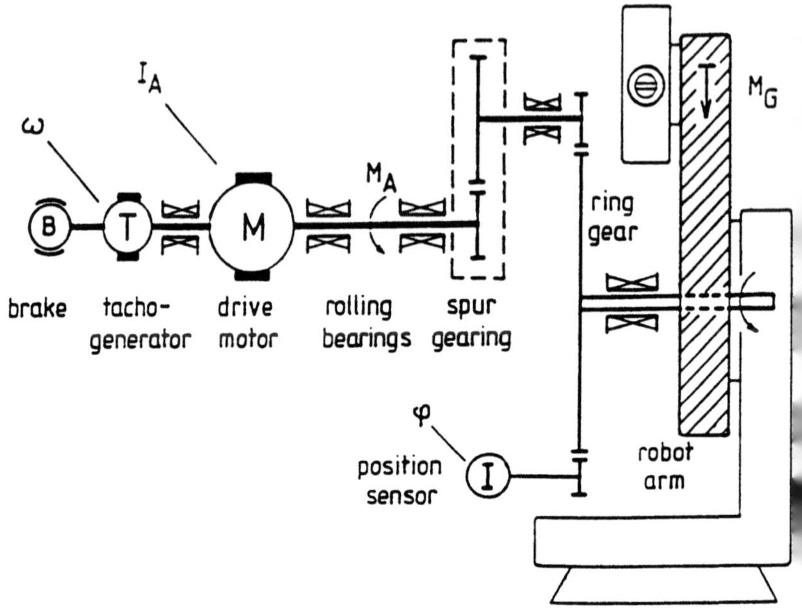

Fig. 3 Scheme of a drive chain (axis) of an industrial robot, indicating the measured quatities

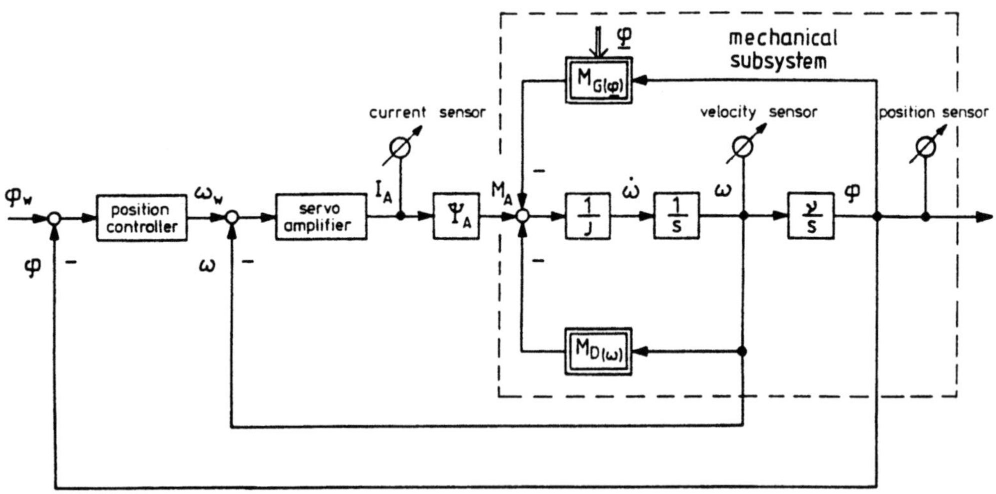

Fig. 4 Block diagram of the model of an industrial robot drive unit in conventional closed loop control

Parameter estimation equations

In order to determine the relevant physical quantities a procedure is proposed based on the continuous-time model Eq. (31) and (32).

The measurements are sampled at discrete time instants $t=kT_0$, $k=t/T_0$. Assumed that e.g. basic axis 3 in considered. Eq. (31) and (32) may be expressed by

$$M_A(k) = J(\underline{\varphi}_0, m_L)\dot{\omega}(k) + M_{D0}\text{sign}(\omega(k)) \tag{33}$$
$$+ M_{D1}\omega(k) + M_{D3}\omega^3(k) + M_G\cos(\varphi(k)) + e(k),$$

where e(k) represents the error between assumed axis model (Fig.4) and real dynamic behaviour of the respective axis. By e(k) stochastic disturbances and slight model inaccuracies are expressed, always present in real application problems . Eq.(33) can be rewritten in vector notation

$$y(k) = \underline{\psi}^T(k)\underline{\theta} + e(k) \tag{34}$$

with $y(k) = M_A(k)$, (35)

$$\underline{\psi}^T(k) = [\dot{\omega}, \text{sign}(\omega), \omega, \omega^3, \cos(\varphi)] \tag{36}$$

$$\underline{\theta}^T = [J, M_{D0}, M_{D1}, M_{D3}, M_G] \tag{37}$$

It becomes obvious, that the θ_j, j = 1...n. (n denotes the number of model parameters of Eq. (37)) directly represent the physical process coefficients p_j. The proportional factors of the sensors are assumed to be known, leading to the determination of the p_j in standardized physical (SI) units.

The acquisition of (not necessarily equidistant) N sets of measurement data (N>n) at different instants yields an overdetermined equation system

$$\underline{Y} = \underline{\Psi}\,\underline{\theta} + \underline{e} \tag{38}$$

Minimization of the quadratic cost function

$$V = \sum_{k=0}^{N} \lambda^{N-k} e^2(k) = \underline{e}^T \underline{W}\,\underline{e} \tag{39}$$

with the weighting matrix $\underline{W}=\text{diag}[\lambda^N, \lambda^{N-1}, \ldots, \lambda, 1]$ leads to the well known least squares Gauss-Markov estimation equation for a set of optimal parameters, see Isermann (1988):

$$\underline{\hat{\Theta}} = [\underline{\Psi}^T \underline{W} \underline{\Psi}]^{-1} \underline{\Psi}^T \underline{W} \underline{Y} = \underline{P}_W \underline{\Psi}^T \underline{Y}_W \tag{40}$$

For the recursive solution, where the matrices become functions of the sampling step k, several efficient algorithms are proposed in e.g. Isermann (1988).

Experimental results

The experimental research was performed with an industrial robot of type R106, see Fig. 5. The 6 rotational axes are actuated by DC-servomotors of high dynamic performance. The incipient fault diagnosis system is implemented on a standard microcomputer of type PC/AT with data acquisition board interfacing analog and digital measurement signals.

The sampling interval T_0 is chosen identically to that of the position controller of 5 msec. Fig. 6 demonstrates the typical time history of the signals in case of a point-to-point movement (PTP) of basic axis 3. The analog measurements are prefiltered by anti-aliasing lowpass-filters at a cutoff-frequency of 100 Hz and then processed by identical digital state variable butterworth filters of 4th order with a cutoff frequency of f_e=50 Hz to generate the acceleration signal $\dot{\omega}$.

For parameter estimation purpose the "DSRFI"-algorithm (Discrete Square Root Filtering in Information Form) is applied as proposed by Biermann (1977) because of its excellent numerical properties. A suitable forgetting factor is λ=0.99.

The nominal process coefficients p_{0j} and their variation coefficients v_{0j} are well known from

- many investigations performed at the industrial robot,
- several research projects on industrial robot controller design, and
- construction data from the manufacturer of this specific device.

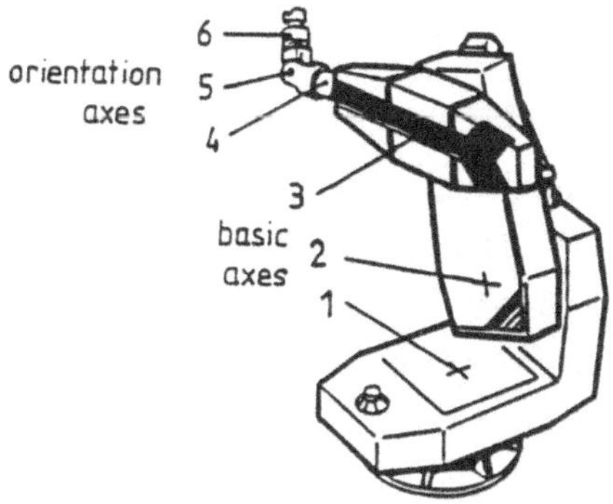

Fig. 5. Industrial robot with 6 rotational axes

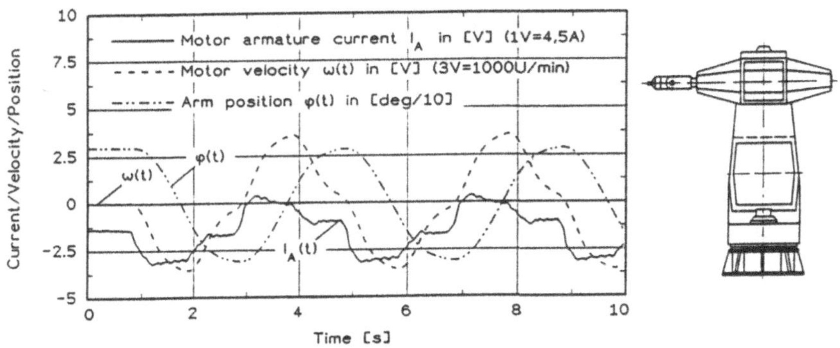

Fig. 6 Typical time history of measurements in case of a periodic point-to-point movement of basic axis 3

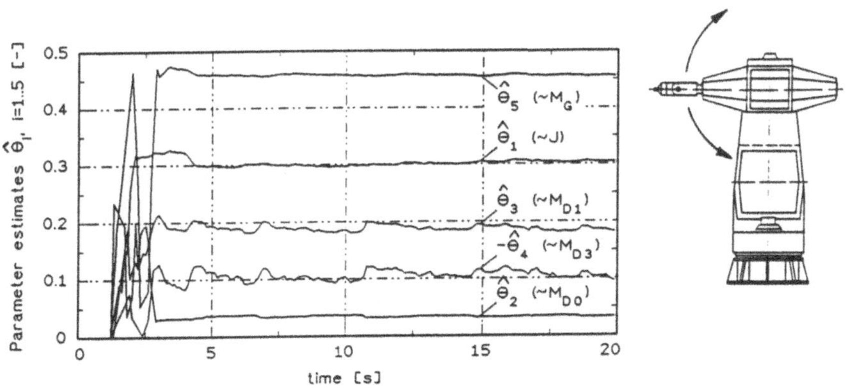

Fig. 7 Convergence of the parameter estimates of basic axis 3

In Freyermuth (1991) and Freyermuth (1990) several mechanical faults (e.g. wrong torque bias of gearing, overheating of drive unit, overload by picked up load or tool, defect of electromagnetic brake a.s.o.) have been subject to detailed experimental investigation applying the proposed method to all robot axes. Pattern of deviations of prcess coefficients caused by these different kind of faults have been recorded and illustrated.

Therefore in the sequel the results of a continuous experiment performed at basic axis 3 will be presented exemplarily to demonstrate fault detection efficiency of the proposed parameter estimation approach. During the cyclic point-to-point movement, Fig. 6, the parameters were estimated recursively, Fig. 7, and stored once at each movement cycle (~ every 2 sec). Then sequentially the following mechanical modifications were performed at the drive chain, as indicated in the gearing scheme of the respective axis, Fig. 8.

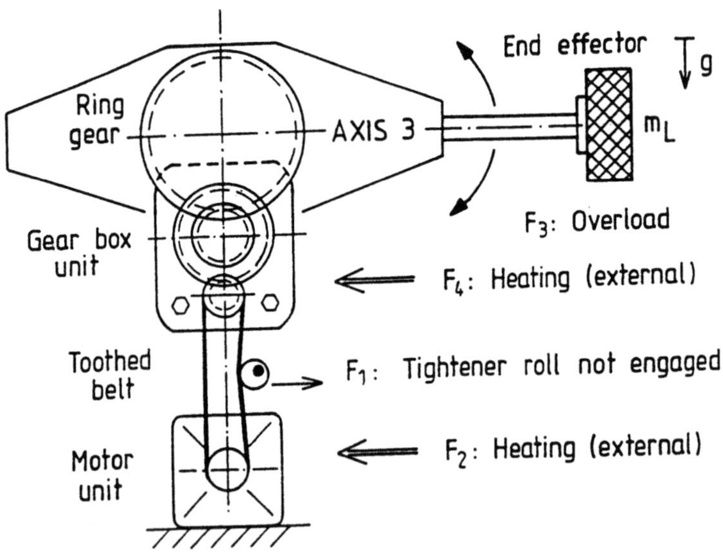

Fig. 8 Gearing scheme of basic axis 3 indicating the components affected by faults F_1 to F_4

F_1 Belt tightener roll not engaged
F_2 Overheating of motor unit (extern.incited)
F_3 Overload by additional mass at end effector
F_3 Overheating of gear box unit (extern. incited)

In industrial environment these faults can be caused e.g. by:

- Assembly or maintenance mistakes (F_1)
- Wear or defect of mechanical elements (F_2)
- Operational mistakes (F_3)
- Wrong environmental conditions (F_4)

Fig. 9 to 12 shows the impact of the different faults on the process coefficients p_j. The respective confidence functions indicate the significance of process coefficient deviations in case of abnormal operating conditions (N_0 denotes the nominal state). The length of the averaging filter with limited memory was chosen m=15.

Values of calculated confidence functions at specific diagnosis instants can be stored in a file and subsequently processed by a knowledge based system, what is described detailed by Isermann et al (1990).

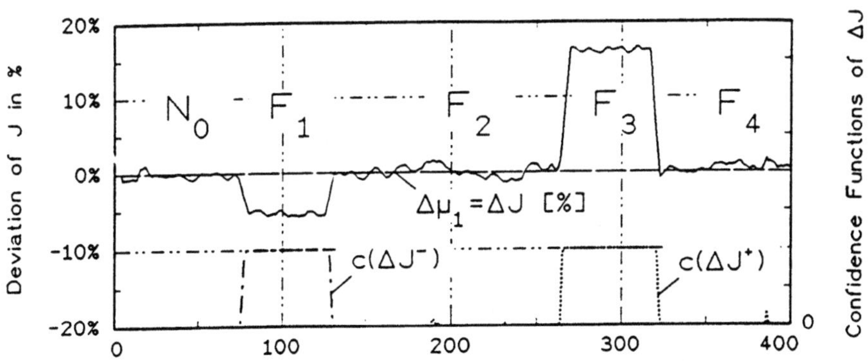

Fig. 9 Behaviour of $\Delta\mu_1 = \Delta J$ (deviation of moment of inertia) in % and respective confidence functions

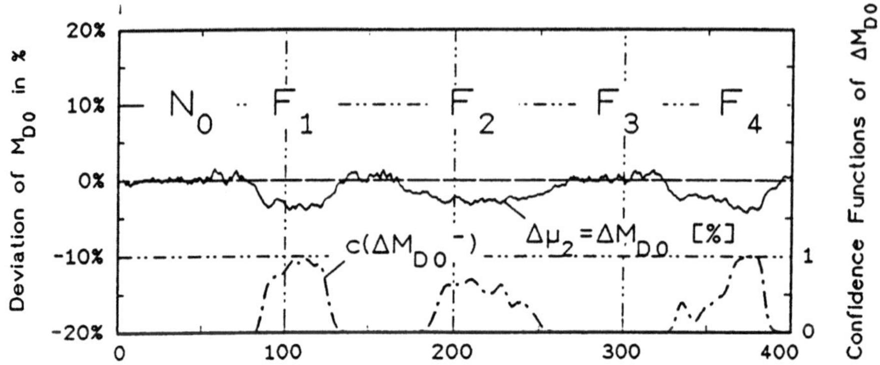

Fig. 10 Behaviour of $\Delta\mu_2$ = ΔM_{D0} (deviation of Coulomb friction coefficient) in % and respective confidence functions

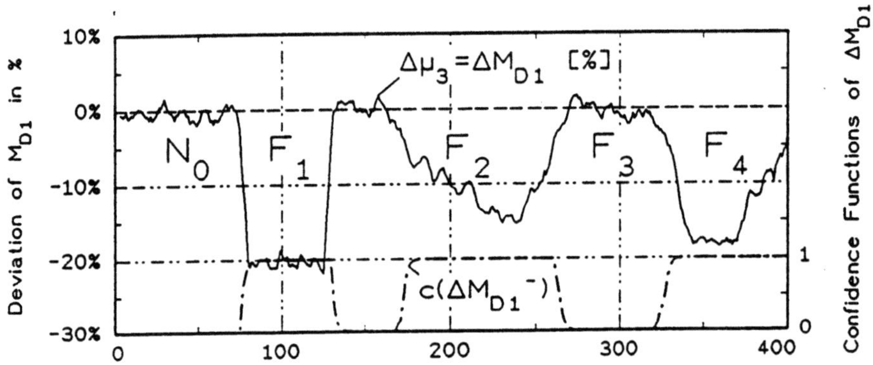

Fig. 11 Behaviour of $\Delta\mu_3$ = ΔM_{D1} (deviation of viscous friction coefficient) in % and respective confidence functions ($\Delta\mu_4$ resembles $\Delta\mu_3$)

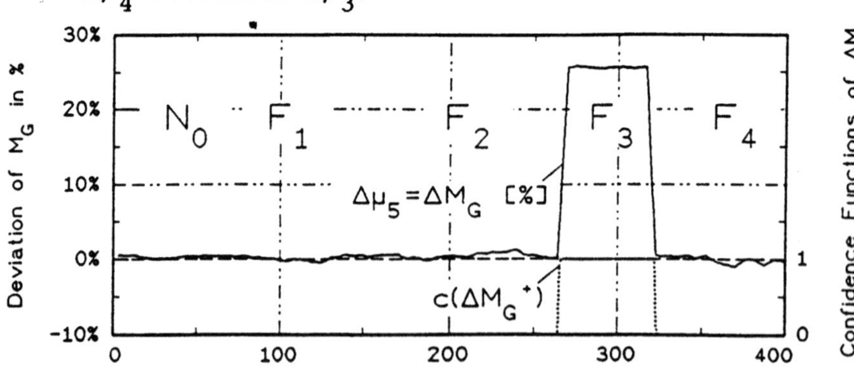

Fig.12 Behaviour of $\Delta\mu_5$ = ΔM_G (deviation of gravitation coefficient) in % and respective confidence functions

Literature

Biermann, G.J., (1977). Factorization methods for discrete sequential estimation. Academic Press, New York.

Dasgupta, S., Anderson, B.D.O. and Kaye, R.I., (1988). Identification of physical parameters in structured systems. Automatica, 24, 217-225.

Eykhoff, P., (1974). System Identification (London: Wiley)

Freyermuth, B., (1990). Modellgestützte Fehlerdiagnose von Industrierobotern mittels Parameterschätzung. Robotersysteme 6, 202-210. Berlin: Springer.

Freyermuth, B., (1991). An approach to model based fault diagnosis of industrial robots. IEEE Intern. Conf. on Robotics and Automation, Sacramento, USA.

Isermann, R., (1988). Identifikation Dynamischer Systeme. Vols. I and II (Berlin: Springer-Verlag).

Isermann, R., (1989) Digital Control Systems. Vol. I, second edition (Berlin: Springer-Verlag).

Isermann, R., (1990) Digital Control Systems. Vol. II, second edition (Berlin: Springer-Verlag).

Isermann, R.,Freyermuth, B., Fuchs, A., Janik, W. Neumann, D., Reiss, Th. and Wanke, P. (1990). Model based fault diagnosis and supervision of machine and drives. Preprints 11th IFAC World Congress on Automatic Control, Vol.7, 1-12, 13.-17. August 1990, Tallinn, USSR.

Nold, S. and R. Isermann, (1986). Identifiability of process coefficients for technical failure diagnosis. 25th IEEE Conference on Decision and Control, Athens.

Ortega, J.M. and W.C. Rheinboldt, (1970). Iterative solution of nonlinear equations in several variables (New York: Academic Press).

Peter, K.H., R. Isermann, (1989). Parameteradaptive PID-control based on continuous time process models. IFAC-Symposium on Adaptive Control and Signal Processing, Glasgow, Scottland.

Peter, K.H., R. Isermann, (1990). Parameter adaptive control based on continuous-time process models. 11th IFAC World Congress, Tallinn (Oxford: Pergamon).

Raksanyi, A., Y. Lecourtier, E. Walter and A. Venot, (1985). Identifiability and distinguishability testing via computer algebra. Mathematical Biosciences, 77, 245-266.

Schumann, A., (1990). Computer algebra - a key technique for computer aided theoretical modelling. 9th IASTED International Conference on Modelling, Identification and Control Insbruck. Austria.

Sorenson, H.W., (1980). Parameter Estimation: Principles and Problems (New York: Dekker).

Specht, R. and R. Isermann, (1988). On-line identification of inertia, friction and gravitational forces applied to an industrial robot. IFAC Symposium on Robot Control (SYROCO 88), Karlsruhe (Oxford: Pergamon Press).

Walter, E., (1982). Identifiability of State Space Models. Springer Lecture Notes in Biomathematics No. 45 (Berlin: Springer Verlag).

Young, P.C., (1984). Recursive Estimation and Time-Series Analysis (Berlin: Springer-Verlag).

Detection and Location of Damage Causing Non-linear System Behavior

H.G. Natke
University of Hannover
Curt-Risch-Institut
Hannover, Germany

J.T.P. Yao
Texas A&M University
Department of Civil Engineering
College Station, TX, USA

1 Introduction - Types and Causes of Faults/Defects

In this paper, systems are defined as existing engineering facilities such as bridges, airplanes, and robots. These systems are subject to (time variant) forces and faults/defects may occur and/or grow during their respective life times.

Faults and defects include voids, inclusions, cracks, fractures, surface distortions, corrosion, sedimentation, reduction of mass by abrasion and wear, increase of mass due to environmental pollution, and material ageing. Special effects include domino-effect (e.g. earthquake followed by fire) and progressive collapse.

Causes are

- lack and reduction of strength due to
  - cumulative damage (e.g., fatigue resulting from vibration)
  - external forces (resonance, accidents, vandalism,...)
  - instability (e.g., selfexcitation)
  - environmental forces (e.g., chemicals)
  - design errors
  - errors in handling and fabrication
  - human errors in construction work
- unexpected forces (extremely large earthquakes, impacts by dropping, explosions etc.)
- defects due to operation (e.g., wear, abrasion)
- corrosion (chemical process) and erosion (e.g., fissure)
- imbalance
- reciprocating imbalance
- fluid structure interaction (e.g. turbulence)
- fluctuating magnetic forces

In the case of stiffness degradation (e.g. due to environmental forces) the resulting stresses may exceed certain limit states. Therefore, a detailed analysis is frequently needed. It is desirable to observe existing systems in order to find symptoms which indicate the presence of faults, to monitor the symptoms, and then to identify the faults. In addition, we need to distinguish between various faults with certain known features and/or patterns. The dynamic behavior of the system is frequently affected by these faults. The effects of non-linearities due to faults of a system are summarized herein. Then they will be evaluated in terms of how they can be used for detection, localization and for assessment. The validated models [1], [2], [3] of the undamaged system under consideration is assumed to be available in such cases.

There are other approaches which are different from dynamic response considerations. For example, one may study propagating (guided stress) waves including energy flow (intensity) which are not treated herein.

2 Effects of Non-linearities

Effects of non-linearities are all those which violate the linearity requirements such as small deformations, proportionalty and homogeneity of structural behavior. In addition, without mentioning any pre-assumptions (for example, with respect to the needed type of forcing), one can look to distortions of Nyquist plots, inequalities of corresponding parts of Fourier and Hilbert transformed functions, (see the enumeration in [4]). A direct search for non-linear behavior can be performed by using scalar indicators like the SIG- function or higher order correlation functions (see also [4]).

An easy way is to apply harmonic forces with sufficiently high amplitudes and looking for higher harmonics. Note that the dynamic behavior of a system with a crack-type fault is not only non-linear but also with additional degrees of freedom [5] due to additional impulse loading and shear forces present at the crack contact surfaces.

To illustrate the many types of apparent non-linearities, several examples [6] will be presented based on finite element calculations with ANSYS (version 4.4A, 1989). Distinctions are made between physical (material), geometrical and structural non-linearities. The latter is due to some type of play (gap), which is described by displacement constraints.

Example 1: Clamped plate with a physical non-linearity.

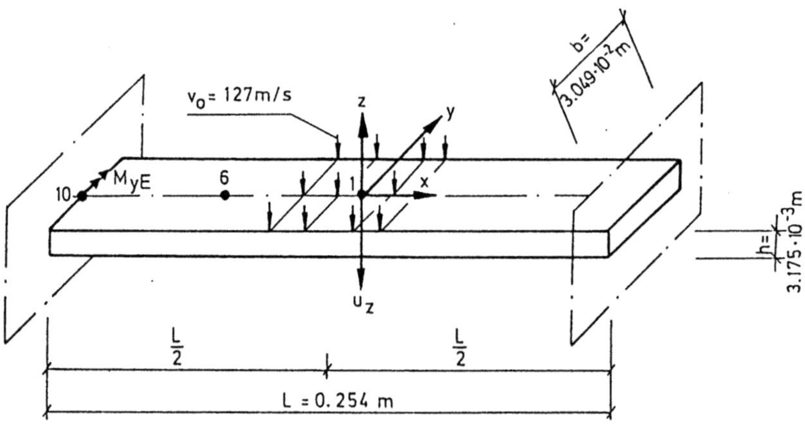

Figure 1 Clamped plate

Given a rectangular plate with two opposite ends clamped and the remaining sides free is subjected to an rectangular impulse applied on the central region (see Fig. 1) with an initial velocity v_0. The vibration of the plate includes geometric (large deflections) and material (elastic- ideal plastic) non-linearities. For comparison a second simulation is carried out with linear material behavior and geometric non-linearity. The duration of the velocity impulse is $0.001s$ with the amplitude v_0 =

127 ms^{-1}. The natural frequencies are those of the associated linear model and the total transients are calculated for a duration of $0.002s$ by using the Newmark numerical integration.

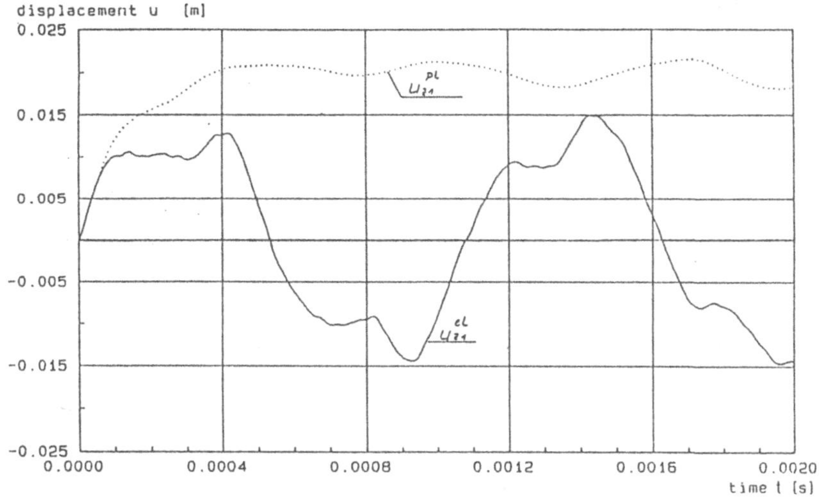

Figure 2 Displacement at the center of the plate

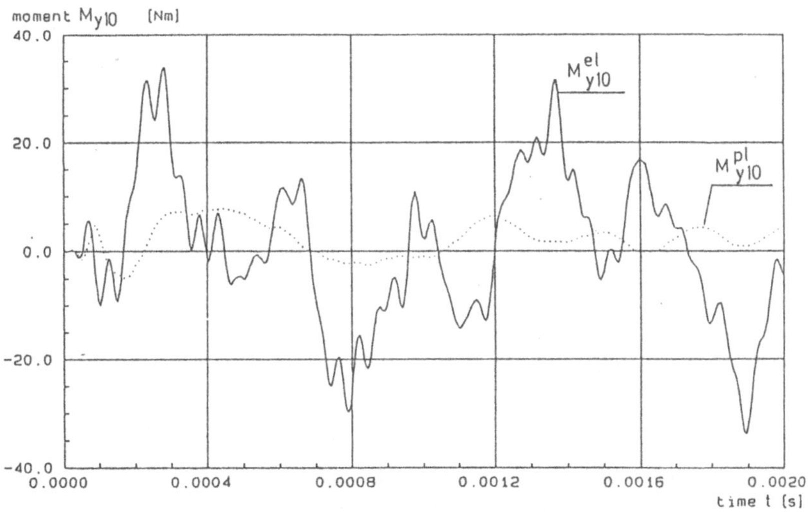

Figure 3 Moment at the fixed end of the plate

The displacement of the plate at the center is shown in Fig. **2** and the moment at the fixed end in Fig. **3** for both cases with *pl* stands for "plastic" and *el* for "elastic". The elastic-plastic response u_{z1}^{pl} indicates little dynamic response within the time interval. These results show that the impulse loading propagate as a transverse wave with severe plastic deformation at the wave front. This significant energy propagation leads to an geometric damping of the system dynamics.

Example 2: Clamped beam with physical and geometrical non-linearities.
The undamaged beam consists of isotropic and elastic material. It is subjected to a transverse rectangular impulse (640 lbs, duration 0.005 s) applied to its midspan. Large deflections and geometric non-linear behavior are assumed.

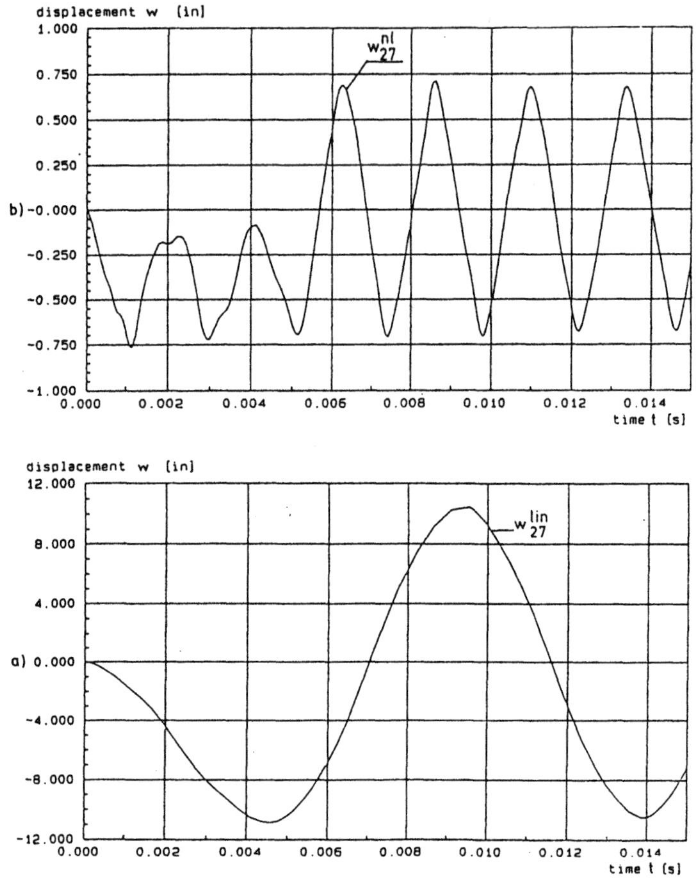

Figure 4 Responses of the beam midspan for small (a) and large (b) deflections

Fig. **4** shows the displacement of the beam midspan $w_{27}^{nl}(t)$ for large deflections and the corresponding response for small deflection theory, w_{27}^{lin}. The maximum displacement of the assumed linear behavior is more than ten times greater than that of the non-linear case demonstrating the importance of performing non-linear analyses. The frequency of the linear response corresponds with the fundamental eigenfrequency of $110,5\ Hz$ while the response frequency of the non-linear calculation is about four times higher. Similar conclusions may be reached from studying the moment at the clamped end of the beam. It can be noticed that the formulae for the non-linear displacement depend on the rotation of the beam axis φ, the trigonometric functions expanded into series yield in even powers in φ.

2.1 GAP2: Now a gap is introduced at midspan underside in order to simulate

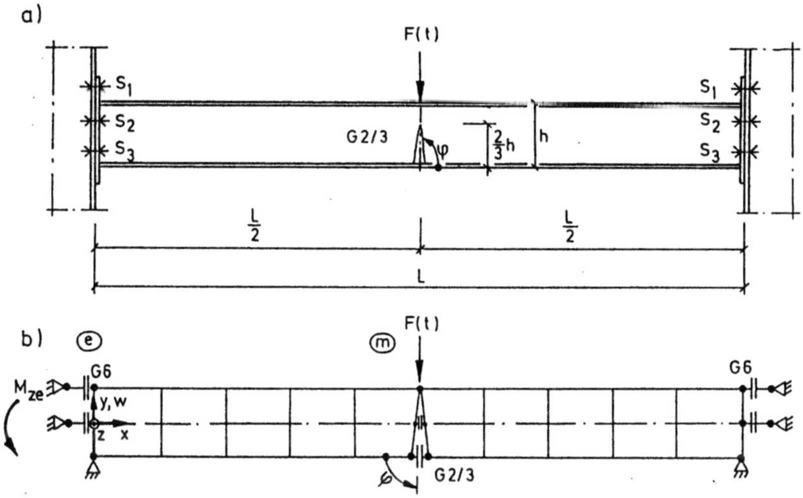

Figure 5 The system and the related model

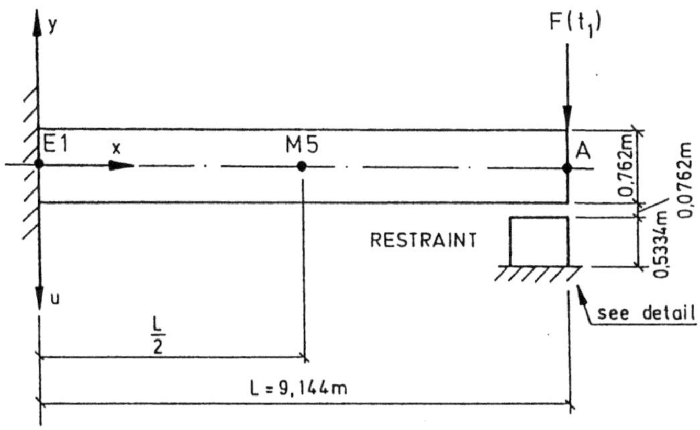

Figure 6 Geometry and boundary conditions

a crack (see Fig. 5 which shows the real system and its FE model). The crack here is assumed to be open during the analysis. The midspan displacements of the undamaged and the damaged beam show that the differences between both cases are more significant in the moments than in the displacements.

2.2 GAP3: If one assumes that the gap introduced in Example 2.1 opens and closes during motion, the results show that the quantities are more sensitive due to the gap than in the prior simulation.

Example 3: Cantilever with displacement constraint, see Fig. 6.

An impulse (2.92 MN, duration 0.116 s) applied the free end of a cantilever beam is considered. It simulates the combination of non-linear material behavior (the

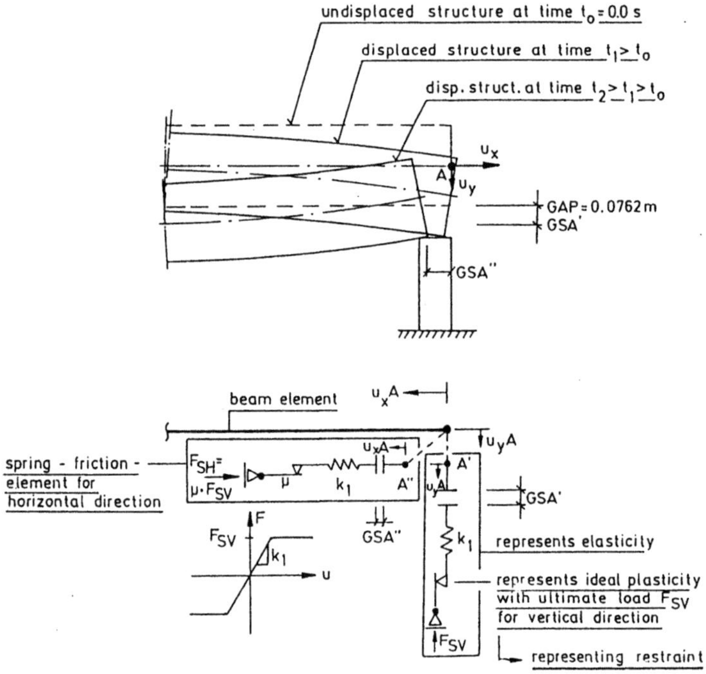

Figure 7 Model of constraint

constraint modelled by a non-linear spring) and structural non-linearities (contact elements, a gap between the cantilever and the restraint).

The cantilever is modelled with beam elements with linear elastic- ideal plastic force deflection curve.

The constraint is modelled as indicated in Fig. **7**. The horizontal friction in the contact surface is also modelled. Figs. **8**,**9** contain the end displacement and the bending moment at the support.

The cases are: *nl* for fully non-linear, *nl1* indicates the elastic behavior of the cantilever and the elastic-plastic behavior of the restraint, *nl2* stands for the elastic-plastic behavior of the cantilever and elastic behavior of the restraint. The figures show clearly what physically happens and that the influence of the cantilever stiffness is superior to the restraint stiffness. The influence of friction in the contact zones is shown by the end displacement in Fig. **10**. Here *nl* designates the frictionless surfaces, *nr1* a smooth surface with $\mu = 0.2$ and *nr2* a rough surface with $\mu = 0.8$.

These examples show for the assumed faults within the chosen systems and excitations that geometric non-linear calculations are necessary for realistic simulations and that in general the resulting non-linear effects can be taken as symptoms. These are often the stresses rather than the displacements, which sometimes are not sensitive enough. This statement is in agreement with the findings in [9]. There exists also examples which show no significant modifications in the displacements and

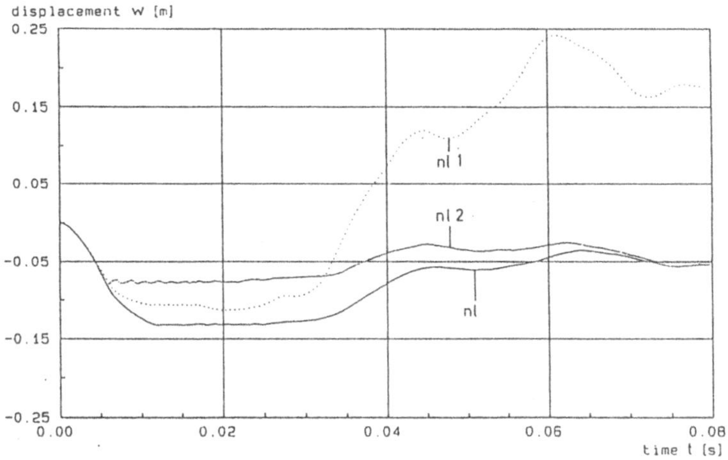

Figure 8 End displacement for physical linear and non-linear analysis

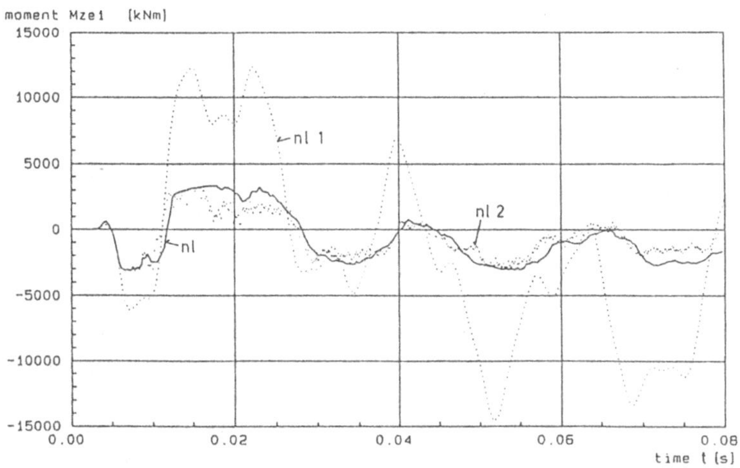

Figure 9 Bending moment at the support for physical linear and non-linear analysis

stresses (dependent on the chosen locations) due to local non-linearities [6].

The following conclusions can be drawn in general: In addition to the general known non-linear characteristics mentioned above

- Finite Element programs exist for PCs with the capability to model and to simulate assumed faults for a particular system.
- In diagnostics it is necessary to study assumed faults in order to obtain
  - their sensitivities with respect to dynamic quantities
  - symptoms, patterns for detection

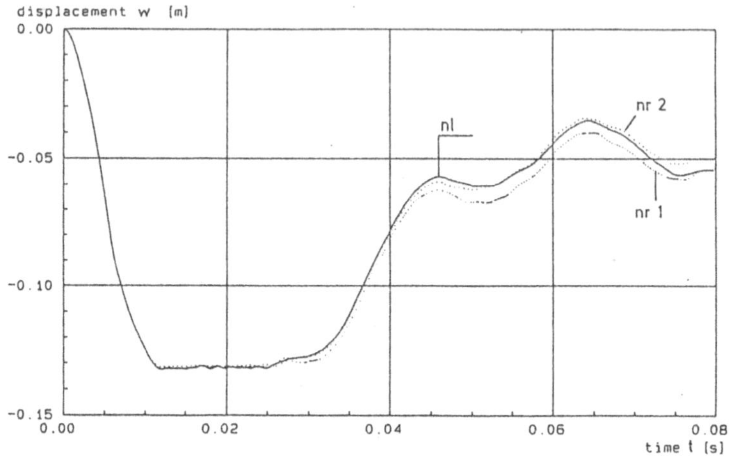

Figure 10 End displacements for various friction levels

- a catalogue of symptoms, features of possible faults
- an assessment for their severity.

3 Damage Detection

Damage and fault detection can be done globally and locally. Global detection is performed with the use of indicator functions such as the MAC [1]. An aim functional (loss function) or the residuals defining an aim functional can also be taken for detection. The problem will be solved by comparing the dynamic response (direct or indirect, e.g. by modal quantities, see [9]) of the undamaged linear system with those of the damaged non-linear system. The question arises as to which symptoms should be chosen. If the undamaged system behaves linearly then damage detection means to discover non-linear behavior. In order to solve this problem many methods are available (see Sect. 2 and [4]). If the undamaged system behaves non-linearly then the detection of faults seems to be much more difficult, because one has to look not only for parameter modifications of a structured non-linear model but also for changes in the structure of the model. Therefore the investigation is restricted to linear undamaged models and the damage introduces some non-linear effects.

The first step of the model-based diagnosis is to improve the initial linear model by system identification methods [1]. The result then will be a mathematical model with known confidence and sufficiently small errors. This model then is chosen as a pattern, and with it possible faults with their effects are studied in order to obtain features (or simpler symptoms) for fault detection and distinction. Measurements of the faulty process and signal processing (error reduction, information condensation) will give the basis for parameter estimation of the prior improved linear model which will result in biased estimates due to the incorrect model structure used [7]. This means modified parameters as compared with the prior validated model and probably not interpretable physically.

A subproblem partly solved is the introduction of a damage measure for detection and assessment [7], [8], [9].

Another method for damage detection which must not be based on the structured identification of the non-linearities at this stage, could be the application of observers, the description of which is not intended to be complete herein. Only one interesting method will be mentioned. It is the detection of non-linear effects through state observers [10], [11]. In the state space domain the non-linear system is described by

$$\dot{x}(t) = Ax(t) + Nn[x(t)] + b(t), \tag{1}$$

$$z(t) = Cx(t). \tag{2}$$

$x(t)$ is the n-dimensional state vector, $z(t)$ the m-dimensional measuring vector, $n(x)$ is a p-vector designating the non-linearities, $n, m, p \in \mathcal{N}$. The matrices A, N, C should be of proper dimensions, $b(t)$ is the external force vector. $n[x(t)]$ is sought, the other quantities are assumed to be known. The estimation of $n(x)$ is done by assuming $n[x(t)] = n(t)$ as an external distortion of the model (1) with the fictive model

$$n[x(t)] \doteq Hv(t), \tag{3}$$

$$\dot{v}(t) = Fv(t) + Gb(t). \tag{4}$$

The exogenous linear model (4) should be of proper dimensions in order to represent or to approximate sufficiently accurate the non-linearity (see Sect. 4.1). $b(t)$ must not be necessarily included in Eq. (4) [11]. The matrices F, G and H must be chosen suitably. For further details see [11]. Simulations even with friction show good results. The choice of Eqs. (3,4) with $G = 0, V = 0$ (integrator model) seems to be in many cases suitable. In a second step a controller can be used for compensation of the non-linearity if one tries to reconstruct the associated linear (undamaged) model.

Fault modelling in the state space is popular in control engineering in connection with observers. However, one has to use some prior knowledge as discussed in Section 4.3.

4 Damage Location

The usage of suitably local residuals (e.g. those of the elements of the physical parameter matrices) will solve the localization of damage directly. This statement holds true if the verified prior linear model will be compared with a validated non-linear model of the damaged system. In order to perform the residuals and to solve the corresponding inverse problems, mathematical models are needed. Some models, their simplifications and parametrizations are discussed in the following.

4.1 Variable Substitution and Coordinate Transformation with Known Model Structure: Linearization

The first step of identifying non-linear systems is the structure (of the model) identification. If the model structure is known, in special cases a variable substitution can result in a linear model [12] for simplification. Also a transformation will result in an (exact) linear model [13]. This method of (exact) linear mathematical representation and decoupling is expensive in computation. Therefore, algebraic computer programs are developed [14], [15] in control engineering. The authors are not aware of practical applications in damage location. The procedure needs the fulfilment of some requirements. However, dynamic models demand only the inclusion of a relatively small number of degrees of freedom, though it seems to be worthwhile to mention it for further investigations. The method of classical linearization as an approximation will not be discussed here, because it is not useful for the purpose of localization based on non-linearities.

4.2 Feedback Procedure for Splitting the Equations of Motion into a Linear and Non-linear Part

Another method can be the feedback procedure for partition the equations of motion into a linear and a non-linear part. A starting point is the method published in [16], [17]. It is a generalized restoring force method. Here a new sight of it is presented.

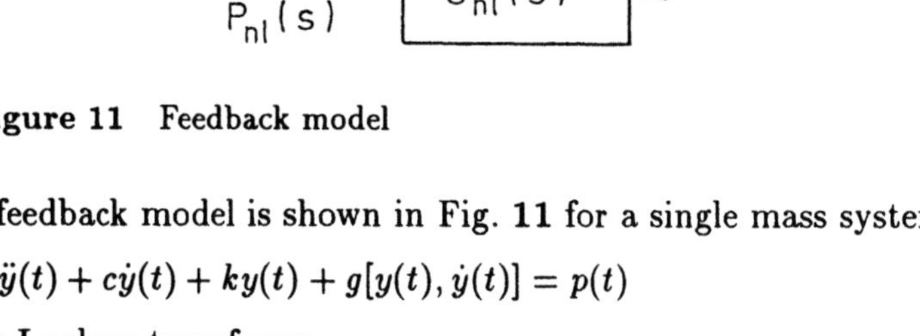

Figure 11 Feedback model

The feedback model is shown in Fig. **11** for a single mass system:

$$m\ddot{y}(t) + c\dot{y}(t) + ky(t) + g[y(t), \dot{y}(t)] = p(t) \tag{5}$$

with its Laplace transform:

$$(s^2 m + sc + k)Y(s) + \mathcal{L}\{g[y(t), \dot{y}(t)]\} = P(s) \tag{6}$$

where the capital letters designating the Laplacian transforms of the corresponding time functions. Defining

$$H_{total}(s) := \frac{Y(s)}{P(s)}, \quad S_{lin}(s) := s^2 m + sc + k, \quad S_{nl}(s) := \frac{\mathcal{L}\{g[y(t), \dot{y}(t)]\}}{Y(s)} \tag{7}$$

it follows that

$$H_{total}(s) = \frac{1}{S_{lin}(s) + S_{nl}(s)}. \tag{8}$$

This is a well-known equation for feedback systems, for example, for servo pick-ups.

An old attempt to partition the excitation into a linear and a non-linear part is given as follows:

$$p(t) = p_{lin}(t) + p_{nl}(t), \tag{9}$$

$$m\ddot{y}(t) + c\dot{y}(t) + ky(t) =: p_{lin}(t), \tag{10}$$

$$g[y(t), \dot{y}(t)] =: p_{nl}(t) \tag{11}$$

(with $y(t)$ denoting the displacement response of the total non-linear system) which [18] has failed to date in default of a practicable criterion. If one assumes that the linear model (10) is known (i.e. the parameters m, c, k) then force measurements (input identification) will estimate $p_{lin}(t)$. The excitation $p(t)$ also is measured, therefore

$$p(t) - p_{lin}(t) = p_{nl}(t) = g[y(t), \dot{y}(t)] \tag{12}$$

will lead to the restoring force surface with input identification of the corresponding linear system. It remains to discuss methods for identification of the correponding linear system and the input identification (see, for example, [19] and the IMAC proceedings).

The generalization to multi-degree of freedom models seems to be not difficult. The estimation using Kalman filtering is described in [16], the approximation of the non-linear part by polynomials in $y(t), \dot{y}(t)$ is mentioned in [16].

4.3 Non-linear Restoring Force Method with Model Structure Identification by a priori Choice

In control theory the fault modelling in the state-space framework is well-known [20]. There the observers are often applied. This formulation modified can be taken for parametrical identification of the restoring forces. The model structure is assumed a priori and validated a posteriori [21]. Another approach uses a library of many types of non-linearities [22].

In the following the basic formula are cited to be used for estimation procedures. The equation of motion in the state space

$$\dot{x}(t) = Ax(t) + b(t) + Fh[x(t), \dot{x}(t), t] + n(t), \tag{13}$$

with the model correction Fh and measuring errors $n(t)$ together with the measuring equation

$$z(t) = Hx(t) + q(t) + m(t) \tag{14}$$

with the measuring matrix H and the vector q describing the sensor faults, $m(t)$ the noise vector, prescribe the correction of the linear model by pre-given parametric models Fh. The parameters of the linear model part, of course, also can be corrected:

$$A^c = A + \Delta A, \; H^c = H + \Delta H. \tag{15}$$

4.4 Non-linear Restoring Force Method with Polynomial Approximation

The non-linear restoring force surface can be approximated by polynomials. This is shown for separabel restoring and damping forces in [4] where in addition to the estimation of the polynomial coefficients a structure identification is performed by estimation of their maximum powers. Due to the ill-posedness of the inverse problems regularization is applied in order to study how to overcome the ill-posedness [23], [24].

The fictitious exclusion of hysteretic behavior (multi-valued function representation) by approximation through polynomials (single-valued functions) is discussed in [25].

Zamirowski [26] discusses the structure identification of non-linear systems by using the prior power estimation as well as the a posteriori estimation using statistical criteria. The modelling of a located fault can be done if it is desired by a linearized state space representation [27].

5 Damage Assessment

Damage assessment can be performed with the use of classification, multi-hypothesis testing, fuzzy logic and/or neural networks. However, in case where validated mathematical models are available, a knowledge base of maximum information and high confidence can be used. Simulations and calculations for cases with known faults may be performed to relate certain features in recorded responses to these particular faults. Therefore the extended identification methodology can be applied as already proposed in [7].

6 Conclusions

In this paper, faults, defects, and causes for damage are discussed. Several examples are given with finite element calculations to obtain responses for non-linear systems with particular faults. Several other methods and approaches are also considered. Results show that it is possible to relate certain features of the response records to particular faults.

Next steps to be taken are

- to solve the problem of fault distinguishability,
- to perform the training and learning phase which results in a trend model and a forecast; and
- to introduce the decision phase based on limit states and certain deterministic or probabilistic methods or fuzzy logic.

References

[1] H.G. Natke, 1992, Einführung in Theorie und Praxis der Zeitreihen- und Modalanalyse - Identifikation schwingungsfähiger elastomechanischer Systeme, Vieweg, Braunschweig, Wiesbaden, 3. Auflage

[2] H.G. Natke and J.T.P. Yao, 1990, System Identification Methods for Fault Detection and Diagnosis, in: Structural Safety and Reliability, Eds. A.H.-S. Ang et al., Proc. of ICOSSAR '89, ASCE New York, 1990, 1387-1393

[3] J.T.P. Yao and H.G. Natke, 1991, Uncertainties in Structural Identification and Control, Internat. Fuzzy Eng. Symposium '91, Yokohama, Japan, Nov. 13-15, 1991, 844-849

[4] H.G. Natke, 1989, Survey on the Identification of Mechanical Systems, in: Road-Vehicle-Systems and Related Mathematics, Ed. H. Neunzert, B.G. Teubner, Stuttgart and Kluwer Academic Publishers, 69-110

[5] O. Abraham, J.A. Brandon, 1991, Modal Behaviour of a Cracked Beam, Symposium EUROMECH 280 on Identification of Nonlinear Mechanical Systems from Dynamic Tests, Oct. 29-31, Lyon, France

[6] L. Meese, 1992, Implementation and Modelling of Structural Damage, Diplomarbeit (Master Thesis) University Hannover, Curt-Risch-Institute

[7] H.G. Natke and J.T.P. Yao, 1988, System Identification Approaches in Structural Safety Evaluation in: Structural Safety Evaluation Based on System Identification Approaches, Eds: H.G. Natke, J.T.P. Yao, Vieweg Braunschweig, Wiesbaden

[8] C. Cempel, H.G. Natke, 1990, Damage Measures and Evolution in Machines and Structures, Research Report CRI-B-1/1990, Curt-Risch-Institute of the University of Hannover

[9] C. Cempel, H.G. Natke and A. Ziołkowski, 1992, Application of Transformed Normal Modes for Damage Location in Structures, in: Structural Integrity Assessment, Ed. P. Stanley, Elsevier Appl. Sci. London, New York, 246-255

[10] P.C. Müller, 1990 , Rechnergestützte Rekonstruktion nichtlinearer Effekte mit Hilfe von Zustandsbeobachtern, in: Dynamische Probleme - Modellierung und Wirklichkeit, Hrg. H.G. Natke et al., Mitteilung CRI-K 1/90 des Curt-Risch-Institutes der Universität Hannover, 471-480

[11] P.C. Müller, 1992, Modellvereinfachung nichtlinearer Systeme, VDI-Berichte Nr. 925, VDI Verlag Düsseldorf, Wiesbaden 161-188

[12] E. Kamke, 1943, Differentialgleichungen, Lösungsmethoden und Lösungen, Akademische Verlagsges. Becker & Erler Leipzig

[13] A. Isidori, 1989, Nonlinear Control Systems, 2nd ed. Springer Berlin, Heidelberg

[14] J. Birk, and M. Zeitz, 1988, Extended Luenberger Observer for Nonlinear Multivariable Systems, Int. J. Control 47, 1823-1836

[15] J. Birk and M. Zeitz, 1991, MACNON - MACsyma Program for NONlinear Systems, Bericht, Institut für Systemdynamik und Regelungstechnik, Universität Stuttgart

[16] G. Renker, 1989, Zur Identifikation nichtlinearer strukturmechanischer Systeme, Mitteilungen aus dem Institut für Mechanik, Ruhr-University Bochum

[17] G. Renker and H. Waller, 1991, Time Domain Identification of Nonlinear Vibrational Systems, in: Structural Dynamics, W.B. Krätzig et al.(Eds.), A.A. Balkema, Rotterdam Brookfield, 299-304

[18] C. Chandivert, 1970, Considerations Relevant to the Vibration Testing of Aeronautical Structures Having Non-ideal Characteristics; Dep. of Supply, Australian Defence Scie. Service, ARLISM 355

[19] H.G. Natke, 1987, Input and Damping Identification within the Frequency Domain, Int. J. of Analytical and Experimental Modal Analysis, 2, 92-98

[20] J. Gertler, 1991, Analytical Redundancy Methods in Fault Detection and Isolation, Survey and Synthesis, in Preprints of IFAC/IMACS- Symposium on Fault Detection, Supervision and Safety for Technical Processes - Safeprocess'91 - Ed. R. Isermann, Baden-Baden, Sept. 10-13,1991, Vol.1, 9-21

[21] Mook, J., 1992, Minimum Model Error Estimation, Paper presented on July 1 within the "Oberseminar für Mechanik" of the University Hannover

[22] Qe, Ling; Meyer, T.J.; Mook, J.; Soong, T.T., 1992, Damage Assessment through Nonlinear Structural Identification; Paper presented at this Int. Workshop, to be published

[23] H.G. Natke, U. Prells, 1991, Contribution to Structure Identification of Nonlinear Mechanical Systems, Symposium EUROMECH 280 on Identification of Nonlinear Mechanical Systems from Dynamic Tests, Oct. 29-31, Lyon, France

[24] H.G. Natke, 1992, On Regularization Methods within System Identification, IUTAM Symposium on Inverse Problems in Engineering Mechanics, May 11-15, Tokyo, Japan

[25] F. Benedettini, D. Capecchi, F. Vestroni, 1990, Nonparametric Models in Identification of Hysteretic Oscillators, DISAT - Pubbl. n. 4, Dic., Dip. di Ing. delle Strutture, delle acque e del Terreno, Univ. dell'Aquila

[26] M. Zamirowski, 1992, Einige zeitdiskrete Parameterschätzmethoden zur Identifikation nichtlinearer mechanischer Systeme, Ph.D. thesis, University of Hannover, Rep. Curt-Risch-Institute CRI-F-1/92

[27] - , 1990, Diagnoseverfahren in der Automatisierungstechnik, VDI- Verlag, Düsseldorf

Nonlinear Problems in Selectively Sensitive Identification of Structures

Yakov Ben-Haim
Faculty of Mechanical Engineering
Technion — Israel Institute of Technology
Haifa, Israel

1 Introduction

In this paper we will formulate and briefly examine several nonlinear problems which arise in the selectively-sensitive identification of dynamic systems. Selective sensitivity is a method for system identification which enables the reduction of high-dimensional estimation problems to a collection of lower dimensional tasks.

The formulation and basic properties of selective sensitivity have been studied in several previous papers [2]–[5], [9]. The idea of sensitivity analysis has found many applications in engineering. The output sensitivity to parameter variation has been related by Aoki and Staley [1] to the Fisher information matrix, and used with the Rao-Cramer inequality for designing inputs for discrete stochastic systems which asymptotically display minimum estimation errors. Haftka and Gürdal [6] use sensitivity analysis in structural optimization.

As in other sensitivity analyses, the method of selective sensitivity depends heavily upon analysis of the output sensitivity to parameter variations. However, our emphasis is on choosing excitations which partition the identification of a large system to a number of lower-dimensional identification problems. This is possible when each parameter to be estimated appears sparsely in the dynamical equations. This is often true in linear elastic systems, in which the physical parameters appear in only a few locations in the dynamic stiffness matrix.

In section 2 we will briefly outline the basic concept of selective sensitivity. In section 3 we will examine the selectively sensitive identification of nonlinear systems. In section 4 we consider selective sensitivity in the context of feedback regulation of the input. In section 5 we consider the question of parameter transformations and their possible utility in simplifying selective sensitivity.

2 Selective Sensitivity for Linear Elastic Systems

Consider an N-dimensional linear elastic structure such as:

$$M\ddot{x}(t) + C\dot{x}(t) + Kx(t) = Hf(t) \tag{1}$$

where M, C and K are inertia, damping and stiffness matrices, respectively. The deflection vector is $x(t)$, the input vector is $f(t)$ and the measurement vector is $y(t)$:

$$y(t) = Gx(t) \tag{2}$$

G and H are real matrices, possibly rectangular, which indicate the locations of the sensors and actuators, respectively. The matrices M, C and K depend on model parameters $p_1, p_2, \ldots$ representing stiffnesses, inertias, geometrical parameters, etc.

The basic question addressed by selective sensitivity is: construct an excitation, $f(t)$, which causes the output to be sensitive to a small subset $\mathcal{I}$ of indices of m parameters of the model, and insensitive to the remaining model parameters. If this can be achieved, then the values of the parameters in $\mathcal{I}$ can be obtained in an m-dimensional estimation.

Let $\phi(s)$ and $\psi(s)$ represent the Laplace transforms of the input and the output vectors, $f(t)$ and $y(t)$ respectively. Then eqs.(1) and (2) can be transformed and combined as:

$$\psi(s) = GF(s)H\phi(s) \tag{3}$$

where $F(s)$ is the transfer matrix (occasionally referred to as the frequency response matrix). Its inverse, $Q(s)$ (sometimes called the dynamic stiffness matrix) is:

$$Q(s) = s^2M + sC + K \tag{4}$$

We now define the output sensitivity to the αth model parameter, p_α:

$$S(p_\alpha) = \left\| \frac{\partial \psi(s)}{\partial p_\alpha} \right\|^2 \tag{5}$$

where $\|\cdot\|$ is the euclidean norm for vectors.

When $S(p_\alpha)$ is large the Laplace-transformed output is sensitive to variations of the model parameter p_α, while if $S(p_\alpha)$ is small then little information about p_α is contained in the measurements. Information about the model parameters is concentrated or filtered in two senses by a selectively sensitive input. First, the excitation is chosen so that $S(p_\alpha)$ is large for one or several parameters and small or vanishing for the rest. Furthermore, the selective sensitivity may differ at different points of the complex plane, so the Laplace transformation can be thought of as a filter for removing undesired information.

We now develop an expression for the sensitivity to p_α. Let us define:

$$A_\alpha(s) = \frac{\partial Q(s)}{\partial p_\alpha} \tag{6}$$

One finds the sensitivity to p_α to be the following hermitian quadratic form in the Laplace transform of the input:

$$S(p_\alpha) = \phi^\dagger \underbrace{H^T[FA_\alpha F]^\dagger G^T G[FA_\alpha F]H}_{D_\alpha} \phi \tag{7}$$

$$= \phi^\dagger D_\alpha \phi \tag{8}$$

where eq.(7) defines the sensitivity matrix, D_α. (Superscript T implies the matrix transpose; superscript $\dagger$ means matrix conjugate transposition.)

The first point to notice is that, since D_α is hermitian and positive semi-definite, it has real non-negative eigenvalues. Furthermore, the extreme values which can be obtained by the sensitivity, normalized to the length of the input vector, $S(p_\alpha)/\phi^\dagger\phi$, are the least and greatest eigenvalues of D_α. These extreme values are obtained by choosing the input vector, $\phi(s)$, proportional to the corresponding eigenvectors of D_α. We know, therefore, how to choose the input so as to enhance or depress the output sensitivity to any particular model parameter.

Let us first qualitatively consider the task of finding a selectively-sensitive input. Any model parameter, p_α, will usually occur at no more than a few positions in the dynamic stiffness matrix, $Q(s)$. Consequently A_α will be a sparse matrix whose rank is much less than its dimension, N, which is large (for a large system). The sensitivity matrix, D_α, being a product of matrices including A_α, is at least as rank deficient as A_α. Each sensitivity matrix, then, has a null space of high dimension. In other words, there are many linearly independent vectors ξ which are orthogonal to any particular sensitivity matrix: $D_\beta \xi = 0$. But no sensitivity matrix is orthogonal to *all* vectors. So, given a vector ϕ not orthogonal to D_α, it will often be orthogonal to many other sensitivity matrices. If we adopt this ϕ as the input, then the output will have non-zero sensitivity to p_α but will have identically zero sensitivity to many (perhaps all) other model parameters. We conclude that it should be possible to find input vectors (in the Laplace domain) which achieve selective sensitivity to a small number of model parameters. Analysis of the null spaces of the sensitivity matrices should lead us to these vectors.

We formalize these conclusions thus: the condition for **selective sensitivity** to the parameters indexed in $\mathcal{I}$ is:

$$S(p_n) = \begin{cases} 0 & \text{if} \quad n \notin \mathcal{I} \\ \text{not zero} & \text{if} \quad n \in \mathcal{I} \end{cases} \tag{9}$$

We can express eq.(9) more explicitly. $S(p_n)$ vanishes if and only if $\partial\psi/\partial p_n$ vanishes. Employing eq.(3) and the relation:

$$\frac{\partial F}{\partial p_n} = -F \frac{\partial Q}{\partial p_n} F = -F A_n F \tag{10}$$

one finds that eq.(9) is equivalent to:

$$GFA_nFH\phi = \begin{cases} 0 & \text{if} \quad n \notin \mathcal{I} \\ \text{not zero} & \text{if} \quad n \in \mathcal{I} \end{cases} \tag{11}$$

One is often justified in assuming that the mass, damping and stiffness matrices are linear in the model parameters [7], [8]:

$$M = \sum_n m_n M_n \,, \quad C = \sum_n c_n C_n \,, \quad K = \sum_n k_n K_n \tag{12}$$

where the matrices M_n, C_n and K_n do not depend on the physical model parameters m_n, c_n and k_n. We will combine these model parameters into a single N_p-dimensional parameter vector, p. Now define matrices $A_n(s)$ from eqs.(4) and (12) so as to express the dynamic stiffness matrix as follows:

$$Q(s) = \sum_{n=1}^{N_p} p_n A_n(s) \tag{13}$$

The $N \times N$ matrices A_n depend on the topology of the structure — what is connected to what — and on the formulation of the finite element representation of the structure, but not on the parameter values themselves. Once the finite element representation of the system has been chosen (not always a trivial matter in itself), the matrices A_n are completely known.

Let us seek an input which causes selective sensitivity to the model parameters whose indices are contained in an index-set $\mathcal{I}$. Our search for the desired excitation will proceed in two stages. We first seek a vector θ satisfying:

$$A_n\theta = \begin{cases} 0 & \text{if} \quad n \notin \mathcal{I} \\ \text{not zero} & \text{if} \quad n \in \mathcal{I} \end{cases} \tag{14}$$

θ is not necessarily[1] the output, but rather just an auxiliary vector with which we seek a selectively sensitive input. We make the important observation that, if the linear sub-system modelling of eq.(13) holds, then solutions of eq.(14) are independent of the model parameters, and depend only on the topology of the structure. Let us assume that a solution θ exists. (This is often the case since the matrices A_n will usually be sparse.) Then we choose the input, ϕ, as a solution of:

$$H\phi = Q\theta \quad \text{or} \quad FH\phi = \theta \tag{15}$$

where Q is the unknown dynamic stiffness matrix and F is its inverse. Such a ϕ will satisfy eq.(11) and yield the desired selective sensitivity, because θ satisfies eq.(14). G does not appear in eq.(15) since θ is not (necessarily) the output. Exploiting the orthogonality properties of θ in eq.(14) and the linear sub-system modelling, eq.(13), eq.(15) becomes:

$$H\phi = \sum_{n\in\mathcal{I}} p_n A_n \theta \tag{16}$$

A vector, ϕ, may exist which satisfies this equation. If so, this vector will also satisfy eq.(11) and thus provide selective sensitivity for model parameters p_n, $n \in \mathcal{I}$.

We summarize our brief formulation of selective sensitivity for linear elastic structures by noting the following basic properties.

1. Formulation of the selectively-sensitive input depends on knowledge of the structure of the model, but not on the values of any model parameters other than the parameters being estimated. This independence is the basis for partitioning the identification into low-dimensional estimation problems. The dependence of the input on the parameters being estimated can be treated in various ways. We will examine one option when we consider the feedback formulation of selective sensitivity in section 4.

2. Formulation of the selectively sensitive input for a linear system, when it exists, can be obtained from the solution of linear algebraic equations.

3. Experience to date with a range of mechanical structures has confirmed the expectation that a large degree of parameter selectivity is attainable. This stems from the fact that physical parameters such as stiffness and geometrical properties such as moments of inertia usually appear in only a few locations in the dynamic stiffness matrix, $Q(s)$. The mathematical consequence is that the matrices A_α characteristically have very low rank.

[1] If G is the identity matrix then θ is precisely the output.

4. The major limitation in the application of selective sensitivity is the need for either a large number of actuators or for mobile actuators which can be moved to different locations on the structure at various stages of the system identification. A large number of sensors is also needed, though this is generally a less severe constraint. Large numbers of accelerometers are routinely employed in current modelling techniques, and laser-interferometric measurement methods promise to expand this virtually without limit.

3 Selective Sensitivity of Non-Linear Systems

The first nonlinear problem we consider is the formulation of a selectively sensitive input for identification of nonlinear model parameters in a general — not necessarily linear — dynamical system.

Consider the finite-dimensional nonlinear system whose state vector is N_D-dimensional and denoted $x(t)$. The dynamics of the system are represented by:

$$\mathcal{L}(x) + \mathcal{N}(x) = Hf(t) \tag{17}$$

where $\mathcal{L}(x)$ is a linear differential operator, $\mathcal{N}(x)$ is a nonlinear N_D-dimensional vector function, $f(t)$ is an N_R-dimensional input vector and H is an $N_D \times N_R$ constant real matrix.

The aim is to design the excitation, $f(t)$, so as to identify the nonlinear part of the model, $\mathcal{N}(x)$, employing knowledge of the linear operator $\mathcal{L}(x)$. The analysis will be based on the concept of selective sensitivity.

The structure of $\mathcal{L}(x)$ is:

$$\mathcal{L}(x) = \sum_{n=0}^{N_L} B_n \frac{d^n x(t)}{dt^n} \tag{18}$$

where each B_n is a real, constant $N_D \times N_D$ matrix. For a linear elastic system the linear part of the model may take the form:

$$\mathcal{L}(x) = M\ddot{x}(t) + C\dot{x}(t) + Kx(t) \tag{19}$$

where M, C and K are inertia, damping and stiffness matrices, respectively.

Now consider a measurement equation which expresses the fact that only N_M measurements are made. The output $y(t)$ is related to the state vector $x(t)$ by:

$$y(t) = Gx(t) \tag{20}$$

where G is a real constant $N_M \times N_D$ real matrix.

The linear part of the model, $\mathcal{L}(x)$, depends on an r_P-dimensional vector p of linear-model parameters. Similarly, $\mathcal{N}(x)$ depends on an r_Q-dimensional vector q of nonlinear model parameters as well as, possibly, on p. We will assume that G and H are independent of the model parameters p and q.

We will denote the Laplace transforms of $y(t)$, $f(t)$ and $\mathcal{N}(x)$ by $\psi(s)$, $\phi(s)$, and $\xi(s)$, respectively.

Now define a "dynamic stiffness matrix" of the linear part of the model, $Q(s)$, in terms of the Laplace transform of $\mathcal{L}(x)$, for zero initial conditions:

$$\mathcal{T}[\mathcal{L}(x)] = Q(s)\mathcal{T}[x] \tag{21}$$

where $\mathcal{T}[\cdot]$ is the Laplace transform operator. $Q(s)$ is a complex $N_D \times N_D$ matrix and, we assume, non-singular for at least one value of s. For example, if $\mathcal{L}(x)$ is given by eq.(19) then $Q(s)$ is defined in eq.(4). In this case $Q(s)$ is invertible provided that s is not an eigenvalue of $Q(s)$.

Now the Laplace transformation of eq.(17) can be expressed:

$$\mathcal{T}[x] = Q^{-1}(s)H\phi(s) - Q^{-1}\xi(s) \tag{22}$$

Employing eq.(20) one obtains the Laplace transform of the output as:

$$\psi(s) = GQ^{-1}(s)H\phi(s) - GQ^{-1}\xi(s) \tag{23}$$

This is not strictly an input-output relation, since the input is ϕ but the state, $x(t)$, appears in both the output $\psi(s)$ and in $\xi(s)$.

We wish to evaluate the sensitivity of the output, $\psi(s)$, with respect to a nonlinear model parameter q_m. We differentiate $\psi(s)$ with respect to q_m, assuming that G, H and $\phi(s)$ are independent of the model parameters but that $Q(s)$, $\psi(s)$ and $\xi(s)$ depend on q. After some manipulations one finds the sensitivity to a nonlinear model parameter q_m is simply:

$$S(q_m) = \left(\frac{\partial\psi}{\partial q_m}\right)^{\dagger}\left(\frac{\partial\psi}{\partial q_m}\right) \tag{24}$$

$$= \xi_m^{\dagger} {Q^{-1}}^{\dagger} G^T G Q^{-1}\xi_m \tag{25}$$

where we define:

$$\xi_m = \frac{\partial\xi}{\partial q_m} \quad , \quad m = 1, \ldots, r_Q \tag{26}$$

Let us also define:

$$V(s) = {Q^{-1}}^{\dagger}(s) G^T G Q^{-1}(s) \tag{27}$$

We note the following points. (1) $S(q_m)$ is real and non-negative. In particular, $S(q_m)$ is positive semi-definite in ξ_m. (2) $V(s)$ is hermitian and positive semi-definite. (3) $V(s)$ is independent of the nonlinear part of the model, q. (4) The derivation of eq.(25) is independent of the structure of the nonlinear part of the model.

The derivation of a selectively sensitive input, when it exists, should be obtainable by exploiting the quadratic structure of eq.(25). This problem has been studied for certain specific polynomial forms for the nonlinear model [3]. Extensions to other nonlinear polynomials and general nonlinear models is needed.

4 Feedback Implementation of Selective Sensitivity

4.1 The Concept

The existence of a vector θ satisfying eq.(14) is central to establishing selective sensitivity to the model parameters indexed in the set $\mathcal{I}$. The matrices A_n depend on the topology of the structure and on the dimension of the dynamic model, but not on the model parameters. Consequently, we can determine whether selective sensitivity for p_α, $\alpha \in \mathcal{I}$, is possible, without knowing the values of any model parameters. However, implementation of the selectively sensitive input depends on solving eq.(16) for ϕ, for which we must know values for p_α, $\alpha \in \mathcal{I}$.

One approach to implementation of a selectively sensitive input is based on realizing that, since we know the desired output but not the required input, we can formulate the problem from a feedback-control perspective. Find a vector, call it $\overline{\theta}$, which satisfies eq.(14). This is the output which would be obtained by a selectively sensitive input constructed, from eq.(16), with the correct values of p_α, $\alpha \in \mathcal{I}$. Now implement a feedback regulator which adapts the values of p_n to cause the output to match $\overline{\theta}$ while the input is constrained by $H\phi = \sum_{n\in\mathcal{I}} p_n A_n \overline{\theta}$. The control parameters are precisely the parameters which we seek to estimate. The idea here is to adapt the input to force the output to a form which has been determined without knowledge of the model parameters; estimation of the desired model parameter results as a by-product of this control procedure.

One difficulty with this idea is that, for dynamic measurements, the input and output are Laplace transform variables. This difficulty is avoided in the following example by considering static loads and deflections.

We will illustrate the feedback-implementation of selective sensitivity by estimating the stiffness at the top floor of a 10-story structure.

4.2 The Model

The dynamic variables and model parameters of the shear-type building model are shown in fig. **1**. N floor units are depicted. The lateral displacement of the nth floor unit is $x_n(t)$, while the resulting shear force in the columns of the nth floor unit is $y_{n-1}(t)$. External lateral loading on the nth floor is represented by $f_n(t)$. The mass of the nth floor is m_n and the stiffness and damping coefficients of the columns are k_n and c_n, respectively. The basic dynamic equations, based on [11], are:

$$f_n + y_n = y_{n-1} + m_n \ddot{x}_n \tag{28}$$

$$y_{n-1} = k_n[x_n - x_{n-1}] + c_n[\dot{x}_n - \dot{x}_{n-1}] \tag{29}$$

There is no ground motion during the integrity tests, so $x_0 = \dot{x}_0 = 0$. In addition, the shear at the top of the building must vanish: $y_N = 0$. The shear forces can be eliminated and these equations can be combined in matrix form as eq.(1). The

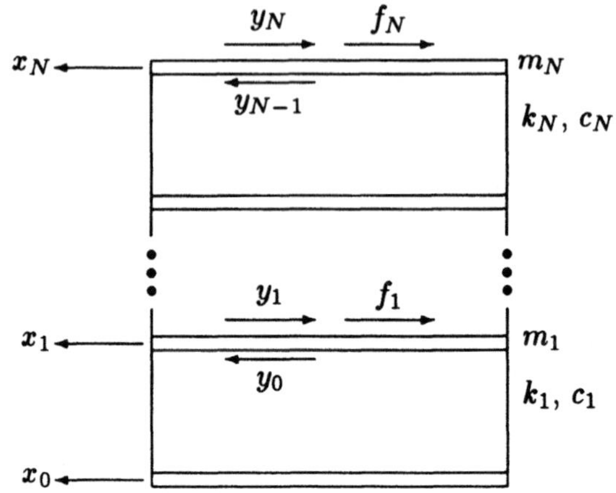

Figure 1 Variables and parameters of the shear-type building.

static load-deflection relation is:

$$Kx = Hf \tag{30}$$

The stiffness matrix $K = [k_{m,n}]$ is tridiagonal, with the following structure:

$$\begin{aligned} k_{n,n} &= k_n + k_{n+1} \\ k_{n,n+1} &= -k_{n+1} \\ k_{n,n-1} &= -k_n \end{aligned} \tag{31}$$

K can be represented in linear sub-model parametrization as in eq.(12). Let e^α represent the N-dimensional standard basis vector: unity in the αth position and zero elsewhere. Then one finds:

$$K_n = \left(e^n - e^{n-1}\right)\left(e^n - e^{n-1}\right)^T \tag{32}$$

4.3 Conditions for Selective Sensitivity to Stiffnesses

We proceed as follows in order to obtain selective sensitivity to the αth stiffness parameter. Using the fact that $\partial K^{-1}/\partial k_n = -K^{-1}(\partial K/\partial k_n)K^{-1}$, one finds that the variation of the output x with the nth stiffness is:

$$\frac{\partial x}{\partial k_n} = -K^{-1}\frac{\partial K}{\partial k_n}K^{-1}Hf = -K^{-1}K_nK^{-1}Hf \tag{33}$$

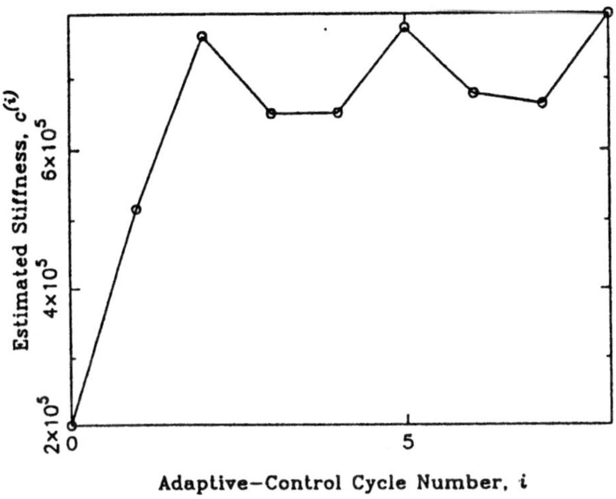

Figure 2 Typical estimation history with feedback-implementation of selectively-sensitive excitation. Estimation terminated after 8 feedback cycles.

The sensitivity to k_n is the inner product of this vector with itself. For selective sensitivity to k_α we require:

$$S(k_n) = \begin{cases} 0 & n \neq \alpha \\ \text{not zero} & n = \alpha \end{cases} \tag{34}$$

Employing eq.(33), this is equivalent to:

$$K_n K^{-1} H f = \begin{cases} 0 & n \neq \alpha \\ \text{not zero} & n = \alpha \end{cases} \tag{35}$$

But $x = K^{-1}Hf$, so a necessary condition for selective sensitivity to the αth stiffness coefficient is that the output satisfy the following relations:

$$K_n x \;=\; 0\,,\;\; n \neq \alpha \tag{36}$$

$$K_\alpha x \;\neq\; 0 \tag{37}$$

Eqs.(37) and (36) are satisfied

$$x = h \sum_{n=\alpha}^{N} \mathrm{e}^n \tag{38}$$

where h is any non-zero constant. Complete selective sensitivity is obtainable for each stiffness coefficient.

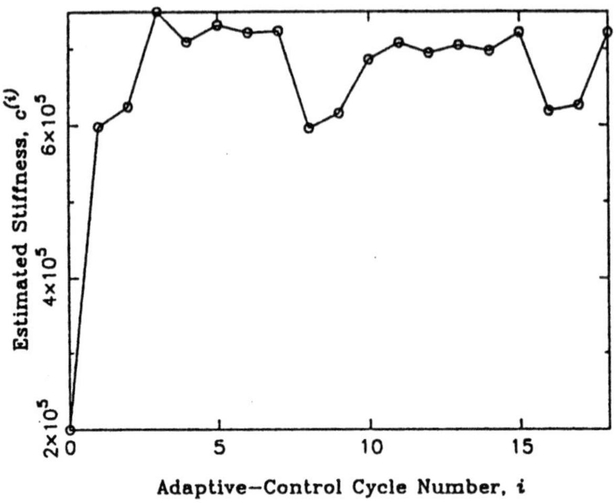

Figure 3 Typical estimation history with feedback-implementation of selectively-sensitive excitation. Estimation terminated after 18 feedback cycles.

Combining eqs.(12), (32) and (38), one finds the static load which is required in order to produce the desired selectively sensitive deflection:

$$Hf = Kx = hk_\alpha \left(\mathrm{e}^\alpha - \mathrm{e}^{\alpha-1}\right) \tag{39}$$

The conclusion is that selective sensitivity for k_α is obtained by statically loading the building at two points only: the $(\alpha-1)$th and αth floors. (When estimating k_1 only one external load is required).

4.4 The Inverse Estimator

We constrain our example to estimation of the stiffness at the top floor. The necessary and sufficient conditions for selective sensitivity to k_N are eqs.(37) and (36) with α These are satisfied by choosing x as in eq.(38):

$$\overline{x} = h\mathrm{e}^N \tag{40}$$

Let c be an initial guess for the value of k_N. Employing eq.(39), letting $H = I$, and recalling that $k_{N+1} = 0$, one sees that the load which must be applied is:

$$f(c) = cK_N\overline{x} = hc\left(\mathrm{e}^N - \mathrm{e}^{N-1}\right) \tag{41}$$

Let γ and ν be zero-mean random vectors corrupting the input and the output, respectively. The measured response to load $f(c)$ is:

$$x(c) \quad = \quad K^{-1}(f(c)+\gamma)+\nu \tag{42}$$

$$= \quad cK^{-1}K_N\overline{x} + K^{-1}\gamma + \nu \tag{43}$$

where K is the unknown stiffness matrix of the structure.

Using the sub-modelling of eq.(12), the matrix K_N can be related to the stiffness matrix K and the stiffness coefficients k_n as:

$$K_N = \frac{1}{k_N}\left[K - \sum_{n=1}^{N-1} k_n K_n\right] \tag{44}$$

Substituting this value for K_N into eq.(43), using eq.(40) and employing the selective-sensitivity conditions: $K_n\overline{x} = 0$ for $n < N$, eq.(43) becomes:

$$x(c) = \frac{hc}{k_N}\mathrm{e}^N + K^{-1}\gamma + \nu \tag{45}$$

where, of course, the value of k_N is unknown and must be determined from the measurements.

The random vector $r = K^{-1}\gamma + \nu$ has zero mean but unknown covariance because K is unknown. Consequently a least-squares technique for estimating k_N from the measurement $x(c)$ would be asymptotically biased. We will use instead an instrumental-variables approach [10]. The inner product of e^N with the measurement vector, eq.(45), is:

$$\mathrm{e}^{N^T} x(c) = \frac{hc}{k_N} + r_N \tag{46}$$

r_N vanishes in the mean, so the expectation of $\mathrm{e}^{N^T} x(c)$ is:

$$\mathrm{E}\left[\mathrm{e}^{N^T} x(c)\right] = \frac{hc}{k_N} \tag{47}$$

which can be used to obtain an asymptotically unbiased estimate of k_N.

4.5 The Feedback Controller-Estimator

We can now formulate the feedback controller-estimator used for adaptive determination of k_N from selectively-sensitive static loads. The estimate of k_N at the ith cycle of the adaptation is $c^{(i)}$. The procedure starts with an initial guess, $c^{(0)}$, at cycle number $i = 0$.

1. At the ith cycle the load is:

$$f^{(i)} = hc^{(i)}\left(\mathrm{e}^N - \mathrm{e}^{N-1}\right) \tag{48}$$

2. The resulting measured deflection is:

$$x^{(i+1)} = \frac{hc^{(i)}}{k_N}\mathrm{e}^N + K^{-1}\gamma + \nu \tag{49}$$

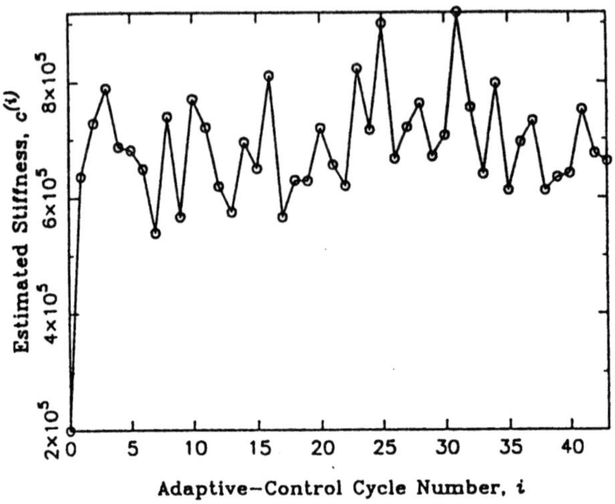

Figure 4 Typical estimation history with feedback-implementation of selectively-sensitive excitation. Estimation terminated after 43 feedback cycles.

This measurement is repeated, with the same input (but different noise vectors), N_M times. These repetitions will be denoted $x^{(i)}(m)$, $m = 1, \ldots, N_M$. The instrumental variable, $\mathrm{e}^{N^T} x$, is averaged over these N_M measurements to estimate its mean:

$$\left\langle \mathrm{e}^{N^T} x \right\rangle = \frac{1}{N_M} \sum_{m=1}^{N_M} \mathrm{e}^{N^T} x^{(i)}(m) \tag{50}$$

3. The control-estimation variable, c, is up-dated as:

$$c^{(i+1)} = \frac{hc^{(i)}}{\left\langle \mathrm{e}^{N^T} x \right\rangle} \tag{51}$$

4. Either increment i and return to step 1 or terminate the feedback-estimation.

4.6 Simulation

We now simulate a feedback-implementation of the selectively-sensitive estimation of k_N. The noise-free input vector is calculated from eq.(41). The input is corrupted by proportional noise, so the elements of the random vector γ are calculated as:

$$\gamma_i = f_i \eta_f \rho_i\,, \quad i = 1, \ldots, N \tag{52}$$

where ρ_i is a standard normal random variable and η_f is an input-noise amplitude factor. The nominal measurement, x^{nom}, is calculated from eq.(42) with $\nu = 0$, and then corrupted by proportional noise by adding the random vector ν calculated as:

$$\nu_i = x_i^{\text{nom}} \eta_x \rho_i' , \quad i = 1, \ldots, N \tag{53}$$

where ρ_i' is a standard normal random variable and η_x is a measurement-noise amplitude factor. If any element of the noisy measurement vector exceeds the safe-deflection limit of 0.18[m], then the test is terminated.

The initial guess of k_N is $c^{(0)} = 2 \times 10^5$ [N/m], quite far from the correct value of 6.871×10^5 [N/m].

In each cycle of the feedback-adaptation, the instrumental variable $x^T e^N$ is averaged N_M times, as explained in step 3 above. $N_M = 10$ in most of the results to be discussed

The feedback loop is required to cycle at least 7 times, and the loop is terminated if 50 cycles have been performed. However, termination is allowed after the ith cycle (earlier than the 50th) if the estimate has stabilized. More precisely, we use the following criterion. Consider the i estimates of k_N which have accumulated since the beginning of the run: $c^{(1)}, \ldots, c^{(i)}$. Ignore the chronological first third of these estimates, and let $\bar{c}_2$ and $\bar{c}_3$ be the average estimates of k_N obtained in the middle and final thirds of these i cycles. Let σ_{23} be the standard deviation of the final two-thirds of the estimates. The adaptive control-estimation is terminated if:

$$\frac{|\bar{c}_3 - \bar{c}_2|}{\sigma_{23}} \leq \varepsilon \tag{54}$$

where $\varepsilon = 0.06$ in our simulations.

Figs. **2**–**4** show typical esti histories, with 1% input and measurement noise ($\eta_f = \eta_x = 0.01$) and $N_M = 10$. The initial guess is rapidly corrected, but the termination time varies considerably from run to run, depending on the input- and measurement-noise sequences.

Table **1** shows further results for various parameter values. The first two columns show the input- and measurement-noise amplitude factors. The third column is the number of repeated measurements to estimate the mean of the instrumental value. The last four columns show results from 50 complete repetitions of the adaptive control-estimation procedure. $\widehat{k}_N$ and σ_{k_N} the mean and standard deviation of these 50 estimates of the stiffness, and k_N represents the actual value of the stiffness of the top floor. Column 4 contains the percent fractional error of the estimate: $100(\widehat{k}_N - k_N)/k_N$. The 5th column is the percent relative error of the sample of runs: $100\sigma_{k_N}/\widehat{k}_N$. The 6th and 7th columns show mean and standard deviations of the cycle-number at termination.

The first three rows of results in table **1** show the performance of the feedback-estimator for various values of input and measurement noise, with the other parameters constant. The percent error of the estimate, column 4, is very sensitive to noise, increasing from 0.009% to 6% as the input and measurement noise increase from 0.5% to 2%. The relative error of the estimate (column 5) also increases with noise, but not as strongly. The mean and standard deviation of the number of cycles

η_f	η_x	N_M	$100\frac{\hat{k}_N - k_N}{k_N}$	$100\frac{\sigma_{\hat{k}_N}}{\hat{k}_N}$	E(i)	σ_i
0.005	0.005	10	0.0087	1.46	22.3	13.1
0.01	0.01	10	0.847	2.68	20.2	10.6
0.02	0.02	10	6.24	6.75	21.8	11.4
0.02	0.02	30	2.00	3.07	21.1	10.3
0.02	0.02	100	1.36	1.68	23.8	13.6
0.02	0.10	10	6.29	5.50	41.5	12.0

Table 1 **Estimates of stiffness at 10th floor, averaged over 50 feedback runs.**

to termination (last 2 columns) is insensitive to noise. Note, however, the large values of σ_i, indicating large variation in cycle-number. Rows 3 to 5 demonstrate the improvement in the estimation with increasing the sample size, N_M, for estimating the mean of the instrumental variable. The 6th row shows that quite large measurement noise ($\eta_x = 0.10$) can be tolerated. Such large noise at the input would lead to enormous estimation errors.

5 Non-Linear Parameter Transformations

The final nonlinear problem associated with selective sensitivity which we will consider is the matter of parameter transformations. Consider a structure whose model parameters are represented by $p^T = (p_1, \ldots, p_N)$. If complete selective sensitivity is not obtainable for these model parameters we inquire: is there a one-one transformation of the parameters p to new model parameters $q^T = (q_1, \ldots, q_N)$ such that selective sensitivity is achievable for each parameter q_n ? When this is possible, we may wish to determine $q_1, \ldots, q_N$ in a sequence of N one-dimensional estimations, and then evaluate $p_1, \ldots, p_N$ by the reverse transformation.

The advantage of this procedure is in reducing the identification to a sequence of one-dimensional estimations. Numerous questions arise, such as propagation of errors in the back transformation, choosing between alternative transformations should more than one be available, and so on. However, we will only demonstrate, by example, that one-one transformations exist which change the selective sensitivity.

Consider the static deflection of the multi-story structure studied in section 4. The stiffness matrix K relates the deflection vector x to the load vector f as in subsection 4.2, with $H = I$. The original model parameters are the stiffness coefficients, k_n. Define new parameters by:

$$q_n = \frac{1}{k_n - k_{n-1}} \tag{55}$$

where we define $k_0 = 0$. The inverse transformation is:

$$k_n = \sum_{i=1}^{n} \frac{1}{q_i} \tag{56}$$

Analysis of the selectivity conditions for these two sets of model parameters reveals that complete selectivity is obtained for the parameter set $k_1, \ldots, k_N$ while incomplete selectivity is obtained for $q_1, \ldots, q_N$. In other words, the "natural" parameters are completely selectable, while the new parameters are not, even though the transformation between them is one-one.

It is a tantalizing hypothesis that "natural" model parameters will always display maximum selectivity. This, however, has not yet been studied. We have, however, shown that selectivity is not necessarily preserved by one-one transformations. Consequently, when the selectivity is incomplete, one may seek a one-one parameter transformation which improves the selectivity.

A question closely related to the improvement of selectivity by parameter transformation is the improvement of sensitivity. Consider a parameter set $p_1, \ldots, p_N$ in which p_1 and p_2 are not separable by a selectively sensitivity excitation. It sometimes occurs (see [9] for an example) that the sensitivities to p_1 and p_2 differ substantially. In order to compare sensitivities to different model parameters, which may have different units, we define the *normalized sensitivity* as:

$$\widetilde{S}(p_n) = \left\| \frac{\partial \psi}{\partial p_n / p_n} \right\|^2 = p_n^2 S(p_n) \tag{57}$$

Suppose that the sensitivity to p_1 is much less than the sensitivity to p_2:

$$\widetilde{S}(p_1) \ll \widetilde{S}(p_2) \tag{58}$$

Since p_1 and p_2 are not separable, the disparity in their sensitivities makes it difficult to estimate p_1, whose sensitivity is low. One then asks for a one-one parameter transformation from p_1, p_2 to a new parameter subset q_1, q_2 in which the sensitivities are more nearly commensurable.

6 Conclusions

We have considered several nonlinear problem areas which arise in the application of selective sensitivity to the identification of dynamical systems.

The basic task addressed by selective sensitivity is to construct an excitation which causes the output to be sensitive to a small number of parameters of the model, and insensitive to the remaining model parameters. When this can be achieved, which for linear elastic systems is quite often, then the model parameters can be obtained in a sequence of low-dimensional estimations.

In considering the selectively-sensitive identification of nonlinear systems we showed that the sensitivity condition, from which the input can be derived, is a quadratic form in a matrix which is independent of the nonlinear part of the model.

This promises to lead to useful input functions, as have already been developed for certain polynomial nonlinear structures.

The basic formulation of selective sensitivity for linear systems leads directly to linear algebraic equations defining the response to a selectively sensitive input. Various schemes have been developed elsewhere for deriving input functions from these conditions on the output. We have shown here that an additional method for implementing a selectively sensitive input is as a feedback-regulator which homes in on the correct output. The adjustment parameters of the regulator are precisely the model parameters whose estimate is desired.

Finally, we have considered the possibility of improving either the parameter selectivity or sensitivity by transforming the original parameter set to a new parameter set. We have examined one example in which a one-one parameter transformation changes the selectivity. This raises the general possibility that when the selectivity is incomplete, one can find a one-one parameter transformation which will improve the selectivity and thereby reduce the dimensionality of the overall system identification.

Acknowledgements

was supported in part by the Fund for the Promotion of Research at the Technion, and by the Technion V.P.R. Fund — E. and M. Mendelson Research Fund.

References

[1] M. Aoki and M. Staley, 1970, On input signal synthesis in parameter identification, *Automatica*, 6: 431–440.

[2] Y. Ben-Haim, 1992, Adaptive Diagnosis of Faults in Elastic Structures by Static Displacement Measurement: The Method of Selective Sensitivity. *Mechanical Systems and Signal Processing*, 6: 85–96.

[3] Y. Ben-Haim, 1993, Identification of certain polynomial nonlinear structures by adaptive selectively-sensitive excitation. To appear in *ASME J. Vibration and Acoustics*.

[4] Y. Ben-Haim, 1993, Adaptive model up-dating by selective sensitivity: Overview and new results. To appear in *International J. of Numerical Methods in Engineering*.

[5] Y. Ben-Haim and U. Prells, 1993, Selective Sensitivity in the Frequency Domain, Part I: Theory. To appear in *Mechanical Systems and Signal Processing*.

[6] R. Haftka and Z. Gürdal *Elements of Structural Optimization*, 3rd edition, Kluwer, Dordrecht, Holland, 1992.

[7] H.G. Natke, D. Collmann, H. Zimmermann, 1974, Beitrag zur Korrektur des Rechenmodells eines elastomechanischen Systems anhand von Versuchsergebnissen, VDI-Berichte nr. 221, pp23–32.

[8] H.G. Natke, 1979, Vergleich von Algorithmen für die Anpassung des Rechenmodells einer schwingungsfä elastomechanischen Struktur an Versuchswerte, *Zeitschrift fü Angewandte Mathematik und Mechanik,* 59: 257–268.

[9] U. Prells and Y. Ben-Haim, 1993, Selective Sensitivity in the Frequency Domain, Part II: Applications. To appear in *Mechanical Systems and Signal Processing.*

[10] T. Söderström 1989, *System Identification,* Prentice Hall, New York.

[11] R.H. Zhang and T.T. Soong, 1992, Seismic design of viscoelastic dampers for structural applications, *ASCE Journal of Structural Engineering,* 118:1375–1392.

Safety Evaluation of Electric Pylons Using Impulse Testing

M.P. Luong
CNRS-LMS Ecole Polytechnique
Palaiseau, France

1 Introduction

Overhead line towers are commonly exposed to dynamic loadings from several sources, including high winds, earthquake ground motions and others. The design task is made quite challenging by the inherent constraints of economics, the demand for extreme reliability and the considerable uncertainty in defining the dynamic loading which the structure must endure. A non-destructive vibratory testing using impulsive tension slacking has been devised for the quality and safety control of overhead line towers of Electricité de France. Several structural control experiments have been conducted on four-legged towers resting on three types of foundations : concrete stepped or pedestal blocks, steel piles and prestressed foundations. Experimental results have given significant vibratory signatures in close connection with the geotechnical response characteristics of both such slender structures and their foundations. This paper presents some significant results of the test program and the safety evaluation of the pylon.

The major difficulties encountered in the application of modern control techniques to structural systems may be listed as follows:

- Active control requires the ability to generate and apply large controlled forces to the structure.
- Modern control theory often leads to feedback control laws, thus requiring on-line measurement or estimation of all the system state variables.
- On-line control requires that both measurement and control be performed in real time.

From a practical standpoint, while the application of large control forces to a structure does not raise insurmountable difficulties, the generation of such forces over sustained periods of time, as necessitated by continuous optimal feedback control theory, may cause the concept of active control to become impracticable. To bypass this possible drawback, the approach under consideration attempts to use pulses of relatively short duration to control the structural system.

The objectives of this vibratory non-destructive testing were :

- to characterize the soil-structure interaction of overhead line towers resting on different types of foundations such as traditional concrete footings, piles or prestressed foundations recently designed in France.
- to verify the geotechnical performance of these foundations subjected to dynamic loading.
- to investigate and develop simple experimental testing in order to obtain readily a vibratory signature for the control and inspection of the mechanical behaviour of the transmission tower.

In several cases, horizontal forces, for instance generated by winds, when loading on slender, high rise and light structures such as electric pylons or transmission line towers, induce onto their foundations both compressive and tensile loadings. Traditional building techniques employ huge heavy concrete blocks, large buried concrete plates or costly pile groups able to resist punching or pulling loads. When subjected to severe and diverse climatological variations or earthquake loadings, the geotechnical behaviour of these slender structures can degenerate leading to failure.

The physical nature of the transmission line system restricted the dynamic excitation alternatives for testing. The classical procedures were sine dwell, sine sweep, fast sine sweep or chirp, random, impulse, etc. The technique in use is based on the release from an initial chosen tension similar to the twang-excitation method [Kemper et al 1981] based on a release from initial displacement of the structure.

This paper presents some significant results of series of slacking tests conducted on overhead line towers of Electricité de France resting on three different types of foundations.

2 Pylon Foundations

The transmission structures subject to vibratory testing involved in our test program are double-circuit, 63 kV, self-supporting, lattice-steel towers. These four-legged towers rest on three types of foundations.

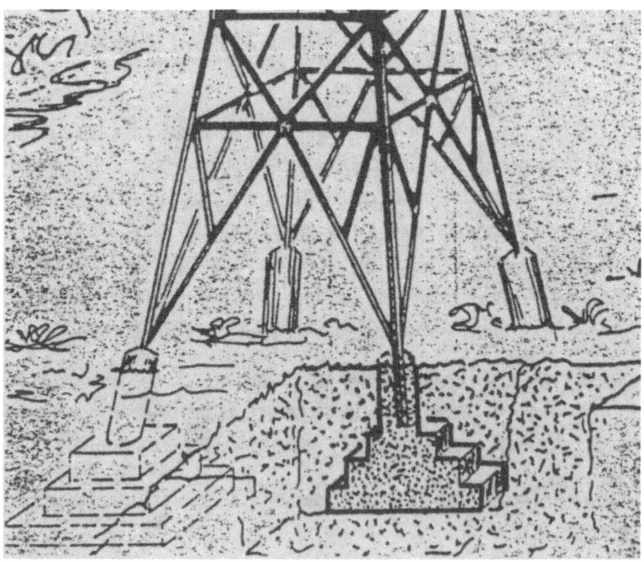

Figure 1 Stepped or pedestal footing for overhead line tower.

Concrete foundation

The stresses of each pylon leg are distributed to a stepped or pedestal concrete footing designed to satisfy the limit total displacement to an acceptable small amount and eliminate differential settlements between parts of a structure as nearly as possible. As the pylon may be subjected to overturning forces, the footing has been designed on the assumption of linear variation of soil pressure and constructed with sufficient resistance to deformation. Stress in the longitudinal direction of the pylon leg is transferred to its pedestal by extending the longitudinal steel into the support (Fig. 1). The stress-transfer bar projects into the base a sufficient compression-embedment distance to transfer the stress in the column bar to the base concrete.

Pile foundation

Piles have been used to resist uplift and overturning moments developed through friction along the sides of the piles and to distribute loads over a sufficiently large vertical area (28 m) of relatively weak soil to enable it to support the loads safely (Fig. 2). The safe friction values to use for the project has been determined by uplift tests. The compressive loads are supported through bearing at the top, friction along their sides, adhesion to the soil, or a combination of these means. Pile foundations are relatively costly. They usually are selected only when soil conditions are such as to make spread or mat foundations impractical or uneconomical.

Figure 2 Pile foundation for overhead line tower.

Prestressed foundation

This new type of foundations (Fig. 3), already in use for overhead line towers, is based on a prestressing technique, applied on the soil foundation in order to harden it, thus ensuring for the foundation a quasi elastic response to dead and live loads with higher deformation modulus and much smaller settlements [Luong 1991].

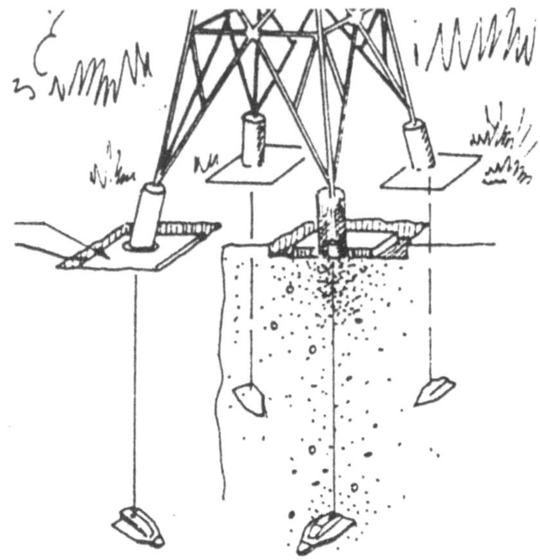

Figure 3 The pylon foundation is stabilized by the prestressing force.

3 Design of Prestressed Foundation

The overhead towers foundations are principally subjected to uplift, compressive loads and/or overturning. They are thus essentially stretched, compressed or bent. These requirements define a loading domain to be imposed on the foundation. This domain determines the magnitude of the recommended prestressing force used firstly to harden the foundation material and secondly to consolidate its stability domain.

To establish the correct design by means of numerical techniques is rather delicate because of the geometrical interaction of neighbouring supports loaded in tension and in compression. This is a three-dimensional configuration difficult to deal with. In addition the self-weight-induced stresses appropriate to the prototype earth structures exert in this case very strong influences [Mandel 1962]. Computational methods are useful only if the essential nature of soil and structure is properly reproduced in the numerical models. Experimental modelling and simulation are therefore vital to a fuller understanding of the problem.

An experimental study has been carried out in centrifuge, in order to determine the basic data required for a clear understanding of the dynamic, vibratory, cyclic and transient responses of pylons resting on their prestressed foundations. This may lead to an optimal and precise design, thus ensuring the safe and economical constructions and also the satisfactory performance of overhead line towers.

An anchoring system is used to load the squared metallic plate resting on the soil. Prestressing forces are applied on the plate by means of four springs. This loading generates a mechanical hardening on the soil foundation. This results in a supporting soil stiffer in appearance, whose mechanical behaviour is quasi elastic.

The response is recorded by acceleration, displacement and force transducers. A force transducer, placed between the electromagnetic exciter and the pylon model, gave the force loading signal. A displacement transducer and three 3D-accelerometers were installed on the anchored plate. The analysis of the results has been focussed on the appearance of non-linear phenomena which announce the beginning of loss of structural stability.

Tests in centrifuge were conducted under forced vibrations at fixed frequency for different force intensities.

The comparative analysis has shown (Fig. 4) that the gravity foundation worked correctly when subjected to the motions of low amplitude, that the pile foundation presented a more stable mechanical behaviour. But when subjected to greater excitory motions, the pile foundation would fail at a given loading threshold. Finally the prestressed foundations using four 160 kN (16 tons) prestressing forces have shown the predicted best mechanical behaviour and this for different intensities of loading excitation.

For sake of ready comparison between the three types of foundations, horizontal, transversal and vertical displacement measurements as recorded in flight have been calculated as for full scale prototype in order to obtain a better subsequent analysis with field tests on actual pylons in service. F_p denotes the prestressing force, F the vibratory amplitude of the applied load, b a characteristic size of the pylon and d the measured displacements. The vibratory motion of the pylon is characterized by bending **B**, torsion **T** and stamping **S**.

Performed tests have permitted the assessment of the role of the prestressing force F_p on the mechanical behaviour of the prestressed foundation. A significant influence of the prestressing force is observed on the amplitude of the excitation within the stability domain.

The experimental results, obtained from centrifuge simulation, show the following principal trends:

a. the domain of mechanical stability of the prestressed soil mass is enlarged with the magnitude of the prestressing force,

b. the rheological behaviour of the prestressed soil mass within the stability domain is quasi elastic,

c. with a small amplitude prestressing force used for the tests (160 kN), the mechanical behaviour of the prestressed foundation proves to be better than the pile group foundation and much more reliable than the buried concrete block foundation.

This centrifuge simulation offers the great advantage of suggesting a ready choice either of excitation force or prestressing force within the stability domain of the prestressed foundation where the mechanical behaviour is quasi elastic.

The study performed in centrifuge reinforces the conceptual predictions of this novel type of foundation. It permits to specify the design method of a prestressed

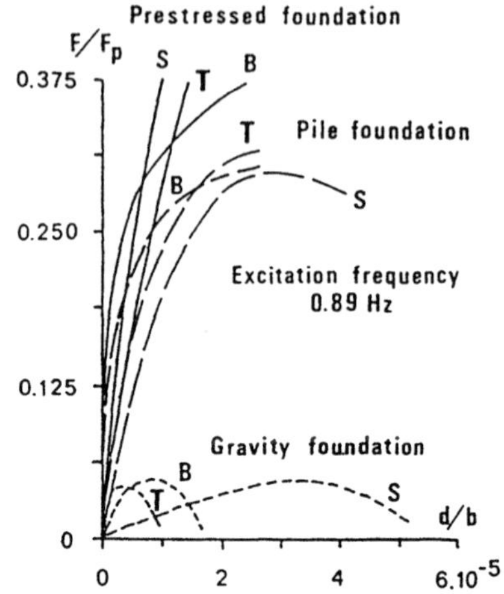

Figure 4 Comparison of the mechanical responses of three types of pylon foundations in centrifuge testing.

foundation that has been effectively installed on several overhead lines of the Electricité de France.

To work up the prestressed foundation for an electric pylon, a concrete slab is laid on the ground and a prestressing force is applied between the anchoring device and the slab.

The prestressing force hardens the soil, thus increasing the elastic soil modulus and decreasing the settlements induced within the foundation by the service loadings. This has been demonstrated on various types of subsoil as shown on the Fig. 5 where E_p and p_l are pressuremeter characteristics.

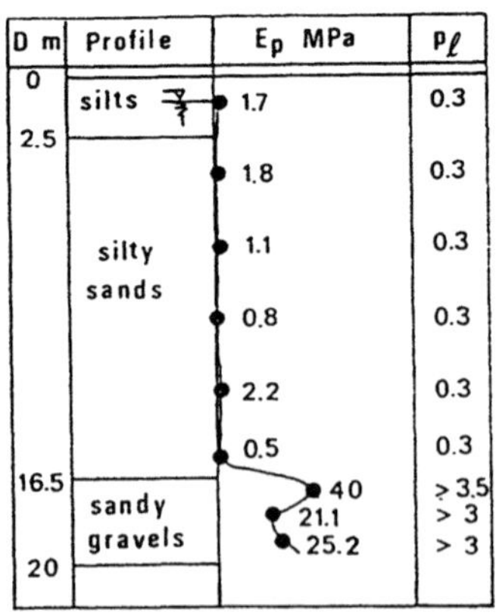

Figure 5 Typical subsoil of the Jonquières–Arles overhead electric line.

4 Non-Destructive Testing oF Soil-Pylon Interaction

An experimental frequency response technique using impulsive slacking from the top of the pylon has been applied for testing the behaviour of the three types of pylon resting on their respective foundations.

Several 3D-accelerometers affixed at different locations on the pylon record the vibratory motions.

A cable was attached on the top of the pylon. It is then fixed in the ground by a small anchoring device, distant from the pylon foundation. A tension system allows to increase the cable tension to a relatively low value for exemple up to 30 kN (Fig. 6).

Figure 6 Tensioning the cable for a slacking test.

An explosive tie allows a sudden release of tension, inducing a rapid excitation on the pylon. The pylon vibrates freely after tension release.

Data reduction of acceleration signals by Fast Fourier Transforms gives the frequency response spectra which are, of course, the vibratory signature of the pylon resting on its prestressed foundation (Fig. 7).

These plots of frequency versus vibrational amplitude in arbitrary units provide vibration signatures of the soil-structure interaction characterizing the mechanical performance of the pylon resting on its prestressed foundations. Of course, this non-destructive test can be applied also to others types of slender structures, because the vibration signature plots pinpoint vibration frequencies and indicate conditions such as poor workmanship, damage in structure or in subsoil. As a periodic maintenance activity, changes in subsequent plots permit early detection and identification of

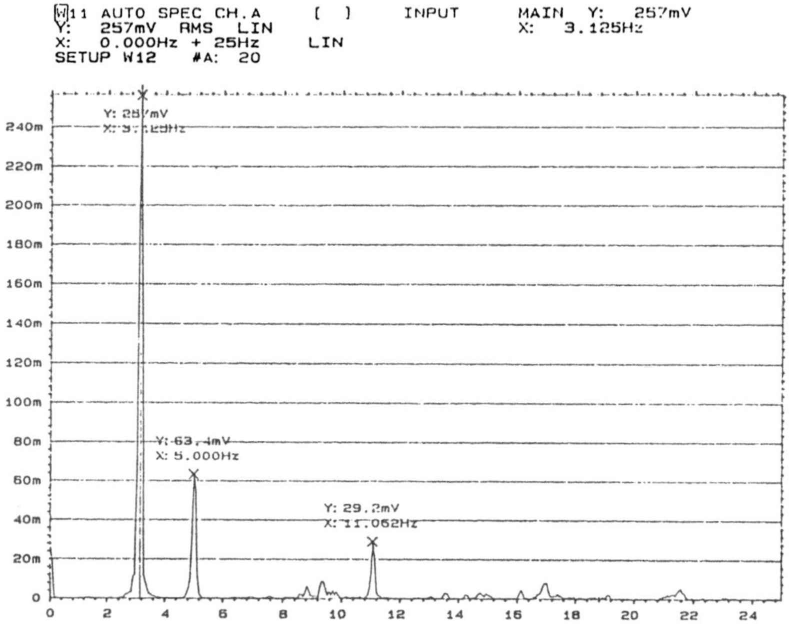

Figure 7 Frequency response of a pylon resting on its prestressed foundation.

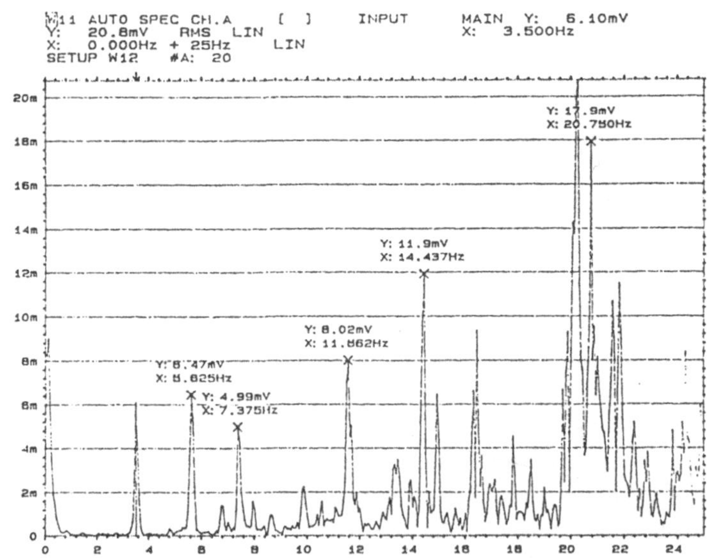

Figure 8 Frequency response of a pylon resting on its concrete stepped footings.

damage. For sake of comparison, the impulsive slacking test has been also conducted on pylons resting on concrete stepped footings (Fig. 8) and on steel pile foundations (Fig. 9).

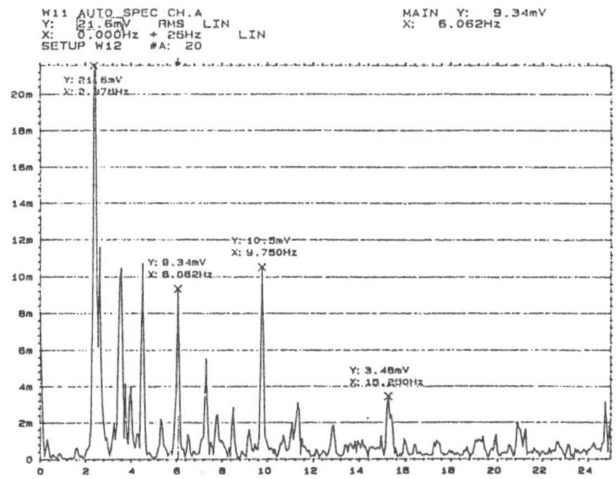

Figure 9 Frequency response of a pylon resting on its steel pile foundations.

5 Specific Data Processing Technique

In many situations, the impulse technique for structural frequency response testing is the simplest and fastest of the various techniques commonly used today. The transient excitation offers good estimates of the required frequency response information. However the nature of the excitation and response signals in the impulse technique requires special signal processing technique.

The frequency response function is defined in terms of the single input/single output system as the ratio of the Fourier transforms of the system output or response y(t) to the system input or excitation x(t).

$$H(f) = Y(f)/X(f)$$

where $Y(f)$ and $X(f)$ respectively denote Fourier transform of system output $y(t)$ and system input $x(t)$.

The only requirement for a complete description of the frequency response function are that the input and output signals be Fourier transformable, a condition that is met by all physically realizable systems, and that the input signal be non-zero at all frequencies of interest.

If the system is nonlinear or time-variant, the frequency response function will not be unique, but it will be a function of the amplitude of the input signal in the case of a nonlinear system and a function of time in the case of a system with time-varying properties.

Linear analysers based on one-dimensional Fourier transforms (FTs) and modal analysis programmes are popular tools used by experimentalists in linear structural dynamics. In this experimental work, an input-output non-parametric approach [Liu & Vinh 1991] has been chosen to portray the nonlinear behaviour of pylon subject to the impulsive slacking test.

Pylons resting on their foundations may be nonlinear structures, a nonlinear functional called Volterra series is used. In this functional, the total response of the system $y(t)$ is decomposed into components of various orders.

$$y(t) = y_1(t) + y_2(t) + \cdots + y_n(t) \tag{1}$$

Each component is defined by the functional

$$y_n(t) = \int_{-\infty}^{\infty} \cdots \int_{-\infty}^{\infty} h_n(\tau_1, \tau_2, \ldots, \tau_n) \prod_{k=1}^{n} x(t - \tau_k) d\tau_k \tag{2}$$

The first order component is decribed by linear convolution

$$y_1(t) = \int_{-\infty}^{\infty} h_1(\tau)\, x(t - \tau) d\tau \tag{3}$$

where $x(\tau)$ denotes the input function and $h_1(\tau)$ the first order impulse response that describes the linear behaviour of the system.

The other components require more than one time variable and a multidimensional signal processing has to be considered. As an example, let us describe the second order component (nonlinear) of the response. From Eq.(2) the following expression can be written :

$$y_2(t_1, t_2) = \int_{-\infty}^{\infty} \int_{-\infty}^{\infty} h_2(\tau_1, \tau_2) x(t_1 - \tau_1) x(t_2 - \tau_2) d\tau_1 d\tau_2 \tag{4}$$

where τ_1 and τ_2 are the two time variables, $h_2(\tau_1, \tau_2)$ denotes the second order impulse response. Within this mathematical framework, two dimensional Fourier transform is appropriate for the study of the behaviour of the structure in the frequency domain.

$$\mathcal{F}_2[h_2(\tau_1, \tau_2)] = \int_{-\infty}^{\infty} \int_{-\infty}^{\infty} h_2(\tau_1, \tau_2) \exp(-j\omega_1 \tau_1 - j\omega_2 \tau_2) d\tau_1 \tau_2 \tag{5}$$

This expression defines the second order transfer function $\mathcal{H}_2$ with two circular frequency variables ω_1, ω_2. $\mathcal{F}$ is the Fourier transform operator.

$$\mathcal{H}_2(\omega_1, \omega_2) = \mathcal{F}_2[h_2(\tau_1, \tau_2)]$$

In this application, a single impulse has been taken as input

$$x(t) = a\delta(t - T) \tag{6}$$

Then from Eq.(1) and Eq.(2)

$$y(t) = \sum_{i=1}^{n} y_i(t) = \sum_{i=1}^{n} \int_{-\infty}^{\infty} \cdots \int_{-\infty}^{\infty} h_i(\tau_1, \tau_2, \ldots, \tau_i) \prod_{k=1}^{i} a\delta(t - T - \tau_k) d\tau_k$$

$$= \sum_{i=1}^{n} a^i h_i(t - T) \tag{7}$$

From the single impulse of m different test magnitudes, Eq.(7) gives :

$$y^{\mathrm{m}}(t) = \sum_{i=1}^{n} a_m^i h_i(t - T) \tag{8}$$

where m indicates m*th* test. The impulse response of first order $h_1(t)$ is the linear impulse response of the system. The occurrence of the nonlinear behaviour is described by the appearance of the impulse responses of higher orders $h_2(t)$, $h_3(t)$, ..., etc.

Under matrix form, Eq.(8) is presented as follows :

$$\{y\} = [a]\{h\} \tag{9}$$

where

$$y = (y^1(t), y^2(t), \ldots, y^{\mathrm{m}}(t))^{\mathsf{T}}$$

$$h = (h_1(t - T), h_2(t - T), \ldots, h_n(t - T))^{\mathsf{T}} \tag{10}$$

$$[a] = \begin{bmatrix} a_1^1 & a_1^2 & \cdots & a_1^n \\ a_2^1 & a_2^2 & \cdots & a_2^n \\ \cdots & \cdots & \cdots & \cdots \\ a_m^1 & a_m^2 & \cdots & a_m^n \end{bmatrix} \tag{11}$$

or

$$\{h\} = [a]^{-1}\{y\} \tag{12}$$

The resolution of this equation permits us to obtain the impulse response of various orders with only one time variable (Fig. 14 and Fig. 15).

For further research, it may be interesting to take a input signal with two impulses, three impulses, ...etc.

6 Validation of the Data Processing Technique

This specific data processing technique has been validated by three slacking tests on a pylon resting on pile foundation.

- A first slacking test with a tension $T = 5\ kN$ has given the following natural frequencies :
  - Bending $N_b = 3.0\ Hz$
  - Torsion $N_t = 5.8\ Hz$
  - Stamping $N_s = 8.6\ Hz$
- A second slacking test with the same tension $T = 5\ kN$ with unscrewed node in the pylon gave the same natural frequencies.
- A third slacking test with the same tension $T = 5\ kN$ with rescrewed node showed the same natural frequencies.

When calculating the quantity NL = (nonlinear energy) / (linear energy),

$$NL = \int_{-\infty}^{\infty} | h_2(\tau) |^2\ d\tau / \int_{-\infty}^{\infty} | h_1(\tau) |^2\ d\tau$$

it can be found respectively :

- $NL_{initial} = 0.97$
- $NL_{unscrewed} = 2.05$
- $NL_{rescrewed} = 0.96$

Thus it can be seen for this case that the ratio NL between the nonlinear energy of order two and the linear energy, is a more sensitive parameter than the natural frequencies.

7 Concluding Remarks

The purpose of these tests was to characterize the dynamic geotechnical properties of transmission line towers in service, especially in case of prestressed foundations. Several non-destructive tests have been carried out on overhead pylons in use with an analysis of their free vibrations. The use of impact excitation together with a Fast Fourier Transform based spectrum analyzer to determine the dynamic characteristics of structures is potentially a very attractive technique. An impact gives excitation across a broad frequency range. It is therefore possible to investigate the whole frequency range of interest in a single test. Experimental determination of

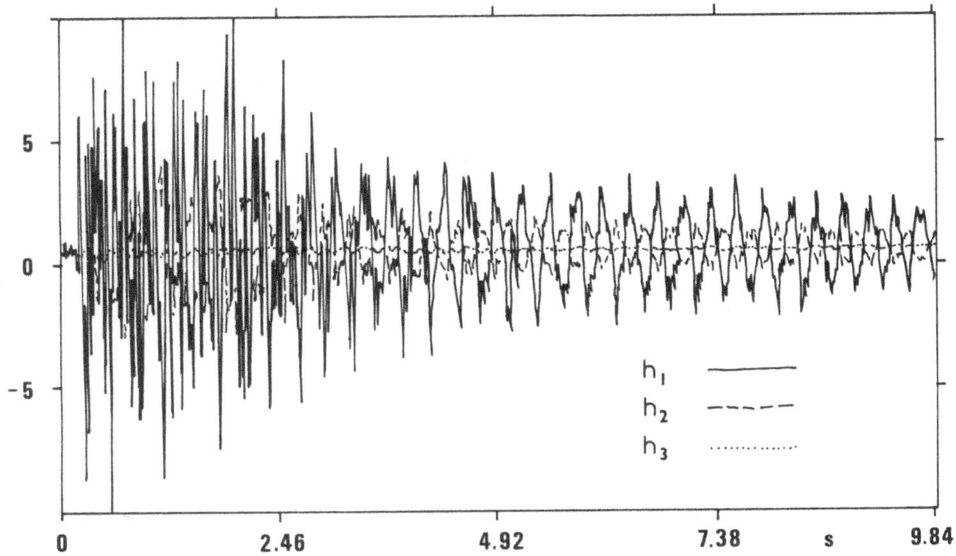

Figure 10 Time response of a pylon resting on its gravity foundation.

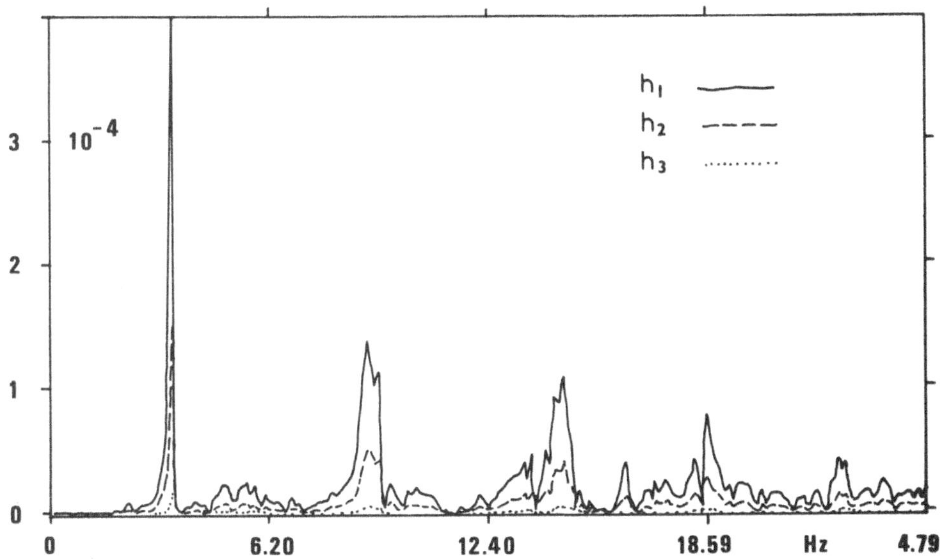

Figure 11 Frequency response of the same pylon subject to an impulse test.

the vibratory signature of such a slender structure resting on its prestressed foundations, subsequent to a sudden release of tension at its summit, reveals to be a very promising technique for integrity control and inspection. This test is very simple and rapidly executed since the need for connecting and aligning a shaker is eliminated. It can be used as routine test. It permits to follow the evolution of the pylon behaviour as regard to fatigue or damage. The control of foundation mechanical behaviour is very useful in cases of earthquake resistant construction and design because it is possible to modify - by changing the prestressing force - the frequency spectrum of the soil-structure response such that there is no interference with the earthquake frequency characteristics.

Acknowledgement - The support of Electricité de France CRTT Sud-Est and the collaboration of Dr H. Liu are gratefully acknowledged.

References

[1] G. Berthomieux, 1976, Modèles tridimensionnels de fondations de pylônes sollicités horizontalement. *Thèse de 3e cycle*, 21-09-1976, Université de Grenoble.

[2] M. Bustamante, A. Frossard, D. Gouvenot and D. Sage, 1986, Nouvelles fondations de supports aériens. *Journées d'Etudes sur les Fondations des Supports Aériens*, ESE, 27-11-1986, Gif-sur-Yvette.

[3] M. Gagneux, 1986, Evolution dans le calcul et choix des fondations des supports aériens HT, THT à l'Electricité de France. *Journées d'Etudes sur les Fondations des Supports Aériens*, ESE, 27-11-1986, Gif-sur-Yvette.

[4] C. Plumelle and J.L. Lapeyre, 1986, Les fondations des supports aériens. *Journées d'Etudes sur les Fondations des Supports Aériens*, ESE, 27-11-1986, Gif-sur-Yvette.

[5] L.Jr. Kemper, R.C. Stroud and S. Smith, 1981, Transmission line dynamic/static structural testing. *J. Struct. Div. ASCE*, 107(ST10), October 1981.

[6] M.P. Luong, 1991, Modeling prestressed foundations for overhead line towers. *Centrifuge 91*, Ko ed., Balkema, Rotterdam.

[7] J. Mandel, 1962, Essais sur modèles réduits en mécanique des terrains - Etude des conditions de similitude. *Revue de l'Industrie Minérale*, No 9, Vol. 44.

[8] M.P. Luong, 1986, Simulation des forces de masse. *Annales de l'ITBTP*, Essais et Mesures No 442, Février 1986.

[9] M.P. Luong, 1988, Signature géotechnique par test du lâcher. *La Revue des Laboratoires d'Essais*, no17, Très Basses Fréquences (TBF) dans l'Environnement - Application aux Transports, Paris, Décembre 1988.

[10] H.G.D. Goyder, 1984, Foolproof for frequency response measurements. *Proc. of 2nd Intern. Conf. on Recent Advances in Structural Dynamics*, Southampton, April 1984.

[11] W.G. Halvorsen and D.L. Brown, 1977, Impulse Technique for Structural Response Testing. *Sound and Vibration*, November 1977, 8-21.

[12] K.A. Ramsey, 1975, Effective measurements for structural dynamics testing : Part I". *Sound and Vibration*, Vol.9, No 11.

[13] H. Liu, 1989, Traitement du signal multidimensionnel et série de Volterra. Application à la dynamique non-linéaire des structures. *Thèse de Doctorat CNAM*, Paris, 28 Avril 1989.

[14] H. Liu and T. Vinh, 1991, Multidimensional signal processing for nonlinear structural dynamics. *Mechanical Systems and Signal Processing*, 5(1), 1991, 61-80.

Damage Assessment through Nonlinear Structural Identification

Ling Ge, T.J. Meyer, D.J. Mook and T.T. Soong
State University of New York at Buffalo
Department of Civil Engineering
Buffalo, NY, USA

Abstract

Motivated by the problem of damage assessment of existing structures, a procedure is developed in this paper for the nonlinear systems identification problem which is based on the minimum model error approach combined with correlation tests and least squares. This procedure enables the form of the damage model to be accurately determined, thus eliminating the requirement that the form be assumed *a priori*. Once the form is determined, the parameters involved in the nonlinear term can be estimated using least squares or other parameter estimation procedures.

This procedure does require cataloging all possible damage models of the structure. However, it does not require accurate models in order to produce accurate state and model error estimates. Additional damage models and combinations of existing models can be added to the correlation portion of the algorithm, virtually eliminating the likelihood that the true damage model terms are not tested.

1 Introduction

After a severe loading episode, such as a strong earthquake, a structure may sustain damage. For the purpose of damage assessment and possible retrofit of the structure, it is desirable to identify severity as well as location of the damage based upon input-output measurements. From the viewpoint of systems identification, it is thus one of nonlinear structure identification in which the form of nonlinearity as well as its associated parameters are to be identified.

Numerous methods for identification of nonlinear systems have been developed in the past two decades. Many of which are reviewed in [2],[12],[15]. Due to complexity of nonlinear systems, systems identification is generally carried out as a two-step procedure; (a) Representing system nonlinearity in a parametric functional form, such as Fourier series, polynomials, or kernel functions and (b) Estimating the parameters contained in the functional form using parameter estimation techniques such as least squares, correlation methods or some other approaches. Some of these procedures are described below.

The direct equation approach is used in [16] where input and output measurements of a dynamic process are expressed by a Fourier Series. System nonlinearity is represented as a sum of polynomials with unknown coefficients. By applying the principle of harmonic balance for different input signals, a set of algebraic equations in these unknown coefficients is set up. Using the method of least squares, these unknown coefficients are estimated. Effects of Fourier series truncation and selection of input signals are studied.

The prediction error model represents the linear submodel by an autoregressive moving average model and the nonlinear part by a Borel function [6],[7]. A non-parametric kernel regression estimation is employed to obtain the final analytical model.

Non-parametric methods generally employ orthogonal polynomials (such as Chebyshev, Legendre or Jacobi) or nonorthogonal polynomials to represent system nonlinearity. Benedettini et al. [1], for example, propose a polynomial model for hysteresis-type nonlinearities in the force, force derivative and velocity space. Combined with a parametric functional form, the identification process is carried out.

The Volterra or Wiener kernel method [2],[3] represents the output as a sum of convolution of inputs and kernel functions known as the Volterra series. The kernel functions which represent model parameters can be estimated by correlation analysis, orthogonal expansion, or frequency domain methods. Although a large class of nonlinear systems can be characterized by the Volterra series, a large amount of computation is required and difficulties associated with identification of the kernel function hinder the attempt to identify practical systems based on this representation.

Yun and Shinozuka [17], among others, use the extended Kalman Filter to linearize the nonlinear model and consider an extended state vector which includes, in additional to the response and its derivative, all the parameters to be identified. Starting from an initial guess, this extended state space is recursively updated as new observations are made available.

All of the techniques outlined above have proved useful in certain applications. However, some of these are sensitive to noise in the measurement data, some require accurate initial conditions, and they often need to assume the nonlinearity form *a priori* or by using a sufficient number of terms in a series to obtain a given accuracy while the physical insight is often lost in the process.

In this paper, an identification algorithm based on a combination of minimum model error (MME) state estimation, correlation technique and least square parameter estimation is employed to identify possible nonlinearities in structural systems, [11],[14]. Given the input measurements, noisy output measurements and an initially assumed model, the state time histories as well as the model error in the initially assumed model are estimated. On the basis of an extensive library which consists of all the possible linear and nonlinear functional forms for a given structural system, correlations of the estimated model error terms with the different functions of the estimated state vector (using the functional forms provided in the library) are compared and the true functional forms are chosen. Finally, least square fit is used to determine the parameters. As shall be shown, a distinct advantage of this approach is that it does not require parameterization of the nonlinear form. Moreover, it does not require accurate models in order to produce accurate state and model error estimates.

2 Identification Procedure

The state-space representation for a multi-degree-of-freedom dynamical system is

$$\dot{\mathbf{x}}(t) = \mathbf{A}\mathbf{x}(t) + \mathbf{B}\mathbf{u}(t) + \mathbf{G}\mathbf{d}(t), \qquad t \in [0, T] \tag{1}$$

where $\mathbf{x}(t)$ is the $n \times 1$ state vector, $\mathbf{u}(t)$ is the $m \times 1$ input vector, and $\mathbf{d}(t)$ is the $q \times 1$ model error vector, called model error term, modeling the unknown nonlinear behavior of the system. The matrices $\mathbf{A}, \mathbf{B}$ and $\mathbf{G}$ are, respectively, the $n \times n$ state matrix for the linear submodel, the $n \times m$ input location matrix, and the $n \times q$ model error coefficient matrix.

The measurement system is assumed to be a discrete-time measurement sequence given by

$$\mathbf{y}(t_k) = \mathbf{g}(\mathbf{x}(t_k), t_k) + \boldsymbol{\xi}_k, \qquad k = 1, 2, \ldots \tag{2}$$

where $\mathbf{y}(t_k)$ is the $p \times 1$ measurement vector at time t_k, which is a function of the state and possible t_k explicitly as represented by the $p \times 1$ vector function $\mathbf{g}$, and $\boldsymbol{\xi}_k$ is the $p \times 1$ measurement noise vector, assumed to be Gaussian distributed with covariance matrix $\mathbf{R}_k$ at time t_k.

2.1 Estimates of State and Model Error

Given the discrete-time measurement sequence $\mathbf{y}(t_k)$, $k = 1, 2, \ldots, t_0 \leq t_k \leq t_f$, the first step in the proposed identification procedure is to determine estimates of $\mathbf{x}(t)$ and $\mathbf{d}(t)$, denoted by $\hat{\mathbf{x}}(t)$ and $\hat{\mathbf{d}}(t)$ respectively, based on the optimal criterion described below.

To estimate the state vector and the model error term, a global cost functional J, consisting of a weighted integral squared model error term plus a weighted sum of the squares of the measurement residuals, is proposed. It has the form

$$\begin{aligned} J = & \sum_{k=1}^{M} [\mathbf{y}(t_k) - \mathbf{g}(\hat{\mathbf{x}}(t_k), t_k)]^T \mathbf{R}_k^{-1} [\mathbf{y}(t_k) - \mathbf{g}(\hat{\mathbf{x}}(t_k), t_k)] \\ & + \int_{t_0}^{t_f} \hat{\mathbf{d}}^T(\tau) \mathbf{W} \hat{\mathbf{d}}(\tau) d\tau \end{aligned} \tag{3}$$

where M is the number of data points and $\mathbf{W}$ is a positive definite weighting matrix. The optimal estimates $\hat{\mathbf{x}}(t)$ and $\hat{\mathbf{d}}(t)$ are obtained by minimizing J.

The justification of choosing this cost functional is as follows: The first term in J penalizes the deviation of the predicted measurements $\mathbf{g}(\hat{\mathbf{x}}(t_k), t_k)$ from the actual measurements $\mathbf{y}(t_k)$; minimization of this term drives the state estimates to the values which, when substituted into the measurement model, predict the actual measurement. The weighting matrix $\mathbf{R}_k^{-1}$ is the inverse of the associated measurement error covariance. Thus accurate measurements (small $\mathbf{R}_k$) are weighted more heavily than inaccurate measurements (large $\mathbf{R}_k$). However, a set of estimates which makes the first term in J a minimum is not the sole objective. Notice that, when we allow larger $\hat{\mathbf{d}}(t)$, we can obtain smaller value for the first term in J. Hence, by adding the second term and minimize J we eliminate the tendency to minimize only the first term. The positive definite matrix $\mathbf{W}$ determines the balance of these two competing terms. One necessary condition for choosing the best weighting matrix $\mathbf{W}$ is that the statistical properties of the estimated measurement noise $\hat{\boldsymbol{\xi}}_k = \mathbf{y}(t_k) - \mathbf{g}(\hat{\mathbf{x}}(t_k), t_k)$ should have the prescribed statistical properties of $\boldsymbol{\xi}_k$.

The minimization of J with respect to $\hat{\mathbf{d}}(t)$, subject to the system equation (Eq. 1), leads to the following two-point boundary value problem (TPBVP) [5]:

$$\left.\begin{aligned}
\dot{\hat{\mathbf{x}}}(t) &= \mathbf{A}\hat{\mathbf{x}}(t) + \mathbf{B}\hat{\mathbf{u}}(t) + \mathbf{G}\hat{\mathbf{d}}(t) \\
\dot{\boldsymbol{\lambda}}(t) &= -\mathbf{A}^T\boldsymbol{\lambda}(t) \\
\hat{\mathbf{d}}(t) &= -\mathbf{W}\mathbf{G}^T\boldsymbol{\lambda}(t) \\
\dot{\boldsymbol{\lambda}}(t_k^+) &= \boldsymbol{\lambda}(t_k^-) + \mathbf{H}_k\mathbf{R}_k^{-1}[\mathbf{y}(t_k) - \mathbf{g}(\hat{\mathbf{x}}(t_k), t_k)]
\end{aligned}\right\} \tag{4}$$

with boundary conditions:

$\hat{\mathbf{x}}(t_0)$ is specified or $\boldsymbol{\lambda}(t_0) = 0$ or combination of the two

$\hat{\mathbf{x}}(t_f)$ is specified or $\boldsymbol{\lambda}(t_f) = 0$ or combination of the two

where $\boldsymbol{\lambda}(t)$ is the $n \times 1$ Lagrange multiplier vector, and

$$\mathbf{H}_k = \left.\frac{\partial \mathbf{g}}{\partial \hat{\mathbf{x}}}\right|_{\hat{\mathbf{x}}(t_k), t_k} \tag{5}$$

The solution for $\hat{\mathbf{x}}(t)$ and $\hat{\mathbf{d}}(t)$ from Eqs. 4 depends on the weighting matrix $\mathbf{W}$. The solution is repeated until a value of $\mathbf{W}$ is found so that the optimal state estimates $\hat{\mathbf{x}}(t_k)$ satisfy the covariance constraint [11]

$$[\mathbf{y}(t_k) - \mathbf{g}(\hat{\mathbf{x}}(t_k), t_k)]^T[\mathbf{y}(t_k) - \mathbf{g}(\hat{\mathbf{x}}(t_k), t_k)] \approx \mathbf{R}_k \tag{6}$$

for $k = 1, \ldots, M$. The model error is thus the minimum adjustment to the model required for the estimated states to predict the measurements with approximately the same covariance as the measurement error.

2.2 Correlation Test

As the second step in the identification procedure, correlation test is employed here to investigate the functional relationship between the model error term and the state trajectory obtained by MME estimation. The estimates of both the model error term and the state vector are variables defined in the time domain. They are not necessarily random but the correlation concept is still valid. The cross-correlation between two discrete-time variables α and β is defined as

$$C(\alpha, \beta) = \frac{\sum_{k=1}^{M}(\alpha(t_k) - \bar{\alpha})(\beta(t_k) - \bar{\beta})}{\sqrt{\sum_{k=1}^{M}(\alpha(t_k) - \bar{\alpha})^2}\sqrt{\sum_{k=1}^{M}(\beta(t_k) - \bar{\beta})^2}} \tag{7}$$

with

$$\bar{\alpha} = \frac{\sum_{k=1}^{M} \alpha(t_k)}{M}, \qquad \bar{\beta} = \frac{\sum_{k=1}^{M} \beta(t_k)}{M} \tag{8}$$

From Schwarz inequality, one has

$$-1 \leq C(\alpha, \beta) \leq 1 \tag{9}$$

Assuming that the functional form between $d_i(t)$ and $\mathbf{x}(t)$ is

$$d_i(t) = a_1 f_1(\mathbf{x}(t)) + a_2 f_2(\mathbf{x}(t)) + \ldots \tag{10}$$

it is obvious that $|C(d_i(t), a_1 f_1 + a_2 f_2 + \ldots)| = 1$. Generally, all the possible function forms $f_1, f_2, \ldots$, can be categorized in a function library. While it is impossible to know exactly the values of $a_1, a_2, \ldots$, it is observed that, when a function f_j does constitute a characteristic part of $d_i(t)$, $C(d_i(t), f_j)$ is still much higher than those associated with non-relevant functional forms [14]. Thus, by testing correlation coefficients for all the functions in the function library, we can obtain the possible true functional forms. The accuracy of the correlation technique depends on the accuracy of the state and model error estimates, and on the completeness of the function library in the sense that the library includes the true functional forms. It can be shown that, since there is always noise in the estimates, the correlation method is robust within some data noise level [14].

2.3 Least Square Fit

As the last step of the identification process, least square fit is used to determine the parameters in the functions identified in the correlation test. Let

$$\hat{d}_i(t) = \mathbf{f}^T(\hat{\mathbf{x}}(t)\mathbf{a} \tag{11}$$

where $\mathbf{a} = [a_1, a_2, \ldots]^T$ is the parameter vector and

$$\mathbf{f}(\hat{\mathbf{x}}(t)) = [f_1(\hat{\mathbf{x}}(t)),\ f_2(\hat{\mathbf{x}}(t)), \ldots]^T.$$

Using the method of least squares, the best estimate of $\mathbf{a}$ can be found by minimizing $\sum_{k=1}^{M} [\hat{d}_i(t_k) - \mathbf{f}^T(\hat{\mathbf{x}}(t_k)\hat{\mathbf{a}}]^2$. The result is

$$\hat{\mathbf{a}} = \left[\sum_{k=1}^{M} \mathbf{f}(\hat{\mathbf{x}}(t_k))\mathbf{f}^T(\hat{\mathbf{x}}(t_k))\right]^{-1} \left[\sum_{k=1}^{M} \mathbf{f}(\hat{\mathbf{x}}(t_k))\hat{d}_i(t_k)\right] \tag{12}$$

Again, there is noise in the estimates $\hat{\mathbf{x}}(t)$ and $\hat{\mathbf{d}}(t)$. The robustness of the least square fit depends on the property of $f_1, f_2, f_3, \ldots$ and the relationships between these functions.

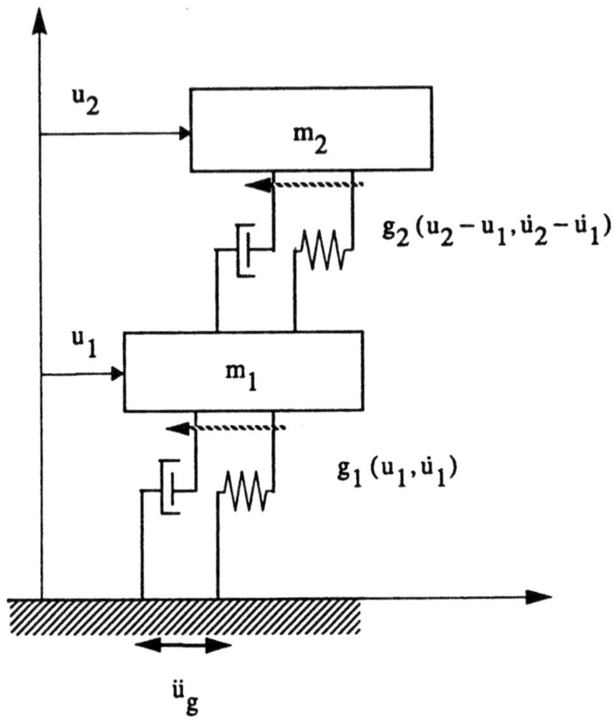

Figure 1 A Two-story structure

3 A Numerical Example

As a simple example, a two-story shear wall structure subjected to ground motion as shown in Fig. 1 is considered here. The equations of motion are

$$\left.\begin{aligned}
&m_1\ddot{u}_1 + g_1(u_1,\dot{u}_1) - g_2(u_2-u_1,\dot{u}_2-\dot{u}_1) = -m_1\ddot{u}_g\\
&m_2\ddot{u}_2 + g_2(u_2-u_1,\dot{u}_2-\dot{u}_1) = -m_2\ddot{u}_g\\
&g_1(u_1,\dot{u}_1) = c_1\dot{u}_1 + \alpha_1 k_1 u_1 + (1-\alpha_1)k_1 z_1\\
&g_2(u_2-u_1,\dot{u}_2-\dot{u}_1) = c_2(\dot{u}_2-\dot{u}_1) + \alpha_2 k_2(u_2-u_1) + (1-\alpha_2)k_2 z_2\\
&\dot{z}_1 = d_1\\
&\dot{z}_2 = d_2
\end{aligned}\right\} \tag{13}$$

where $u_1(t)$ and $u_2(t)$ are, respectively, the displacements of the first and second floors relative to the ground, $\ddot{u}_g(t)$ is the ground acceleration, m_1 and m_2 are the floor masses, g_1 and g_2 are the restoring forces which include the unknown nonlinear terms $d_1(t)$ and $d_2(t)$, k_1 and k_2 are the initial stiffnesses, and c_1 and

c_2 are the damping coefficients. For simplicity, it is assumed that $m_1 = m_2 = m$, $c_1 = c_2 = c$, $k_1 = k_2 = k$, $\alpha_1 = \alpha_2 = \alpha$. Denoting $\omega = \sqrt{k/m}$, $2\zeta = \sqrt{c/k}$, $x_1 = u_1$, $x_2 = \dot{u}_1$, $x_3 = u_2$, $x_4 = \dot{u}_2$, $x_5 = z_1$ and $x_6 = z_2$, the state-space equation for this system is

$$\begin{bmatrix} \dot{x}_1 \\ \dot{x}_2 \\ \dot{x}_3 \\ \dot{x}_4 \\ \dot{x}_5 \\ \dot{x}_6 \end{bmatrix} = \begin{bmatrix} 0 & 1 & 0 & 0 & 0 & 0 \\ -2\alpha\omega^2 & -4\zeta\omega & \alpha\omega^2 & 2\zeta\omega & -(1-\alpha)\omega^2 & -(1-\alpha)\omega^2 \\ 0 & 0 & 0 & 1 & 0 & 0 \\ \alpha\omega^2 & 2\zeta\omega & -\alpha\omega^2 & 2\zeta\omega & 0 & (1-\alpha)\omega^2 \\ 0 & 0 & 0 & 0 & 0 & 0 \\ 0 & 0 & 0 & 0 & 0 & 0 \end{bmatrix} \begin{bmatrix} x_1 \\ x_2 \\ x_3 \\ x_4 \\ x_5 \\ x_6 \end{bmatrix} + \begin{bmatrix} 0 \\ -\ddot{u}_g \\ 0 \\ -\ddot{u}_g \\ d_1 \\ d_2 \end{bmatrix} \tag{14}$$

The measurement system is

$$\begin{bmatrix} y_1(t_k) \\ y_2(t_k) \\ y_3(t_k) \\ y_4(t_k) \end{bmatrix} = \begin{bmatrix} 1 & 0 & 0 & 0 & 0 & 0 \\ 0 & 1 & 0 & 0 & 0 & 0 \\ 0 & 0 & 1 & 0 & 0 & 0 \\ 0 & 0 & 0 & 1 & 0 & 0 \end{bmatrix} \begin{bmatrix} x_1(t_k) \\ x_2(t_k) \\ x_3(t_k) \\ x_4(t_k) \\ x_5(t_k) \\ x_6(t_k) \end{bmatrix} + \begin{bmatrix} \xi_1 \\ \xi_2 \\ \xi_3 \\ \xi_4 \end{bmatrix}, \quad k = 1, 2 \ldots \tag{15}$$

where $\boldsymbol{\xi} = [\xi_1, \xi_2, \xi_3, \xi_4]^T$ represents noise in the measurements of relative displacements and relative velocities. The noise level is defined as

$$\sqrt{\mathrm{Var}(\xi_i)} \Big/ \sqrt{\frac{\sum_{k=1}^{M} y_i^2(t_k)}{M}}, \qquad i = 1, 2, 3, 4 \tag{16}$$

The function library in this case consists of four possible cases when only hysteresis-type nonlinearities are included. They are: Case 1: Both $d_1(t)$ and $d_2(t)$ consist of linear parts only, corresponding to no-damage condition, Case 2: $d_2(t)$ includes nonlinearity, corresponding to second floor being damaged, Case 3: $d_1(t)$ includes nonlinearity, corresponding to first floor being damaged, and Case 4: Both $d_1(t)$ and $d_2(t)$ include nonlinearity, corresponding to both floors being damaged.

The true model is taken to be Case 2 with $d_1(t)$ and $d_2(t)$ given by

$$\left.\begin{aligned} d_1(t) &= x_2 \\ d_2(t) &= x_2 - x_4 - 50(|x_2 - x_4|x_6 + (x_2 - x_4)|x_6|) \end{aligned}\right\} \tag{17}$$

The input ground acceleration is shown in Fig. 2, and the noise in the measurement system is $\sqrt{\mathrm{Var}(\xi_i)} = 0.01$ (noise level is approximately 10%). Figures 3-6 show the measurement data. The state and model error estimates, obtained using the MME approach, are compared with the true state and the true model error (based on the simulation results) in Figs. 7-14. These figures show that the estimates for $x_1(t)$, $x_2(t)$, $x_3(t)$ and $x_4(t)$ are very good (in the sense that the noise is reduced to about 1% of the noise in the measurement data), while the noise in the estimates for $x_5(t)$, $x_6(t)$, $d_1(t)$ and $d_2(t)$ is not negligible.

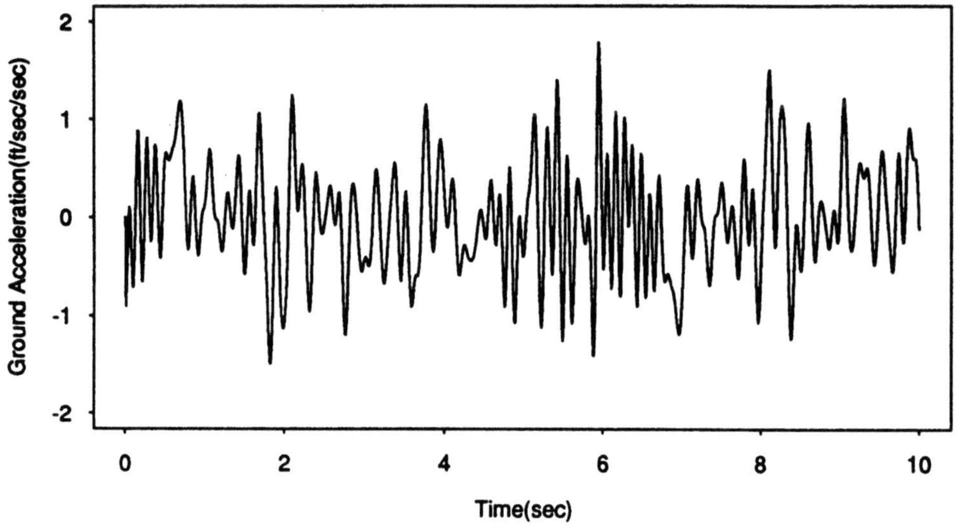

Figure 2 Ground Acceleration $\ddot{x}_g(t)$

The correlation test result is listed in Table 1. From the table we conclude that $\hat{d}_1(t)$ correlates well with the linear function, while $\hat{d}_2(t)$ correlates well with the hysteresis function. So the identified functional forms chosen by the correlation test are

Table 1 Correlation tests

	$\hat{d}_1(t)$	$\hat{d}_2(t)$
Case 1	0.91	0.24
Case 2	0.91	0.98
Case 3	0.09	0.24
Case 4	0.09	0.98

$$\left.\begin{aligned} d_1(t) &= x_2 \\ d_2(t) &= x_2 - x_4 + a(|x_2 - x_4|x_6 + (x_2 - x_4)|x_6|) \end{aligned}\right\} \tag{18}$$

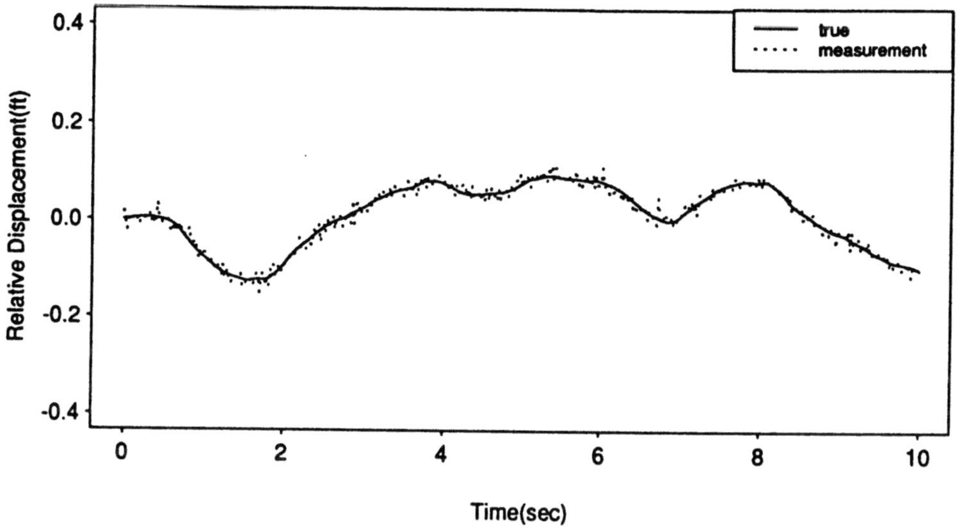

Figure 3 $u_1(t)$ Measurements

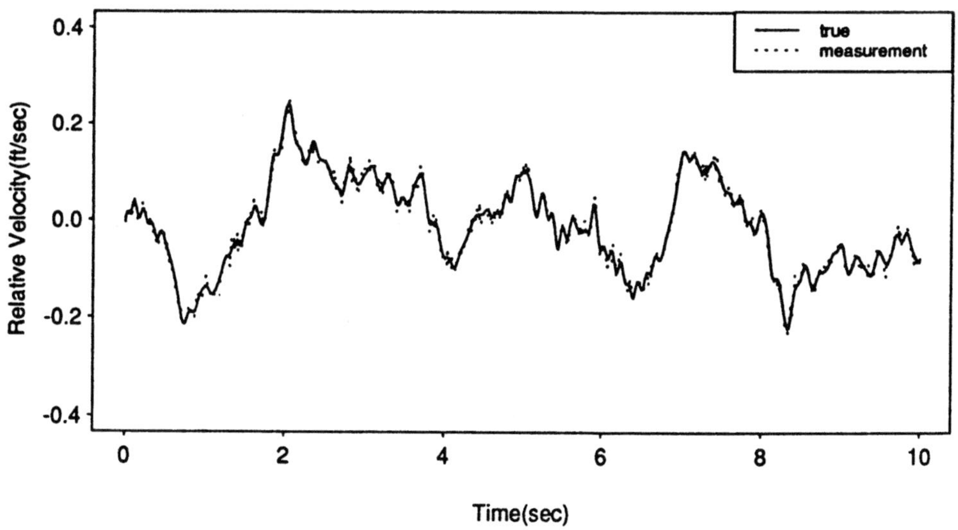

Figure 4 $\dot{u}_1(t)$ Measurements

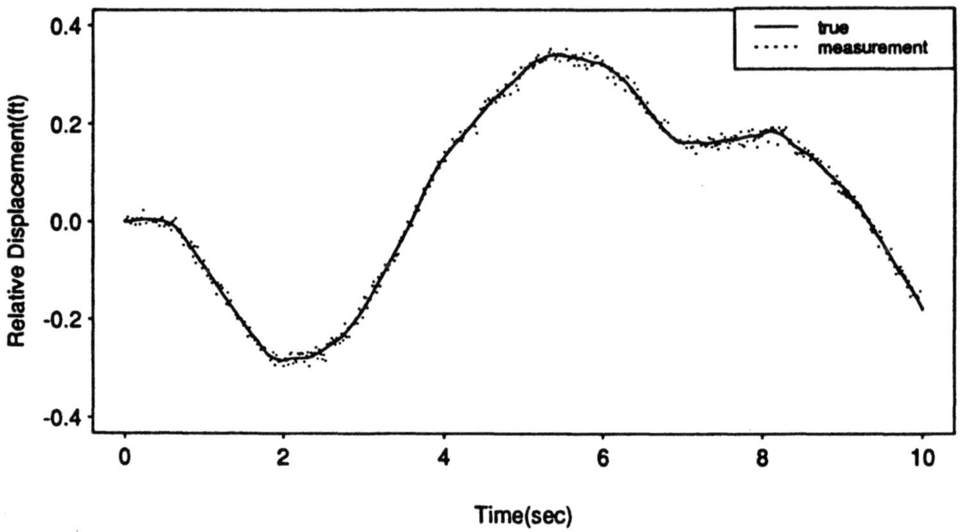

Figure 5 u_2(t) Measurements

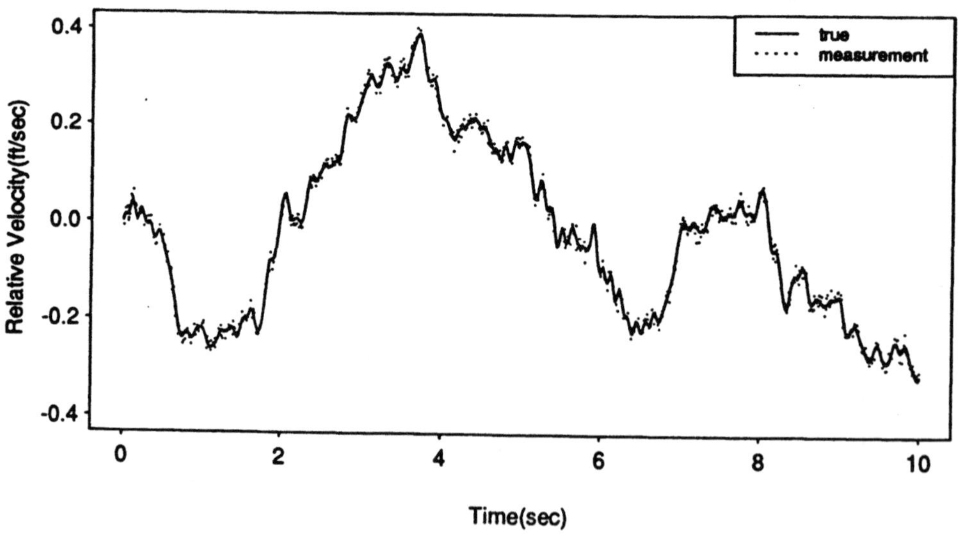

Figure 6 $\dot{u}_2(t)$ Measurements

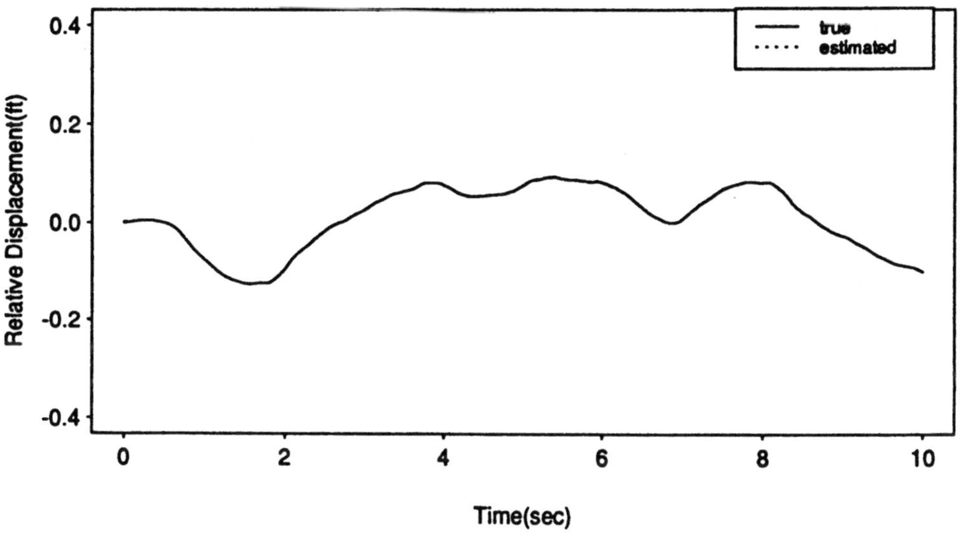

Figure 7 True value and estimate $\hat{u}_1(t)$ of $u_1(t)$

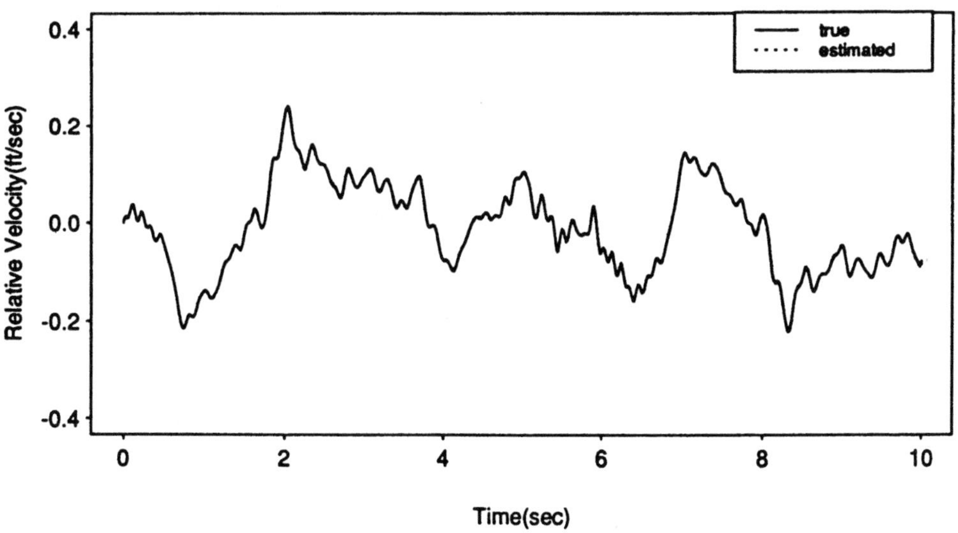

Figure 8 True value and estimate $\hat{\dot{u}}_1(t)$ of $\dot{u}_1(t)$

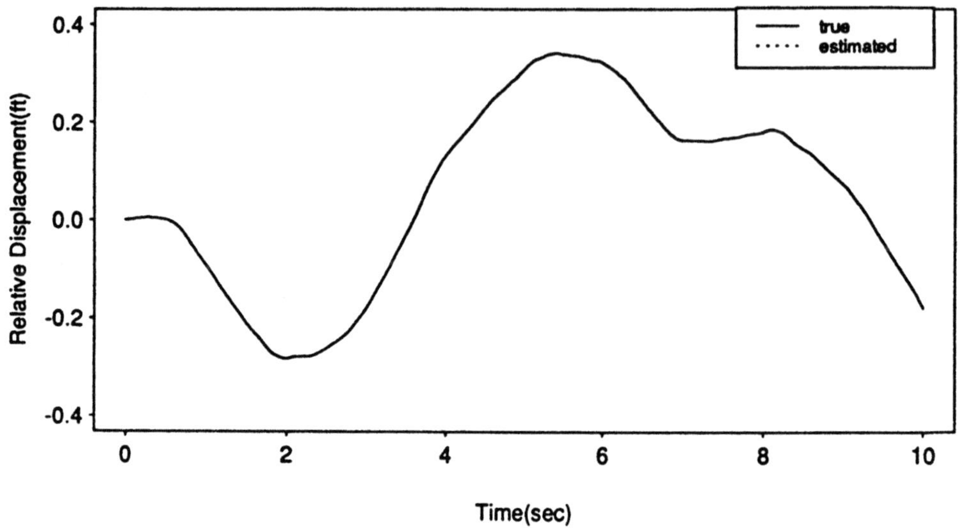

Figure 9 True value and estimate $\hat{u}_2(t)$ of $u_2(t)$

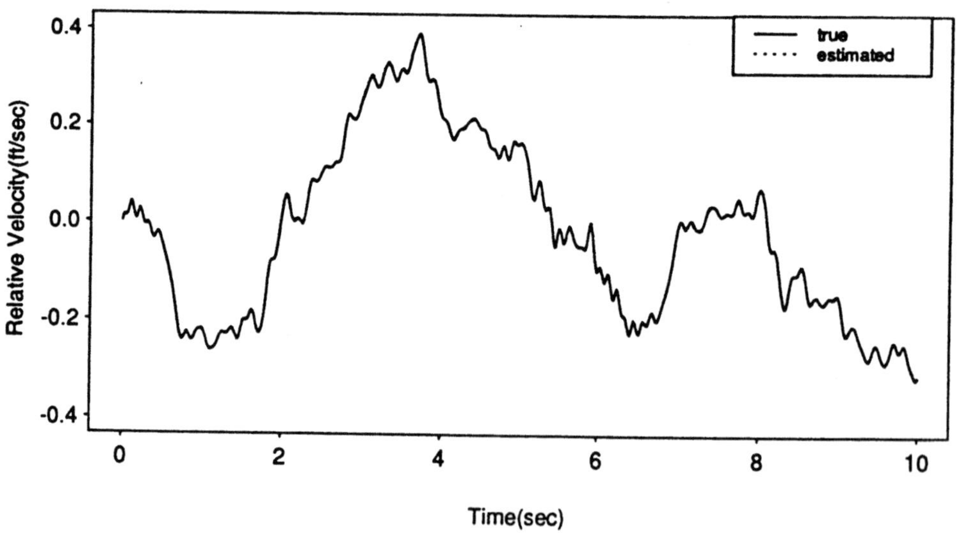

Figure 10 True value and estimate $\hat{\dot{u}}_2(t)$ of $\dot{u}_2(t)$

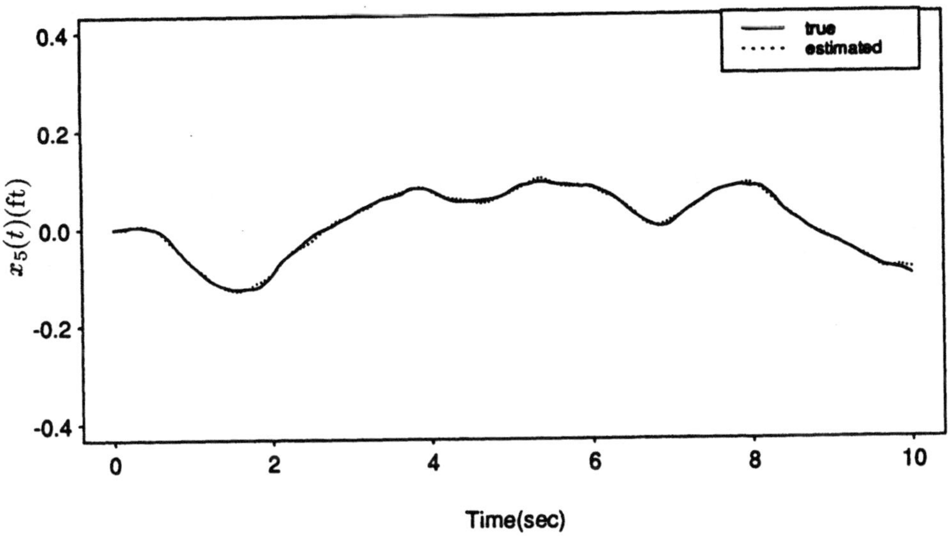

Figure 11 True value and estimate $\hat{x}_5(t)$ of $x_5(t)$

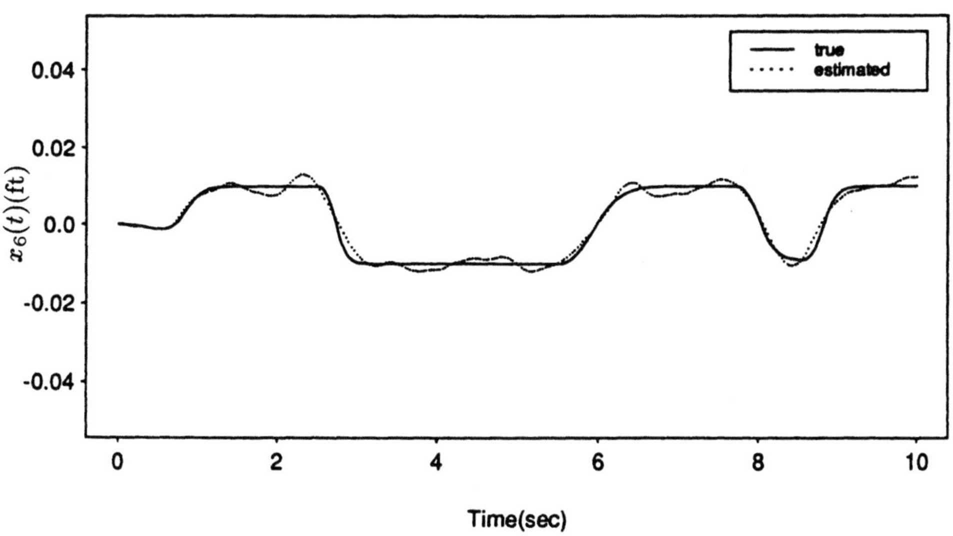

Figure 12 True value and estimate $\hat{x}_6(t)$ of $x_6(t)$

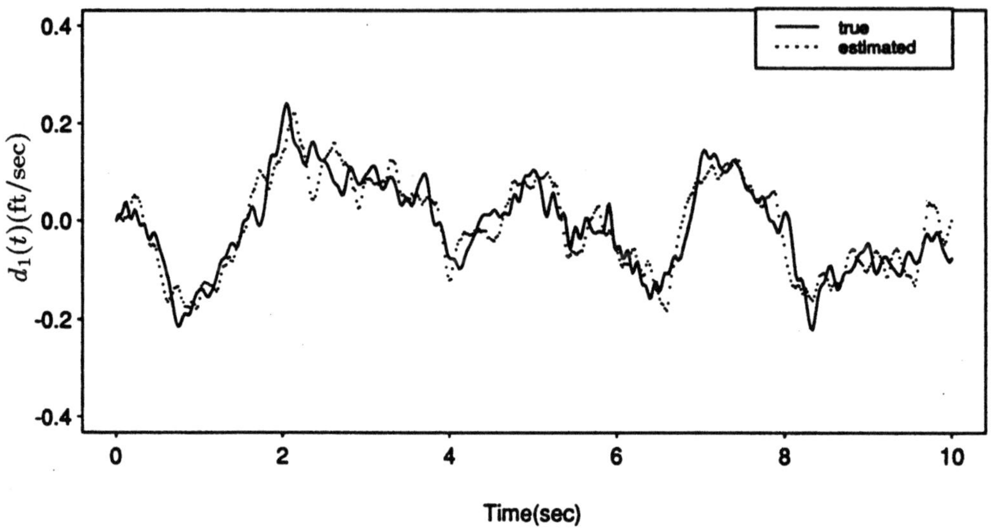

Figure 13 True value and estimate $\hat{d}_1(t)$ of $d_1(t)$

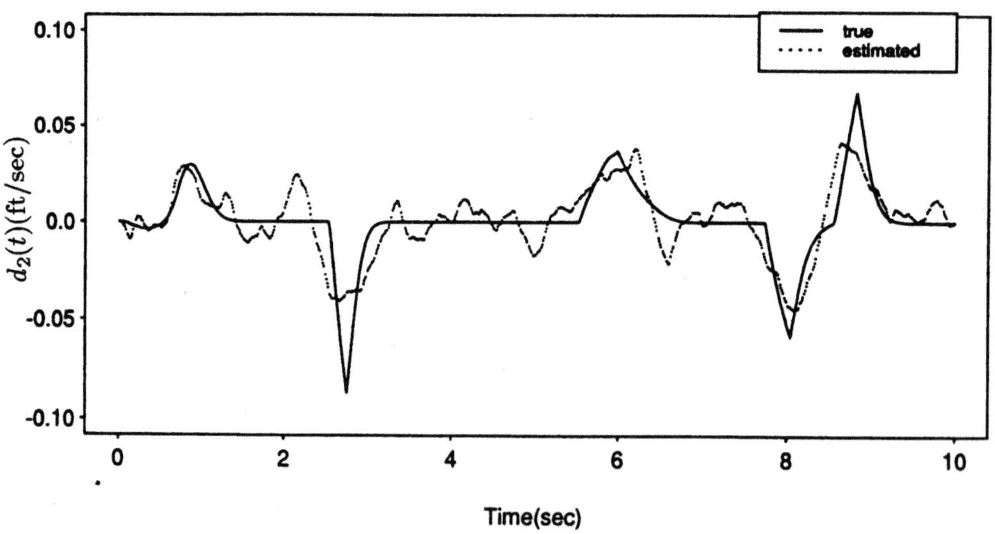

Figure 14 True value and estimate $\hat{d}_2(t)$ of $d_2(t)$

After the functional forms have been chosen via the correlation test, the least square curve fit gives $\hat{a} = -48.33$, which is close to the true value $a = -50.00$.

4 Concluding Remarks

An identification algorithm based on a combination of MME approach, correlation test and least square curve fit is proposed for the purpose of structural damage assessment. In comparison with existing identification techniques for nonlinear systems, a distinct advantage associated with this procedure is that the forms of the nonlinear terms do not need to be assumed *a priori*. It, however, does require cataloging all possible damage models for the structure under consideration, which may vary for different structural types and under different loading conditions. The construction of a comprehensive damage model library thus represents a major task in its application to structural damage assessment. However, numerical experimentation has shown that, even with a very large number of functions in the library, the proposed estimation procedure is efficient and robust in face of uncertainties.

The numerical example shows that estimates of $d_1(t)$ and $d_2(t)$ are more sensitive to the noise in the measurements; this is a consequence of the fact that $d_1(t)$ and $d_2(t)$ are not components of the state vector but rather their derivatives. The accuracy with which the nonlinear error terms can be estimated thus dictates the measurement noise level that can be tolerated in producing accurate identification results. However, as Table 1 shows, even with significant noise present in the estimates of the error terms, correlation tests provide unambiguous identification of the correct nonlinear functions from a class of possible functional forms in the library.

5 References

[1] Benedettini, F., Capecchi, D. and Vestroni, F., 1990, Non-parametric Models in Identification of Hysteretic Oscillators, DISAT Publ. No. 4, Universitá dell'Aquilla, Italy.

[2] Billings, S.A., 1980, Identification of Nonlinear systems - a Survey, IEE Proc., Vol. 127, Pt. D, No. 6, 272-284.

[3] Fakhouri, S.Y., 1980, Identification of the Volterra Kernels of Nonlinear Systems, IEE Proc., Vol. 127, Pt. D, No. 6, 296-304.

[4] Ghanem, R.G., Cavin, H. and Shinozuka, M., 1991, Experimental Verification of a Number of Structural System Identification Algorithms, Technical Report NCEER-91-0024, National Center for Earthquake Engineering Research, Buffalo, NY.

[5] Geering, H.P., 1976, Continuous Time Optimal Control Theory for Cost Functionals including Discrete State Penalty Terms, IEEE Trans. A.C., Vol. AC-21, 863-869.

[6] Goodwin, G.C. and Payne, R.L., 1977, Dynamic System Identification: Experiment Design and Data Analysis, Academic Press, New York, NY.

[7] Greblick, W. and Pawlack, M., 1987, Hammerstein System Identification by Non-parametric Regression Estimation, Int. J. Control, Vol. 45, No. 1, 343-354.

[8] Juang, J.N. and Pappa, R.S., 1985, An Eigensystem Realization Algorithm (ERA) for Modal Parameter Identification and Model Reduction, AIAA Journal of Guidance, Control and Dynamics, Vol. 8, No. 5, 620-627.

[9] Lin, C.C., Soong, T.T. and Natke, H.G., 1990, Real-time System Identification on Degrading Structures, ASCE Journal of Engineering Mechanics, Vol. 116, No. 10, 2258-2274.

[10] Meyer, T.J., 1992, Nonlinear Identification Package, Department of Mechanical and Aerospace Engineering, State University of New York at Buffalo, Buffalo, NY.

[11] Mook, D.J. and Junkins, J.L., 1987, Minimum Model Error Estimation for Poorly Modeled Dynamic Systems, J. guidance, Vol. 11, No. 3, 256-261.

[12] Natke, H.G., Juang, J.N. and Gawronski, W., 1988, A Brief Review of the Identification of Nonlinear Mechanical Systems, Proceedings 6th IMAC, Orlando, FL, 1569-1574.

[13] Osborne, M.R., 1969, On Shooting Methods for Boundary Value Problems, Journal of Mathematical Analysis and Applications, Vol. 27, 417-433.

[14] Stry, G.I. and Mook, D.J., 1991, Correlation Techniques in Robust Nonlinear System Realization/Identification, AAS/AIAA Spaceflight Mechanics Meeting, Houston, TX.

[15] Ting, E.C., Chen, H.S.J. and Yao, J.T.P., 1978, System Identification, Damage Assessment and Reliability Evaluation of Structures, Technical Report No. CE-STR-78-1, Purdue University, Lafayette, IN.

[16] Yasuda, K., Kawamura, S. and Watanabe, K., 1988, Identification of Nonlinear Multi-Degree-of-Freedom Systems, JSME International Journal, Vol. 31, No. 1, 502-509.

[17] Yun, C-B and Shinozuka, M. 1980, Identification of Nonlinear Structural Dynamic Systems, Journal of Structural Mechanics, Vol. 8, No. 2, 187-203.

The Identification of Non-linear Systems with Statistically Equivalent Polynomial Systems[1]

J.H. Zhang
Xi'an Jiaotong University
Department of Engineering Mechanics
Xi'an, P.R. China

H.G. Natke
University of Hannover
Curt-Risch-Institut
Hannover, Germany

Y. Qiu
Xi'an Jiaotong University
Department of Engineering Mechanics
Xi'an, P.R. China

X.N. Zhang
Xi'an Jiaotong University
Department of Engineering Mechanics
Xi'an, P.R. China

[1]Supported by the Chinese Nature Science Foundation

1 Introduction

Volterra (or Wiener) series models of non-linear systems have been studied by many authors, e.g. Rugh[1], Tomlinson[2]. By means of the Multi-dimensional Frequency Response Function (MFRF) based on the Volterra (or Wiener) kernel the non-linear dynamical systems can be analysed in the frequency domain. However, there are some crucial problems concerning the estimation and application of MFRFs[3]. The first is the expense of estimating MFRFs by direct methods, which impels us to study the parameter estimation methods. Billings[4] has developed the estimation method of MFRF by means of the NARMAX model. The polynomial approach for estimating MFRF is discussed in detail in the paper. From the point of view of the estimation of MFRF, the NARMAX model looks more convenient because the output relates directly to the input. However, from the point of view of the discussion on dynamics behaviour, there are more advantages with the parameter identification of the polynomial non-linear model. This is because it is used together with the equation of motion of the dynamic system. Natke and Zamirowski[5] discussed the method of structure identification for the class of polynomials within mechanical systems, which laid a foundation for the parameter estimation method of MFRF. The second crucial problem is that the difficulties arise from using MFRF to describe the non-linear properties of the system because the multi-frequency is without a physical meaning. In the second part of this paper an attempt is made to decrease these difficulties. While the standardized formula of the parameter estimation is given with the response and the excitation, emphasis is given to the study of the statistical equivalent 3rd order polynomial system, the spectral structure of the response expressed with MFRF is analysed in detail, and the Extended Transfer Functions (ETF) are defined, which are only functions of one-dimension frequency. In addition, the effectiveness of the statistical equivalence is analysed. The advantages of the method used here are that the MFRF of the polynomial non-linear system has the theoretical analytical expression, and by studying a non-linear system particular properties with extended transform functions avoid the difficulty of graphing the MFRF.

2 Polynomial System

The general mathematical model of a non-linear dynamic system with N degrees of freedom can be expressed with the second order differential equation, that is

$$M\ddot{y} + g(\dot{y}, y, t) = x(t) \tag{1}$$

where

M mass matrix
g non-linear function vector
y displacement vector
x exciting force vector

If g is not an explicit function of t and can be represented as

$$g(\dot{y}, y) = C(\dot{y}) + K(y) + D(\dot{y}, y) \tag{2}$$

where

$$\begin{aligned} C(\dot{y}) &= C_1\dot{y} + C_2(\dot{y}\otimes\dot{y}) + C_3(\dot{y}\otimes\dot{y}\otimes\dot{y}) + \cdots \\ &= C_1\dot{y} + C_2\dot{y}^{(2)} + C_3\dot{y}^{(3)} + \cdots \end{aligned} \tag{3}$$

$$\begin{aligned} K(y) &= K_1 y + K_2(y\otimes y) + K_3(y\otimes y\otimes y) + \cdots \\ &= K_1 y + K_2 y^{(2)} + K_3 y^{(3)} + \cdots \end{aligned} \tag{4}$$

$$D(\dot{y}, y) = \sum_{i=1}^{n}\sum_{j=1}^{n} D_{ij} y^{(i)} \otimes \dot{y}^{(j)} \tag{5}$$

the system is called a time-invariant Polynomial Non-linear System (PNS) for y and $\dot{y}$, where $\otimes$ is a Kronecker product and $y^{(2)}$ is defined as $y\otimes y$.
If

$$C_n \equiv 0, n \neq 1 \ and \ D_{ij} \equiv 0, i, j \forall I,$$

the system is defined as a Polynomial Stiffness System (SPNS).
If

$$K_n = \begin{cases} \neq 0, & n = 1 \\ \equiv 0, & n \neq 1 \end{cases} \qquad and \ D_{ij} \equiv 0, i, j\forall I,$$

the system is defined as a Polynomial Damping System (DPNS).
If

$$K_n = \begin{cases} \neq 0, & n = 1. \\ \equiv 0, & n \neq 1. \end{cases} \qquad and \ C_n \equiv 0, n \neq 1; D_{ij} \neq 0, \exists i, j\forall I,$$

the system is defined as a Cross-Polynomial System (CPNS).
If

$$D_{ij} \equiv 0, i, j\forall I,$$

the system is defined as a separable system.

2.1 MFRF with single input

SPNS

The equation describing the polynomial stiffness system with a single input is

$$\begin{cases} M\ddot{y} + C_1\dot{y} + K(y) = \mathbf{1}x(t) \\ y(0) = \dot{y}(0) = 0 \end{cases} \tag{6}$$

where $\mathbf{1} = [1, 0, \ldots, 0]^T$, $x(t)$ is a scalar input.
According to the Volterra series expansion, the displacement vector $y(t)$ may be formulated as

$$\begin{aligned} y(t) &= \sum_{n=1}^{\infty} \int_{-\infty}^{\infty} \cdots \int_{-\infty}^{\infty} h_n(t_1, \ldots, t_n) \prod_{i=1}^{n} x(t - t_i) dt_i \\ &= y_1(t) + y_2(t) + \cdots + y_n(t) + \cdots \end{aligned} \tag{7}$$

where[1]

$$y_n(t) = \int_{-\infty}^{\infty} h_n(t_1, \ldots, t_n) \prod_{i=1}^{n} x(t - t_i) dt_i \tag{8}$$

In the expression $h_n(t_1, \ldots, t_n)$ is known as the Volterra kernel. If $h_i = 0$, $i \neq n$, and $h_n \neq 0$, $y(t) = y_n(t)$, and the system described by Eq. 8 is referred to as an nth-order homogeneous Volterra system.
The Fourier transform of $h_n(t_1, \ldots, t_n)$

$$H_n(\omega_1, \omega_2, \ldots, \omega_n) = \int_{-\infty}^{\infty} h_n(t_1, \ldots, t_n) e^{[-j\Sigma_{i=1}^{n}\omega_i t_i]} dt_1 \ldots dt_n \tag{9}$$

is called the nth-order MFRF of the system. If Eq. 8 is written in the form

$$y_n(t_1, \ldots, t_n) = \int_{-\infty}^{\infty} h_n(\sigma_1, \ldots, \sigma_n) \prod_{i=1}^{n} x(t_i - \sigma_i) d\sigma_i \tag{10}$$

then

$$y_n(t) = y_n(t_1, \ldots, t_n) \mid_{t_1 = t_2 = \cdots = t_n = t} \tag{11}$$

If Eq. 10 is transformed into the frequency domain,we get

$$Y_n(\omega_1, \ldots, \omega_n) = H_n(\omega_1, \ldots, \omega_n) \prod_{i=1}^{n} X(\omega_i) \tag{12}$$

where $Y_n(\omega_1, \ldots, \omega_n)$ and $X(\omega_i)$ are Fourier transforms of $y_n(t_1, \ldots, t_n)$ and $x(t_i)$ respectively.
If the exciting force is assumed as $\alpha x(t)$, α is constant, then the equation is

$$\begin{cases} M\ddot{y}_\alpha + C_1\dot{y}_\alpha + K(y_\alpha) = \mathbf{1}\alpha x(t) \\ y_\alpha(0) = 0 \\ \dot{y}_\alpha(0) = 0 \end{cases} \tag{13}$$

and the response follows to

$$y_\alpha(t) = \alpha y_1(t) + \alpha^2 y_2(t) + \cdots + \alpha^n y_n(t) + \cdots \tag{14}$$

When substituting it into Eq. 13, the coefficients of corresponding powers of α are necessarily equated, since this differential equation and the equation for the initial state must hold for all α. It follows that

$$\begin{cases} M\ddot{y}_1 + C_1\dot{y}_1 + K_1 y_1 = \mathbf{1}x(t) \\ y_1(0) = 0 \\ \dot{y}_1(0) = 0 \end{cases} \tag{15}$$

$$\begin{cases} M\ddot{y}_2 + C_1\dot{y}_2 + K_1 y_2 = -K_2 y_1^{(2)} \\ y_2(0) = 0 \\ \dot{y}_2(0) = 0 \end{cases} \tag{16}$$

[1] The multiple integral symbol is still expressed with $\int$ for simplicity.

$$\begin{cases} M\ddot{y}_3 + C_1\dot{y}_3 + K_1 y_3 = -K_3 y_1^{(3)} - K_2(y_1 \otimes y_2 + y_2 \otimes y_1) \\ y_3(0) = 0 \\ \dot{y}_3(0) = 0 \end{cases} \quad (17)$$

....

This is a sequence of linear equations. From Eq. 15 the response $y_1(t)$ is

$$y_1(t) = \int_0^t h_1(t_1)x(t-t_1)dt_1 \quad (18)$$

In this expression $h_1(t)$ is the first column of the one-dimensional impulse response function matrix $h(t)$.
In the frequency domain it is

$$Y_1(\omega) = H_1(\omega)X(\omega) \quad (19)$$

where $Y_1(\omega)$ and $X(\omega)$ are Fourier transforms of $y_1(t)$ and $x(t)$ respectively. $H_1^k(\omega) = H_1(\omega)$ is called the first order MFRF, where the superscript k represents the SPNS, and it is the first column of the one-dimensional frequency response function matrix

$$H(\omega) = [-\omega^2 M + j\omega C_1 + K_1]^{-1} \quad (20)$$

Now, in Eq. 16 the input is already known, namely

$$\begin{aligned} y_1^{(2)}(t_1,t_2) &= y_1 \otimes y_2 \\ &= \int_0^{t_1}\int_0^{t_2} h_1(\tau_1) \otimes h_1(\tau_2)x(t_1-\tau_1)x(t_2-\tau_2)d\tau_1 d\tau_2 \\ y_1^{(2)}(t) &= y_1^{(2)}(t_1,t_2)\,|_{t_1=t_2=t} \end{aligned} \quad (21)$$

which shows the parallel multiplication connection of two systems that are the same. In the frequency domain we get

$$Y_1^{(2)}(\omega_1,\omega_2) = H_1(\omega_1) \otimes H_1(\omega_2)X(\omega_1)\cdot X(\omega_2) \quad (22)$$

The system can be shown in Fig. 1 in the frequency domain. Based on Fig. 1, we obtain

$$\begin{aligned} Y_2(\omega_1,\omega_2) &= H(\omega_1+\omega_2)[-K_2 H_1(\omega_1)\cdot X(\omega_1) \otimes H_1(\omega_2)\cdot X(\omega_2)] \\ &= H(\omega_1+\omega_2)[-K_2 H_1(\omega_1) \otimes H_1(\omega_2)]\cdot X(\omega_1)X(\omega_2) \end{aligned} \quad (23)$$

Comparing the expression with Eq. 12, the 2nd-order MFRF is

$$H_2^k(\omega_1,\omega_2) = -H(\omega_1+\omega_2)K_2 H_1(\omega_1) \otimes H_1(\omega_2) \qquad \in C^{N\times 1} \quad (24)$$

With the same procedure, from Eq. 17 we can get the 3rd-order MFRF

$$\begin{aligned} H_3^k(\omega_1,\omega_2,\omega_3) &= -H(\omega_1+\omega_2+\omega_3)\{K_3 - K_2[I \otimes (H(\omega_2+\omega_3)\cdot K_2)] \\ &\quad -K_2[(H(\omega_1+\omega_2)K_2) \otimes I]\}H_1(\omega_1) \otimes H_1(\omega_2) \otimes H_1(\omega_3) \qquad \in C^{N\times 1} \end{aligned} \quad (25)$$

where I is the nth-order unit matrix.

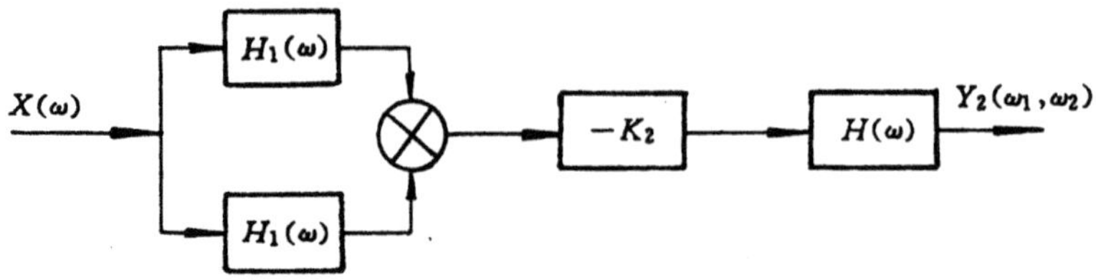

Figure 1 Block diagram of the system

The formulas of higher order MFRF can be obtained in similar procedures. The derivation is tedious but not difficult.

From the above discussion we can deduce: MFRF of the polynomial stiffness system with single input are formulated in structural form, with a one-dimensional frequency response function $H(\omega)$.

DPNS

The equation describing the polynomial damping system with single input is

$$\begin{cases} M\ddot{y} + C(\dot{y}) + K_1 y = \mathbf{1}x(t) \\ y(0) = 0 \\ \dot{y}(0) = 0 \end{cases} \tag{26}$$

In this equation $C(\dot{y})$ can be represented with Eq. 3. With the same approach as SPNS, we have

$$\begin{cases} M\ddot{y}_1 + C_1\dot{y}_1 + K_1 y_1 = \mathbf{1}x(t) \\ y_1(0) = 0 \\ \dot{y}_1(0) = 0 \end{cases} \tag{27}$$

$$\begin{cases} M\ddot{y}_2 + C_1\dot{y}_2 + K_1 y_2 = -C_2\dot{y}_1^{(2)} \\ y_2(0) = 0 \\ \dot{y}_2(0) = 0 \end{cases} \tag{28}$$

$$\begin{cases} M\ddot{y}_3 + C_1\dot{y}_3 + K_1 y_3 = -C_3\dot{y}_1^{(3)} - C_2(\dot{y}_1 \otimes \dot{y}_2 + \dot{y}_2 \otimes \dot{y}_1) \\ y_3(0) = 0 \\ \dot{y}_3(0) = 0 \end{cases} \tag{29}$$

....

When these equations are compared with Eq. 15 ~ Eq. 17, they are completely similar, except that y_i, K_i is replaced by $\dot{y}_i$, C_i on the right side of the equations. By virtue of the fact that Eq. 20 is a one-dimension displacement frequency response function, the one-dimension velocity frequency response is $j\omega H(\omega)$. Therefore with a similar procedure we have [6]

$$H_1^c(\omega) = H_1(\omega) \tag{30}$$

$$H_2^c(\omega_1,\omega_2) = \omega_1\omega_2 H(\omega_1+\omega_2)[C_2 H_1(\omega_1)\otimes H_1(\omega_2)] \qquad \in C^{N\times 1} \tag{31}$$

$$\begin{aligned} H_3^c(\omega_1,\omega_2,\omega_3) \;=\; & -(j\omega_1)(j\omega_2)(j\omega_3)H(\omega_1+\omega_2+\omega_3)\{C_3- \\ & C_2[I\otimes(H(\omega_2+\omega_3)\cdot C_2)] - C_2[(H(\omega_1+\omega_2)\cdot \\ & C_2)\otimes I]\}H_1(\omega_1)\otimes H_1(\omega_2)\otimes H_1(\omega_3) \qquad \in C^{N\times 1} \end{aligned} \tag{32}$$

where superscript c represents DNPS.
As with the polynomial stiffness system with single input we come to the same conclusion: MFRF for the polynomial damping system is indicated by a one-dimension frequency response function $H(\omega)$ in structural form.

CPNS

The following refer to the approaches for SPNS or DPNS, MFRFs of CPNS

$$H_1^d(\omega) = H_1(\omega) \tag{33}$$

$$H_2^d(\omega_1,\omega_2) = -(j\omega_2)H(\omega_1+\omega_2)[D_{11}H_1(\omega_1)\otimes H_1(\omega_2)] \tag{34}$$

$$\begin{aligned} H_3^d(\omega_1,\omega_2,\omega_3) = {} & H(\omega_1+\omega_2+\omega_3)\{D_{11}[(j\omega_2)(j\omega_3)^2(I\otimes H(\omega_1+\omega_2)\cdot \\ & (I\otimes D_{11}) + (j\omega_2)(j\omega_3)(H(\omega_1+\omega_2)\otimes I)\cdot(D_{11}\otimes I)] - (j\omega_3)D_{21} \\ & -(j\omega_2)(j\omega_3)D_{12}\}H_1(\omega_1)\otimes H_1(\omega_2)\otimes H_1(\omega_3) \end{aligned} \tag{35}$$

where superscript d represents the CPNS.

ADDITION THEOREM

The addition theorem of MFRFs of PNS can be expressed as

$$H_1(\omega) = H_1^k(\omega) = H_1^c(\omega) = H_1^d(\omega) \tag{36}$$

$$\begin{aligned} H_n(\omega_1,\omega_2,\ldots,\omega_n) \;=\; & H_n^k(\omega_1,\omega_2,\ldots,\omega_n) + H_n^c(\omega_1,\omega_2,\ldots,\omega_n) + \\ & H_n^d(\omega_1,\omega_2,\ldots,\omega_n) \qquad (n=2,3,\ldots) \end{aligned} \tag{37}$$

2.2 Characteristics of MFRF of PNS

1. Recursive formulation
   The nth-order MFRF can be expressed by first-order to (n-1)th-order MFRFs. That is, the higher order MFRFs can be all expressed only by the first order MFRF.

2. Relationship between H_n and K_i, C_i, D_{ij}
   The nth-order MFRF is independent of the coefficients of the higher order exponential terms than n, i.e. H_n is independent of K_i, C_i, $(i,j>n)$ and $D_{ij}(i+j>n)$.

3. Principal diagonals
The principal diagonal of MFRF is $H(\omega_1, \ldots, \omega_n) \mid_{\omega_1 = \ldots = \omega_n}$. It refers to the harmonic response for PNS from the point of view of output and $\omega_1 = \ldots = \omega_n = \omega_0 = constant$ $(n \geq 2)$ is with regard to a constant frequency ($= \omega_0$) input from the point of view of input.

4. Constant frequency response subdomain.
From the inverse Fourier transform

$$y_n(t) = \frac{1}{(2\pi)^n} \int Y_n(\omega_1, \ldots, \omega_n) e^{j(\omega_1 + \cdots + \omega_n)t} d\omega_1 \ldots d\omega_n \tag{38}$$

the constant frequency response subdomain is

$$\sum_{i=1}^{N} = \omega_0 = const. \tag{39}$$

5. Pole frequency set
<u>Definition</u>: for nth-order MFRF, $H_n(\omega_1, \omega_2, \ldots, \omega_n)$, and the nth-order pole frequencies are the solutions of any n of the following equations

$$\begin{array}{ll} \omega_i = \omega_{pk} & i = 1, \ldots, n \quad k = 1, \ldots, N \\ \omega_i + \omega_j = \omega_{pk} & i = 1, \ldots, n \\ & j = 1, \ldots, n \;\; and \; j \neq i \;\; k = 1, \ldots, N \\ \ldots\ldots & \\ \omega_{i_1} + \omega_{i_2} + \ldots + \omega_{i_{n-1}} & i_1, i_2, \ldots, i_{n-1} = 1, \ldots, n \\ \quad = \omega_{p_k} & and \; i_1 \neq i_2 \neq \ldots \neq i_{n-1} \quad k = 1, \ldots, N \\ \sum_{i=1}^{n} \omega_i = \omega_{p_k} & k = 1, \ldots, N \end{array} \tag{40}$$

where N is the number of degrees of freedom, and $\omega_{pk}(k = 1, \ldots, N)$ are the natural frequencies of the accompanied linear system with $H(\omega)$. In general, only the positive and zero solutions are studied.

<u>Corollary</u>: the nth-order pole frequencies are only related to the first order pole frequencies.

For example,
The **1st** order pole frequencies for $H_1(\omega)$ are

$$\omega = \omega_{pk} \qquad k = 1, \ldots, N$$

The **2nd** order pole frequencies for $H_2^K(\omega_1, \omega_2)$ of KPNS, from

$$H_2^k(\omega_1, \omega_2) = H(\omega_1 + \omega_2)[-K_2 H_1(\omega_1) \otimes H_1(\omega_2)] \tag{41}$$

and by virtue of the fact that $H_2^k(\omega_1, \omega_2)$ is represented by $H(\omega)$, are closely related to ω_{pk}, i.e.

$$\begin{cases} \omega_1 + \omega_2 = \omega_{pk} \\ \omega_1 = \omega_{pk} \qquad k = 1, \ldots, N \\ \omega_2 = \omega_{pk} \end{cases} \tag{42}$$

All ω_1 and ω_2, calculated with two frequencies of Eq. 40, are pole frequencies of $H_2^k(\omega_1, \omega_2)$ and form a set of pole frequencies. Each equation in Eq. 42 indicates a line in (ω_1, ω_2) Cartesian coordinates. An example of a system with two degrees of fredom is shown in Fig. 2. The magnitudes of $H_2^k(\omega_1, \omega_2)$ on the lines look like ridges of a mountain so we call them ridge lines. The lines represented by Eq. 42 can intersect in the (ω_1, ω_2) plane. The values of ω_1 and ω_2 in the cross-points are approximately equal to the values of ω_1 and ω_2 at peaks of $\mid H_2^k(\omega_1, \omega_2) \mid$. The cross-points are therefore called the 2nd order poles. The frequency values of the 2nd order pole can be calculated by two equations in Eq. 42.

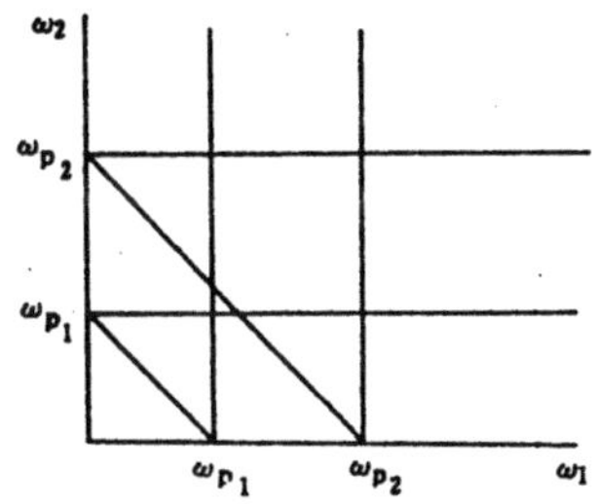

Figure 2 Distribution of 2nd order pole frequencies

As far as the polynomial damping system is concerned, because $H_1^c(\omega) = H_1(\omega)$ and the forms of its MFRFs whose orders are greater than 1 are also similar to the polynomial stiffness system, it follows that the analysis programme of poles for the polynomial damping system is the same as the polynomial stiffness system, and that the fundamental properties obtained are identical. However, it is different, in that $H_n^c(\omega_1 \ldots\ldots \omega_n) = 0$, $n > 1$, at $\omega_i = 0$ for the polynomial damping system. Therefore, the nth-order pole frequencies of $H_n^c(\omega_1 \ldots\ldots \omega_n)$, $n > 1$ disappear when $\omega_i = 0$.

Finally, the pole analysis of MFRF for a polynomial system is similar to the above consideration because of the addition theorem.

2.3 Numerical Example

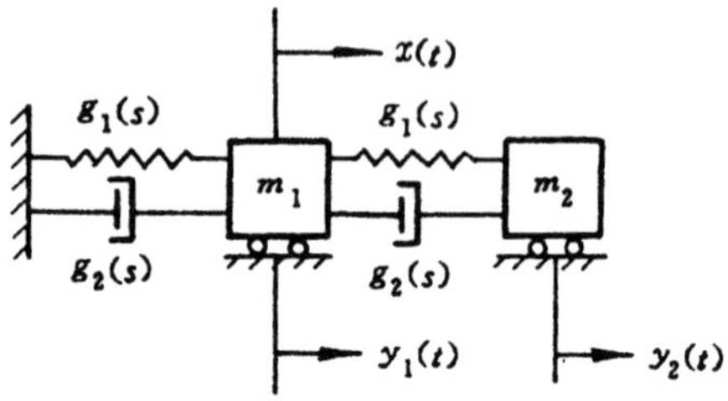

Figure 3 Non-linear M-K system

A spring mass system with two degrees of freedom is shown in Fig. 3. $m_1 = m_2 = 1$. The restoring force function is

$$g_1(s) = k_1 s + k_2 s^2 + k_3 s^3 \tag{43}$$

with $k_1 = 10^4$, $k_2 = 10^8$, $k_3 = 10^{10}$, where s is the deformation of the spring. The damping is assumed as a polynomial

$$g_2(\dot{s}) = c_1 \dot{s} + c_2 \dot{s}^2 + c_3 \dot{s}^3 \tag{44}$$

where, $c_1 = 1$, $c_2 = 10^4$, $c_3 = 10^8$, $\dot{s} = \frac{ds}{dt}$, and the system is considered as a zero initial state.
Now the MFRF of SPNS, DPNS and separable PNS will be discussed as follows.

SPNS

Let $c_2 = c_3 = 0$, the motion equation of the system is

$$M\ddot{y} + C_1\dot{y} + K_1 y + K_2 y^{(2)} + K_3 y^{(3)} = \mathbf{1}x(t) \tag{45}$$

where

$$M = \begin{bmatrix} 1 & 0 \\ 0 & 1 \end{bmatrix} \qquad C_1 = \begin{bmatrix} 2c_1 & -c_1 \\ -c_1 & c_1 \end{bmatrix}$$

$$K_1 = \begin{bmatrix} 2k_1 & -k_1 \\ -k_1 & k_1 \end{bmatrix} \qquad K_2 = \begin{bmatrix} 0 & k_2 & k_2 & -k_2 \\ k_2 & -k_2 & -k_2 & k_2 \end{bmatrix}$$

$$K_3 = \begin{bmatrix} 2k_3 & -k_3 & -k_3 & k_3 & -k_3 & k_3 & k_3 & -k_3 \\ -k_3 & k_3 & k_3 & -k_3 & k_3 & -k_3 & -k_3 & k_3 \end{bmatrix}$$

$$\mathbf{1} = [1, 0]^T \qquad y = [y_1\ y_2]^T \tag{46}$$

From Eq. 20

$$H(\omega) = [-\omega^2 M + j\omega C_1 + K_1]^{-1} \tag{47}$$

by means of the theory of the linear system, the 1st order pole frequencies calculated easily are

$$\omega_{p1} = 61.8 \qquad \omega_{p2} = 161.8$$

From Eq. 47

$$H_1^k(\omega) = H_1(\omega) = \left\{ \begin{array}{c} H_{11}(\omega) \\ H_{21}(\omega) \end{array} \right\} \tag{48}$$

$H_{11}(\omega)$ and $H_{21}(\omega)$ are two elements in the first column of $H(\omega)$.
The formula of the 2nd order MFRF is

$$H_2^k(\omega_1, \omega_2) = H(\omega_1 + \omega_2)[-K_2 H_1(\omega_1) \otimes H_1(\omega_2)] \tag{49}$$

The magnitudes of the elements of $H_2^k(\omega_1, \omega_2)$ are shown in Fig. 4 (a) and (c).
The ridge line equations are

$$\left\{ \begin{array}{l} \omega_1 + \omega_2 = \omega_k \\ \omega_1 = \omega_k \qquad\qquad k = 1, 2 \\ \omega_2 = \omega_k \end{array} \right. \tag{50}$$

Solving the equation system formed with any two of the above equations yields 2nd order pole frequencies as follows

(61.8,0) (161.8,0) (0,61.8) (0,161.8)
(61.8,61.8) (161.8,61.8) (61.8,161.8) (161.8,161.8)
(61.8,100) (100,161.8)

Moreover, the peak locations shown by the contour of the magnitude of the element of $H_2^k(\omega_1, \omega_2)$ can demonstrate that the 2nd order pole frequencies calculated are correct, as is shown in Fig. 4 (b) and (d).

DPNS

Let $k_2 = k_3 = 0$, the motion equation is

$$M\ddot{y} + C_1\dot{y} + C_2\dot{y}^{(2)} + C_3\dot{y}^{(3)} + K_1 y = \mathbf{1}x(t) \tag{51}$$

where M,K_1 and C_1 are same as in Eq. 46, and

$$C_2 = \left[\begin{array}{cccc} 0 & c_1 & c_2 & -c_2 \\ c_2 & -c_2 & -c_2 & c_2 \end{array} \right]$$

$$C_3 = \left[\begin{array}{cccccccc} 2c_3 & -c_3 & -c_3 & c_3 & -c_3 & c_3 & c_3 & -c_3 \\ -c_3 & c_3 & c_3 & -c_3 & c_3 & -c_3 & -c_3 & c_3 \end{array} \right]$$

The 1st order MFRF and pole frequencies are identical with SPNS.The 2nd order MFRF is

$$H_2^C(\omega_1, \omega_2) = -\omega_1\omega_2 H(\omega_1 + \omega_2)[-C_2 H_1(\omega_1) \otimes H_1(\omega_2) \tag{52}$$

Due to the fact that the linear parts of Eq. 45 and Eq. 51 are identical the same ridge line equations and the same 2nd order pole frequencies of $H_2^c(\omega_1, \omega_2)$ exist for DPNS and SPNS, except that the 2nd order pole frequencies on ω_1 and ω_2 axes vanish.
For separable PNS, the addition theorem is employed to get MFRFs. The pole frequencies are calculated as easily as some SPNS.

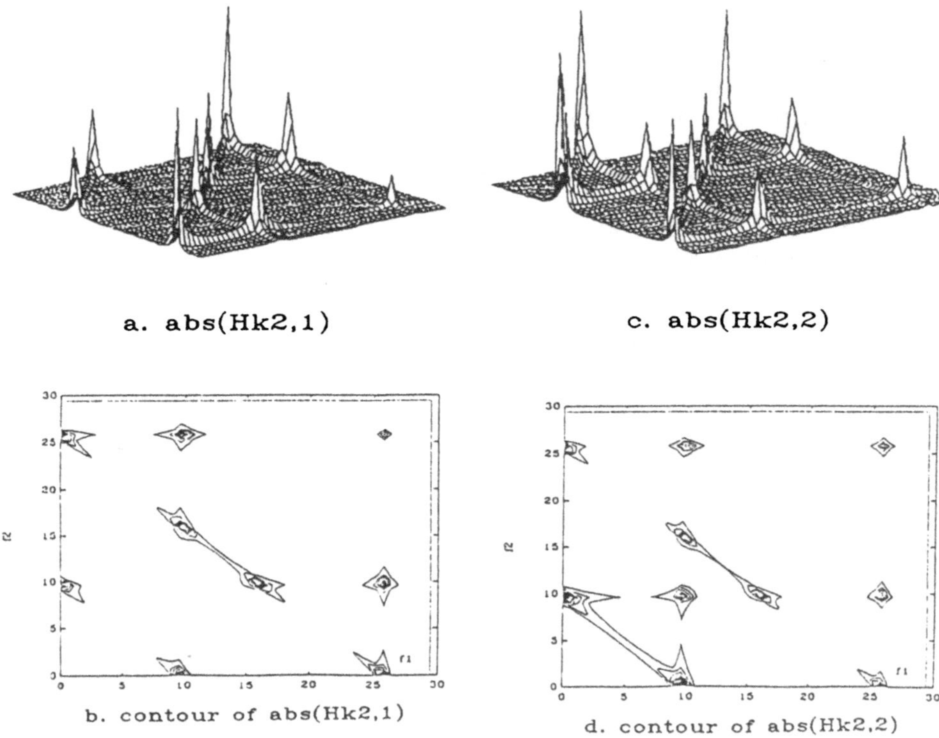

Figure 4 2nd-order MFRF of the polynomial stiffness system

3 Statistical Equivalence Approach

3.1 Equivalent Equation

For a time-invariant non-linear dynamic system with a single degree of freedom Eq. 1 can be rewritten as

$$m\ddot{y} + g(y, \dot{y}) = x(t) \tag{53}$$

where the exciting force $x(t)$ can be limited to be a stationary Gaussian stochastic process in the following discussion, and $E[x(t)] = 0$. The response $y(t)$ is also assumed to be a stationary stochastic process, but is not a Gaussian process in general, and $E[y(t)] \neq 0$. A separable system, $g(y, \dot{y}) = g_c(\dot{y}) + g_k(y)$,can be equivalent with an (n,l) order separable polynomial system

$$m\ddot{y} + \sum_{i=1}^{n} c_i \dot{y}^i + \sum_{j=1}^{l} k_j y^j = x(t) \tag{54}$$

where $c_i(i = 1, \ldots, n)$ and $k_j(j = 1, \ldots, l)$ are to be estimated coefficients. The error function defined in terms of the equation errors is

$$\varepsilon(t) = \hat{x}(t) - (m\hat{\ddot{y}} + \sum_{i=1}^{n} c_i \hat{\dot{y}^i} + \sum_{j=1}^{l} k_j \hat{y^j}) \tag{55}$$

where the superscript $\hat{}$ represents the measured values of responding variables. In order to get a simpler standard form of the parameter estimation equation, let

$$
\begin{array}{ll}
C^{(n)} = [c_1, \ldots, c_n]^T & ---damping\ coefficient\ vector \\
K^{(l)} = [k_1, \ldots, k_l]^T & ---stiffness\ coefficient\ vector \\
\alpha = \begin{bmatrix} C^{(n)} \\ K^{(l)} \end{bmatrix} & ---estimated\ parameter\ vector \\
Y = [\hat{y}, \hat{y^2}, \ldots, \hat{y^l}]^T & ---measured\ displacement\ power\ vector \\
\dot{Y} = [\hat{\dot{y}}, \hat{\dot{y}^2}, \ldots, \hat{\dot{y}^n}]^T & ---measured\ velocity\ power\ vector \\
z = \begin{bmatrix} \dot{Y} \\ Y \end{bmatrix} & ---state\ power\ vector
\end{array}
$$

The superscripts n and l are the order of the damping polynomial and the stiffness polynomial respectively. Eq. 55 is abbreviated to

$$\varepsilon(t) = \hat{x} - (m\hat{\ddot{y}} + \alpha^T z) \tag{56}$$

With $E[\varepsilon^2]$ being minimized, the estimation equation of the parameter is

$$M_{zz}\alpha = M_{\hat{x}z} - mM_{\hat{\ddot{y}}z} \tag{57}$$

where

$$M_{zz} = E[z\ z^T], \quad M_{\hat{x}z} = E[\hat{x}\ z], \quad M_{\hat{\ddot{y}}z} = E[\hat{\ddot{y}}\ z]$$

Because $E[y\dot{y}] = 0$ and $E[\dot{y}\ddot{y}] = 0$, some elements in M_{zz} and $M_{\hat{\ddot{y}}z}$ become zero when the measuring error is neglected. It is easy to prove that M_{zz} is a symmetry matrix. If it is written as

$$M_{zz} = \begin{bmatrix} M_{\dot{Y}\dot{Y}} & M_{\dot{Y}Y} \\ M_{Y\dot{Y}} & M_{YY} \end{bmatrix} \tag{58}$$

where $M_{\dot{Y}\dot{Y}} = E[\dot{Y}\dot{Y}^T]$, $M_{Y\dot{Y}} = M_{\dot{Y}Y}^T = E[Y\dot{Y}^T]$, $M_{YY} = E[YY^T]$, are all symmetry matrixes.

When $n = 1$ and $l = 1$, the statistically equivalent linear system is obtained. If the measurement noises responding to every measured varable are mutually uncorrelated, the equivalent damping coefficient and equivalent stiffness coefficient can be worked out from

$$c_1^{(1)} = E[x\dot{y}]/E[\hat{\dot{y}^2}] \qquad k_1^{(1)} = [E[xy] - E[\ddot{y}y]]/E[\hat{y^2}]$$

where the subscript (1) represents linear equivalence. If the measurement noises are neglected, $E[\hat{\dot{y}^2}] = E[\dot{y}^2]$ and $E[\hat{y^2}] = E[y^2]$ and it becomes a well known result. The parameter estimation Eq. 57 can also be employed for a non-separable polynomial system only taking $d = \begin{bmatrix} C^{(n)} \\ K^{(l)} \\ D \end{bmatrix}$ and $z = \begin{bmatrix} \dot{Y} \\ Y \\ Y \otimes \dot{Y} \end{bmatrix}$ instead of $\alpha = \begin{bmatrix} C^{(n)} \\ K^{(l)} \end{bmatrix}$ and $z = \begin{bmatrix} \dot{Y} \\ Y \end{bmatrix}$

3.2 Extended Transfer Function

Many engineering problems can be analysed approximately with the 3rd order statistically equivalent polynomial system, whose MFRFs can be written in analytical forms as in Section 2. Although all of the higher order MFRF can be expressed with 1st order MFRFs, $H(\omega)$, it is difficult to put this method into practice because the multi-dimensional frequency is without specific physical meaning. If the transfer characteristic can be expressed with one dimension frequency just as is done in a linear system, it will be convenient to use the developed spectral analysis and modal analysis techniques in the linear system. Thus the Extended Transfer Function (ETF) can be defined with the relationship between the power spectral density of the response and that of the excitation to describe the particular characteristics of a non-linear system. If the effect of higher order MFRF is neglected, the power spectral density of the response can be decomposed into the following form [7]

$$S_{yy}(\omega) = S_0 + S_1(\omega) + 2S_2(\omega) + 6S_3(\omega) \tag{59}$$

Where

$$S_0 = \mathbb{H}_0^2 \delta(\omega) \tag{60}$$

$$\mathbb{H}_0 = \int_{-\infty}^{\infty} H_2(v, -v) S_{xx}(v) dv \tag{61}$$

$$S_1 = \mid \mathbb{H} \mid^2 S_{xx}(\omega) \tag{62}$$

$$\mathbb{H}(\omega) = H_L(\omega) + 3H_N(\omega)$$

$$\mathbb{H}_L = H(\omega)$$

$$\mathbb{H}_N = \frac{1}{2\pi} \int_{-\infty}^{\infty} H_3(\omega, v, -v) S_{xx}(v) dv$$

$$S_2 = \frac{1}{2\pi} \int_{-\infty}^{\infty} \mid H_2(\omega - v, v) \mid^2 S_{xx}(v) dv \tag{63}$$

$$\begin{aligned} S_3 \quad = \quad & \tfrac{1}{(2\pi)^2} \textstyle\int_{-\infty}^{\infty} \int_{-\infty}^{\infty} \mid H_3(\omega - v_1 - v_2, v_1, v_2) \mid^2 \\ & S_{xx}(\omega - v_1 - v_2) S_{xx}(v_1) S_{xx}(v_2) dv_1 dv_2 \end{aligned} \tag{64}$$

where $\delta(\omega)$ is the Dirac δ function, $S_{xx}(\omega)$ is a PSD of the excitation. Here the interpretations of each term are given in the following. A similar analysis has been done by Zhang and Qiu[3] with harmonic excitation.

Response mean

$\mathbb{H}_0$ is the response mean. Even though the excitation mean is zero, the response mean is not zero in general.

Home-frequency response

S_1 is the home-frequency response, i.e. the bandwidth of the response is the same as that of excitation. $\mathbb{H}$ is called the Home-frequency Transfer Function(HTF), which consists of two parts: $\mathbb{H}_L(\omega)$ and $\mathbb{H}_N(\omega)$. $\mathbb{H}_L(\omega)$ is the Linear Transfer Function(LTF), which is only related to $c_1^{(3)}$ and $k_1^{(3)}$, and the different frequency components of the excitation do not produce a coupled response. $\mathbb{H}_N(\omega)$ is called the Non-linear home-frequency Transfer Function(NTF), which is not only related to $c_1^{(3)}$ and $k_1^{(3)}$, but also to $c_3^{(3)}$ and $k_3^{(3)}$, as well as to the excitation. The two frequency components of the excitation can couple to result in the response. Hence, $\mathbb{H}_N(\omega)$ will be further decomposed. Let the excitation be stationary narrow-band white noise with Gaussian distribution i.e.

$$S_{xx}(\omega) = \begin{cases} \frac{B}{2(\Omega_2-\Omega_1)} & \Omega_1 \leq |\omega| \leq \Omega_2 \\ 0 & other \end{cases} \tag{65}$$

thus

$$\mathbb{H}_N(\Omega) = \frac{B/2\pi}{\Omega_2 - \Omega_1} \int_{\Omega_1}^{\Omega_2} H_3(\Omega, v, -v)dv \tag{66}$$

Although the Ω in this equation can have an arbitrary value, the discussion is performed only on $\Omega_1 \leq |\Omega| \leq \Omega_2$ because it is a home-frequency response. If $\Omega_1 \to \Omega$ and $\Omega_2 \to \Omega$, the above integral becomes

$$\mathbb{H}_N(\Omega) = \frac{B}{2\pi} H_3(\Omega, \Omega, -\Omega) \tag{67}$$

Thus $H_3(\Omega, \Omega, -\Omega)$ is the self -gain of the narrow-band random excitation, and then

$$\mathbb{H}_{Na}(\omega) = H_3(\omega, \omega, -\omega) \tag{68}$$

and is called the non-linear Self-gain Transfer Function(STF). If the excitation is assumed to be composed of two separated narrow-band white noises, i.e.

$$S_{xx}(\omega) = \begin{cases} \frac{B_1}{2(\Omega_2-\Omega_1)} & \Omega_1 \leq |\omega| \leq \Omega_2 \\ \frac{B_2}{2(\Omega_4-\Omega_3)} & \Omega_3 \leq |\omega| \leq \Omega_4 \qquad \Omega_3 > \Omega_2 \\ 0 & other \end{cases} \tag{69}$$

When Ω satisfies $\Omega_1 \leq \Omega \leq \Omega_2$, and $\Omega_2 \rightarrow \Omega_1$, $\Omega_4 \rightarrow \Omega_3$, there will be

$$\mathbb{H}_N(\Omega) = \frac{B_1}{2\pi} H_3(\Omega, \Omega, -\Omega) + \frac{B_2}{2\pi} H_3(\Omega, \Omega_3, -\Omega_3) \tag{70}$$

The first term describes the self-gain of the response responding to a narrow-band random excitation expressed as Eq. 68. The second term describes the response caused by the cross effect between the narrow-band random excitation with different frequencies, hence the term

$$\mathbb{H}_{Nc}(\Omega, \omega) = H(\Omega, \omega, -\omega) \tag{71}$$

is called the non-linear Cross gain Transfer Function(CTF).

Order-2 response

S_2 is the response caused by the 2nd order MFRF, and is called the order-2 response. It comes out at a new frequency due to the coupling(or modulation) between the excitation components with different frequencies. It should be noted that this response is different from the non-linear cross-gain transfer function. The latter is the home-frequency response, and the former is the response outside the frequency band of excitation. If S_2 is written into the discrete form

$$\frac{\Delta\omega}{2\pi} \sum_{i=-\infty}^{+\infty} \mid H_2[(k-i)\bigtriangleup\omega, i\bigtriangleup\omega] \mid^2 S_{xx}(i\bigtriangleup\omega) S_{xx}((k-i)\bigtriangleup\omega) \tag{72}$$

the above interpretation is very easy to understand: a narrow-band response with a central frequency $k\bigtriangleup\omega$ is produced by the modulation between one narrow-band excitation component with a central frequency $i\bigtriangleup\omega$ and another with a central frequency $(k-i)\bigtriangleup\omega$. If $i = \frac{k}{2}$, $k - i = \frac{k}{2}$, i.e. the modulation between $S_{xx}(\frac{k}{2}\bigtriangleup\omega)$ and itself results in the double frequency response, which is mainly determined by $\mid H_2(\omega, \omega) \mid^2$ apart from being related to the excitation. In order to make a comparison between the different non-linear systems, choose the white noise as the comparison standard, and then $H_2(\omega, \omega)$ can be used to describe the double frequency response characteristic of the system, hence the term

$$\mathbb{H}_2(\omega) = H_2(\omega, \omega) \tag{73}$$

is called the Double frequency Transfer Function(DTF).
In general, the spectrum of S_2 is very rich. The term

$$\mathbb{H}_{\Sigma_2}(\omega) = \int_{-\infty}^{\infty} \mid H_2(\omega - v, v) \mid^2 dv \tag{74}$$

is an order-2 modulation Transfer Function(DMTF). If the bandwidth of the excitation is $[\Omega_1, \Omega_2]$, the bandwidth of the response produced by $\mathbb{H}_{\Sigma_2}(\omega)$ is apparently $[0, 2\Omega_2]$.

Order-3 response

S_3 is the response produced by the coupling among the three frequency components of the excitation. Analogous to the analysis for the order-2 response, it can be called the Tri-frequency Transfer Function(TTF).

$$\mathbb{H}_3(\omega) = H_3(\omega, \omega, \omega) \tag{75}$$

and the Tri-Modulation Transfer Function(TMTF)

$$\mathbb{H}_{\Sigma_3}(\omega) = \int_{-\infty}^{\infty} \int_{-\infty}^{\infty} \mid H_3(\omega - v_1 - v_2, v_1, v_2) \mid^2 dv_1 dv_2 \tag{76}$$

If the bandwidth of the excitation is $[\Omega_1, \Omega_2]$, the highest frequency of the order-3 response is $3\Omega_2$.
For the sake of clarity, the extended transfer functions defined above are gathered in Table 1. Note that their dimensions are different.

3.3 Effectiveness check

We do not usually know the non-linear structure of a general non-linear system, but according to the above statistically equivalent method, the non-linear identification can be done approximately with the help of the order-3 polynomial or the order-2 polynomial or the linear system. The degree of the approximatation determines the effectiveness of the method. The following methods can be used to examine the effectiveness.

(1) Examination with a normalized factor
Define the normalized factor to be

$$Zh_c^{(n)} = [\mid c_n(\sigma_y)^n \mid / \sigma_x] / \sum_{i=1}^{n} [\mid c_i(\sigma_y)^i \mid / \sigma_x] \tag{77}$$

$$Zh_k^{(l)} = [\mid k_l(\sigma_y)^l \mid / \sigma_x] / \sum_{j=1}^{l} [\mid k_j(\sigma_y)^j \mid / \sigma_x] \tag{78}$$

where σ_y, $\sigma_{\dot{y}}$, and σ_x are the root mean square values of the response displacement, the velocity and the excitation respectively. The method for examining the effectiveness is that if $Zh_k^{(l-1)} < [Zh_k]$ and $Zh_k^{(l)} < [Zh_k]$, where $[Zh_k]$ is the permitted error index, the order-(l-2) statistically equivalent polynomial stiffness system is effective. Similarly, while $Zh_c^{(n-1)} < [Zh_c]$ and $Zh_k^{(n)} < [Zh_c]$, where $[Zh_c]$ is the permitted error index, the order-(n-2) statistically equivalent damping system is effective.
If the system is really a polynomial system, the normalized factor can be used to determine the order of the polynomial, and the estimated parameters independent of the excitation.

(2) Examination with ETF
The effectiveness of the statistically equivalent order-3 polynomial system can be examined with the properties of ETF. The highest frequency of the response of

an order-3 system is three times that of the excitation, double that for an order-2 system, and the equivalent linear system only has the home-frequency response. The comparison between the effectiveness of the equivalent linear system and that of an order-3 system can be determined by the comparison between $\mathbb{H}_L(\omega)$ and $\mathbb{H}_N(\omega)$. The comparison between the double frequency response, tri-frequency response and home-frequency response can be used to distinguish the effectiveness of an order-2 system and an order-3 one.

3.4 Examples

Example 1. Duffing equation

$$\ddot{y} + 2\pi\dot{y} + (20\pi)^2 y + 1.5 \cdot 10^{10} y^3 = x(t) \tag{79}$$

Table 1 Extended Transfer Function

home frequency response	LTF	$\mathbb{H}_L(\omega) = H(\omega) = (-m\omega^2 + j\omega c_1 + k_1)^{-1}$
	NTF	$\mathbb{H}_N(\omega) = \frac{1}{2\pi} \int H_3(\omega, v, -v) S_{xx}(v) dv$
		STF — $\mathbb{H}_{Na}(\omega) = H_3(\omega, \omega, -\omega)$
		CTF — $\mathbb{H}_{Nc}(\Omega, \omega) = H_3(\Omega, \omega, -\omega)$
	HTF	$\mathbb{H}(\omega) = \mathbb{H}_L(\omega) + 3\mathbb{H}_N(\omega)$
modulation response	DTF	$\mathbb{H}_2(\omega) = H_2(\omega, \omega)$
	DMTF	$\mathbb{H}_{\Sigma_2}(\omega) = \int \mid H_2(\omega - v, v) \mid^2 dv$
	TTF	$\mathbb{H}_3(\omega) = H_3(\omega, \omega, \omega)$
	TMTF	$\mathbb{H}_{\Sigma_3}(\omega) = \int\int \mid H_3(\omega - v_1 - v_2, v_1, v_2) \mid^2 dv_1 dv_2$

Note:
LTF —linear transfer function
NTF —non-linear home-frequency transfer function
HTF(LTF and NTF) —home-frequency response transfer function
DTF —double frequency transfer function
DMTF($\subset$DTF) —order-2 modulation transfer function
TTF —tri-frequency transfer function
TMTF($\subset$TTF) —order-3 modulation transfer function

is an order-3 polynomial system, where $x(t)$ is a stationary Gaussian white noise on $[0, 100HZ]$, and $E[x(t)] = 0$ as well as $E[x(t)^2] = 1$. The responses $y(t)$, $\dot{y}$, and $\ddot{y}$ are calculated with the numerical integral for the sample of $x(t)$, which are refered to as the numerical simulation of the measured data. This example is used to: (a) examine the statistically equivalent parameter method, (b) exhibit the effect of the order of the polynomial, (c) study the effect of the measuring noise.
The estimated parameters are shown in Table 2, from which the validity of the parameter estimating method and the effectiveness of the examining method stated in this paper are proved. The changes of the orders of c_n and k_l do not affect each other, and the effect of the measuring noise is bigger from a numerical result. The estimated parameters for the given order of the polynomial depend on the intensity of the excitation, but a suitable order can make the estimated parameters independ of that.

Example 2. A quadratic damping system is

$$\ddot{y} + \beta\dot{y} + y + v \mid a + \dot{y} \mid (a + \dot{y}) = x(t) \tag{80}$$

where $\beta = 0.1$, $v = 1$, $a = 0.3$, $x(t)$ is a stationary random Gaussian white noise with the bandwidth $[0.04\ 0.12Hz]$. The estimated parameters of the equivalent polynomial are shown in Table 3. It is diffcult to determine the order of the statistically equivalent system only from the results in Table 3, but the effectiveness of the statistic equivalence can be analysed with the power spectral density of response and the characters of the ETF shown in Fig. 5. The curve-3 calculated with the order (2,1) statistically equivalent system is nearest to the curve-2 that is the numerical solution of Eq. 79. The fact that the band width of the response is approximately twice the excitation shows that the order (2,1) statistically equivalent polynomial system is checked to be valid.

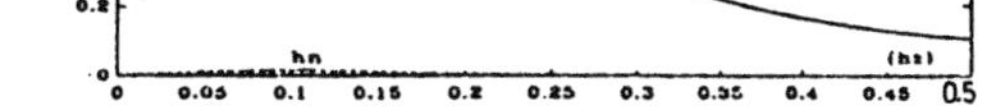

Figure 5 (a) PSD of resp. and excit. (b) HTF of order(2,1) sys.

Table 2 Results of parameter estimation

order of c_n	order of k_l	c_i	$Zh_c^{(n)}$	k_j	$Zh_k^{(l)}$
1	1	6.3395	1	7.700×10^3	1
1	2	6.3394	1	7.6994×10^3	
				9.1345×10^3	0.1748
1	3	6.2830	1	3.9478×10^3	
				-0.0886	
				1.5×10^{10}	0.9992
1	5	6.2830	1	3.9478×10^3	
				1.0472	
				1.5×10^{10}	
				-3.0609×10^6	
				1.5476×10^{10}	3.2875×10^{-4}
2	3	6.2830		3.9478	
		5.9102×10^{-5}	1.5684×10^{-6}	-0.2731	
				1.5×10^{10}	

Note: $\sqrt{E[y_2]} = 3.1901 \times 10^{-4}, \sqrt{E[\dot{y}_2]} = 0.0278$, average time=20.48s

Table 3 Estimated parameters

order of c_n	order of k_l	c_i	k_j
1	1	1.0008	0.9197
2	1	1.1039	0.9711
		0.5017	
		0.6765	
3	1	0.6652	0.9864
		0.5628	
3	2	1.0960	0.9990
		0.4635	0.0315
		0.6780	
2	2	0.6310	1.0089
		0.5524	0.0258
		0.6781	0.9921
3	3	0.6237	0.0308
		0.5505	0.0054
		0.6235	
		1.0667	1.0098
5	2	0.8099	0.0155
		-0.4392	
		-0.2391	

Note:

$E[y] = -0.2695$

$E[\dot{y}] = -0.001$

$\sqrt{E[y^2] - (E[y])^2} = 1.1089$

$\sqrt{E[\dot{y}^2] - (E[\dot{y}])^2} = 0.5896$

4 Remarks

1. MFRF, the generalization of the frequency response function of a linear system can be used to analyse the characteristics of the non-linear system in the frequency domain.

2. For the polynomial non-linear system, the analytical formulae of MFRFs are derived and are shown that MFRFs are expressed only with the FRF of the linear system corresponding to the non-linear system. A series of properties of MFRF are also discussed.

3. An advantage of the parameter estimation method of MFRF is its small computation cost. The statistically equivalent method and the validity check methods proposed in this paper can be used to get the statistically equivalent polynomial system of a non-polynomial non-linear system, which may generalize the MFRF theory of the polynomial non-linear system in practice.

4. Based on the 3rd-order statistically equivalent polynomial, the ETF defined in the paper is of clear physical meaning and convenient to apply in practice.

5. There is no experience of MFRF or ETF is applied to the detection, location and assessment of structural damage.

References

[1] Rugh, W.J., Nonlinear System Theory — the Volterra/Wiener Approach, 1981.

[2] Gifford, S.J. ,G.R.Tomlinson, Recent Advances in the Application of Functional Series to Non-linear Structures, J. of Sound and Vibration (1989) 135 (2), 289–317.

[3] Zhang,J.H., Y.Qiu, Generalized Spectral Analysis for Non-linear Systems, in Chinese, Structure & Environment, No. 3,1992,pp1-10.

[4] Billings, S.A. and K.M.Tsang, Spectral Analysis for Non-linear System, Part 1 : Parametric Non-linear Spectral Analysis, Mechanical Systems and Signal Processing (1989) 3(4) 319–339. Part 2: Interpretation of Non-linear Frequency Response Functions, Mechanical Systems and Signal Processing (1989)3(4) 341–359. Part 3: Case Study Examples, Mechanical Systems and Signal Procession (1990)4(1),3–21.

[5] Natke, H.G. ,and M.Zamirowski, On methods of structure identification for the class of polynomials within mechanical systems, ZAMM. Z.Angew, Math. Mech. 70(1990)10, 415–420.

[6] Zhang,X.N., J.H.Zhang, MFRF of Polynomial Non-linear Systems, In Chinese, Acta Aeronautica Et Astronautica Sinica, will be published.

[7] Bendat J.S., Non-linear System Analysis and Identification from Random Data, JOHN WILEY & SONS, 1990.

Damage Assessment from the Dynamic Response of Deteriorating Structures

F. Brancaleoni
Universitá "G. D'Annunzio"
Pescara, Italy

D. Spina
Universitá "La Sapienza"
Roma, Italy

C. Valente
Universitá "G. D'Annunzio"
Pescara, Italy

1 Introduction

The demand for the residual life determination of a structure is a topic growing in importance, due to the high rehabilitation costs in the case of late damage detection. To this end condition monitoring techniques reveal an essential tool for planning repair or strenghtening and, as long as non destructive evaluation (NDE) methods are applicable, they are usually prefereable with respect to those requiring more or less extensive sampling. Among the NDE methods special attention seems to be paid to dynamic testing, because of its execution speed and capability to work with ambient vibrations [1],[2]. The paper subject pertains to this latter field and aims to present a monitoring technique devoted to provide a synthetic index of the structural state, whose value, if critical, must address towards more refined and oriented investigations. Within this context, it has found advantageous not to try to identify the remaining load carrying capacity directly, but to split the problem into two stages. Initially the dynamic behaviour of the structure is exploited and global damage symptoms are introduced in order to identify the damage effects at least in an overall sense; subsequently further problem dependent quantities, responsible for the damage origin and growth, are derived from the damage indicators and used for safety assessment purposes. Since this latter stage corresponds to a conventional design process, the critical point in the safety assessment problem is reduced to the identification of the dynamic properties of the structure. To this end, signal analysis procedures able to deal directly with the dynamic responses arising from almost arbitrary nonconservative, nonlinear and time - variant systems are used. A time domain approach is followed in either case and the Hilbert transform or the Wigner distribution is used depending whether the system allows for the uncoupling of the vibrating contributions or not. Accuracy and reliablity of the two approaches are discussed by the help of some numerical simulations, whereas their effectiveness is tested against an actual case study.

2 Signal Analysis Procedures

In a signal analysis based identification procedure the primary damage symptom to be detected is the frequency change in the recorded motion. This section provides therefore the general theoretical framework of two suitable signal processing techniques whose outcomes will be used later for identification purposes. A time domain approach is followed in either case and the only dynamic system response is analysed for identification purposes.

2.1 Time domain Hilbert transform

The Hilbert transform (HT) is a linear operator that does not move the definition domain of the transformed function [4], therefore it allows for the dual processing of a signal either in the frequency domain or in the time domain [3]; nevertheless the latter case looks more promising [3] and will be therefore followed hereafter.

The HT specialised to time signals reads:

$$\tilde{x}(t) = \mathcal{H}x(t) = \int_{-\infty}^{+\infty} d\tau \frac{x(\tau)}{\pi(t-\tau)} = \frac{1}{\pi t} * x(t) \tag{1}$$

Yet, the physical meaning of the $\mathcal{H}$ operator is better appreciated if one replaces integral with algebraic operations by of a couple of Fourier transforms ($\mathcal{F}$), so that (1) reads:

$$\tilde{x}(t) = \mathcal{F}^{-1}\mathcal{H}\mathcal{F}x(t) = \mathcal{F}^{-1}\mathcal{H}X(\omega) = \mathcal{F}^{-1}\tilde{X}(\omega) \tag{2}$$

$$\tilde{X}(\omega) = \mid X(\omega) \mid e^{-i[\Theta(\omega)+\Phi(\omega)]} \tag{3}$$

$$\Phi(\omega^-) = +\pi/2; \quad \Phi(\omega^+) = -\pi/2 \tag{4}$$

Eqns. (2),(3) and (4) prove that the mechanical equivalent of the mathematical action of the $\mathcal{H}$ operator corresponds to a filter action on the original signal that leaves the magnitude $\mid X(\omega) \mid$ unchanged but shifts the phase $\Theta(\omega)$ of each frequency components by $\pm\pi/2$ depending whether they are negative or positive frequencies. This latter result in conjunction with the modulation property of the HT [4] can be rephrased by stating that when dealing with single component modulated waveforms:

$$x(t) = A(t)\sin\theta(t) \tag{5}$$

the transformed signal results a quadrature version of the original signal, provided that $A(t)$ is a smooth, slowly varying function of time [4]:

$$\tilde{x}(t) = A(t)\cos\theta(t) \tag{6}$$

The combined use of (5) and (6) allows to find out a very simple expression for the computation of the envelope function $A(t)$ (that is the energy distribution in time) and of the sought frequency content via the instantaneous phase $\theta(t)$:

$$A(t) = \sqrt{x^2(t) + \tilde{x}^2(t)} \tag{7}$$

$$\theta(t) = \tan^{-1}[x(t)/\tilde{x}(t)] \tag{8}$$

On the contrary, when multiple component signals are concerned, the simple expression (6) cannot be recovered and more sophisticated analysis tools must be substituted for the HT.

2.2 Time domain Wigner distribution

Unlike the HT, bilinear time - frequency representations (BTFR) combine time and frequency domain analyses such that both temporal and spectral characteristics (i.e. joint energy - frequency distribution vs. time) of the processed signal can be simultaneously displayed. Therefore, despite the number of components, it reveals a simple matter to analyse how the signal energy divides over each frequency.

From a general viewpoint a time domain BTFR can be given in the form [7]:

$$T_x(t,\omega) = \int_{-\infty}^{+\infty} dp_1 \int_{-\infty}^{+\infty} dp_2 \Big[K_T(t,\omega,p_1,p_2)\mathcal{X}(t,p_1)\Big] \tag{9}$$

where p_1 and p_2 are, for the moment, two unspecified integration variables. Each class of the general BTFR (9) is then generated by the particular choice of the signal to be processed (that is the signal itself, or some its preprocessing); in the present case it has been selected the following "correlation" form [7]:

$$\mathcal{X}(t,\tau) = a\Big[t + (\frac{1}{2} + \alpha)\tau\Big]a^*\Big[t - (\frac{1}{2} - \alpha)\tau\Big] \quad \text{where} \quad p_1 = \tau \tag{10}$$

where $*$ stands for conjugate and $a(t)$ is the analytic signal given by [4]:

$$a(t) = x(t) + i\tilde{x}(t) \tag{11}$$

Each member of the selected class (10) is, in turn, generated by the particular choice of the transformation kernel K_T whose form defines also the properties of that particular representation. In order to match some important physical requirements it is convenient to select K_T in such a way that the marginal property (12) and the shift invariance property (13):

$$\int\int_{-\infty}^{+\infty} dt d\omega T_x(t,\omega) = E_x \tag{12}$$

$$T_y(t,\omega) = T_x(t - t_0, \omega - \omega_0) \quad \text{with} \quad y(t) = x(t - t_0)e^{i\omega_0 t} \tag{13}$$

turn out to be satisfied. In fact, properties (12) and (13) can be tought as the minimum set of requirements in order that T_x provides an "undistorted" representation of the information embedded into $x(t)$. In particular, property (12), stating that the total energy content E_x of $x(t)$ must be conserved, allows for interpreting T_x as an "undistorted" energy distribution over the (t,ω) plane (i.e. energy is neither fed nor removed but possibly only spread by T_x); whereas property (13), stating that a time or a frequency shift in $x(t)$ will result in an equal shift in T_x, allows for interpreting the (t,ω) plane as an "undistorted" domain for T_x (i.e. no scaling of the adopted measure for t and ω due to T_x).

Among the possible choices for K_T that satisfy simultaneously (12) and (13) the simplest form is (δ = Dirac delta function):

$$K_T(t,\omega,\tau,p_2) = \delta(t - p_2)e^{-i\omega\tau} \quad \text{where} \quad p_2 = \text{dummy var.} \tag{14}$$

so that, inserting (10) and (14) in (9), one gets the specialized form ($\alpha = 0$):

$$\mathcal{W}_x(t,\omega) = \int_{-\infty}^{+\infty} d\tau \Big[a(t + \frac{\tau}{2})a^*(t - \frac{\tau}{2})\Big]e^{-i\omega\tau} \tag{15}$$

that is known as the Wigner Distribution (WD) of $x(t)$ [9].

In the case of single component modulated waveforms (5) the WD performs as well as the HT and the resulting representation reads:

$$\mathcal{W}_x(t,\omega) = 2\pi A(t)^2 \delta(\omega - \omega_d) \tag{16}$$

that shows how the WD concentrates all the signal energy around the sought frequency ω_d, see eqn. (22). But, far more important, the same feature shown by (16) is maintained also in the case of multiple component signals allowing for the isolation of each single vibrating contribution in $x(t)$. However resolution problems can arise in the case of close frequencies due to the cross-terms introduced by the bilinearity of the WD [9]. This drawback can easily be shown if the WD is applied to the superposition of two single component signals $x_1(t)$ and $x_2(t)$:

$$\mathcal{W}_{x_1+x_2}(t,\omega) = \mathcal{W}_{x_1}(t,\omega) + \mathcal{W}_{x_2}(t,\omega) + Re[\mathcal{W}_{x_1 x_2}(t,\omega)] \tag{17}$$

where the third term on the right hand side accounts for the "spurious" cross-terms appearance.

2.3 Computational aspects

When performing numerical computations the following aspects must be taken into account: (i) truncation errors (weighting with a time window function); (ii) discretization (sampling) and (iii) noise.

These topics have been studied in [6] for the HT and in [9] for the WD, in conjunction with linear systems. For a detailed discussion the interested reader is referred to the above references; hereafter it is worth to underline that problems (i) and (ii) can be neglected for practical purposes, whereas the identification process shows insensitive to problem (iii) if zero mean white noise is added to the signal.

3 Identification Method

The identification method relies on the basic assumption to split the safety assessment problem into two simpler and sequential problems:

- limit the monitoring technique to estimate only those global parameters (first lower mode shapes, frequencies and related stiffnesses) that act as damage indicators in detecting, quantifying and possibly localizing the damage;

- establish correlation tables to deduce from the damage indicators the related quantities responsible of damage growth (prestressing force, statical redundant forces redistribution, etc.) that act as safety indicators in measuring the reduced distance of the stress state from the limit safe surface.

The above splitting reveals fundamental in reducing the complexity of the posed problem. In fact the correlation tables can be constructed once and for all, according to the given structural problem, by means of conventional design methods. The safety assessment problem can then be considered as a separate and specific appendix of the general problem whose solution therefore depends on a proper damage identification, item to which the attention will be focused in the following.

When analysing the response signal of nonstationary systems, the primary quantity to be displayed is the time varying frequency content of the recorded motion. Consequently a joint time - frequency representation must be provided and, according to the signal processing tools previously introduced, two identification paths will be discussed:

- HT processing whenever some filtering process is applicable;
- WD processing if vibrations coupling must be retained.

3.1 System identification via single component signals

A problem oriented filtering technique of the model updating type will be discussed later in connection with the case study. For the present purposes, it is assumed that a set of single component signals has been obtained, via some available method that allows for uncoupling, and that the unknown constitutive function of the equivalent set of single degree of freedom systems to be identified is initially given in the implicit form $f(\dot{x}, x)$.

However it has found advantageous to reduce the problem to a separate identification of the dissipating $f_d(\dot{x})$ and restoring mechanisms $f_r(x)$ that contribute to $f(\dot{x}, x)$. The assumed splitting is suggested by the observation that the modulating $A(t)$ and vibrating $\theta(t)$ features of the response signal are respectively connected to the dissipating and restoring mechanisms of the system. Since the former can be computed by the separate eqns. (7) and (8), the same separation applies to the constitutive system function that is therefore reduced to the split form:

$$f(\dot{x}, x) = f_d(\dot{x}) + f_r(x) = C_{eq}\dot{x} + f_r(x) \tag{18}$$

where it is understood that in the case of nonseparable nonlinearities (e.g. hysteretic systems) the uncoupling is forced by the introduction of an equivalent viscous damping mechanism C_{eq} capable of supplying the same amount of the actual dissipated energy inside the same processed time window.

In view of eqn. (18), the identification procedure develops as follows: first the dissipating $f_d(\dot{x})$ and restoring $f_r(x)$ mechanisms are detected by the analysis of the decaying (for stable systems) rate of $A(t)$, eqn. (7), and of the time varying frequency content $\theta(t)$, eqn. (8); then, once an explicit functional form for the constitutive models $f_d(\dot{x})$ and $f_r(x)$ is known, a quantitative identification is carried out via a parameter estimate of the coefficients of the two functions.

In the case of linear systems both the signal and system constitutive functions $A(t)$, $\theta(t)$ in (1) and $f_d(\dot{x})$, $f_r(x)$ in (17) are known in explicit form:

$$A(t) = A_0 e^{-t/\tau} \qquad \theta(t) = \omega_d t + \theta_0 \tag{19}$$

$$f_d(\dot{x}) = c\dot{x} \qquad f_r(x) = kx \tag{20}$$

where $\tau = 2m/c$ is the system time constant, ω_d the damped angular frequency and k the undamped stiffness. It reveals then immediate to estimate the concerned coefficients by the use of (7) and (8) within the expressions (19) and (20):

$$c = -2m\left[\frac{d}{dt}logA(t)\right] \tag{21}$$

$$\omega_d = \frac{d}{dt}\theta(t) \tag{22}$$

from which it is seen that c is given by the slope of the straight line $A(t)$ when plotted in a semilog scale, whereas ω_d is given by the intercept on the $\dot{\theta}(t)$ axis when plotted in linear scale [6]. Note also that a clear representation of the constitutive relation can be achieved through the following relation:

$$k = \frac{f_r[x(t)]}{x(t)} = m[\omega_d^2 + \tau^{-2}] \tag{23}$$

provided that the mass m is a known constant parameter [3].

In the case of nonstationary systems the signal and system constitutive functions are not known in advance, further and more important the modulation property does not hold so that (6) is no longer theoretically valid. Nevertheless, due to its simplicity, one can still retain the same procedure available for linear systems forcing the validity of (6). The envelope function $A(t)$ and the instantaneous phase $\theta(t)$ can then still be used through (7) and (8) that are, in turn, no longer identically satisfied and therefore cause some degree of distortion in the computed functions; yet it has been found [5] that the errors introduced by the above assumption cause only local perturbations having oscillatory character whose wavelength agrees with the system instantaneous frequency.

Consequently, the identification process can still be based on the analysis of both the envelope and phase functions and, since their representations are typical for each type of nonconservative or nonlinear mechanism, it is possible to find out: (i) the actual dissipating force f_d, by analysing the $A(t)$ graph; (ii) the time varying frequency content of the system by analysing the changes in the time derivative of $\theta(t)$ and (iii) the actual restoring force looking at f_r in (23) as a function of $A(t)$ instead of $x(t)$:

$$k(t) = \frac{f_r[x(t)]}{x(t)} = m\dot{\theta}(t) - \left[C_{eq}\frac{\dot{A}(t)}{A(t)} + m\frac{\ddot{A}(t)}{A(t)}\right] \tag{24}$$

In order to minimize the errors in (24), now not negligible, $k(t)$ is computed at the discrete set of time instants in which $\tilde{x}(t)$ vanishes (i.e. $\theta(t) = 0$), assumption that allows also for the assumed substitution $A(t) = x(t)$. This has been found not to be sufficient [3]. and the correction terms in square brackets account for the violation of the identities (7) and (8). Note further that the same correction terms apply also in the case of a linear system if the undamped stiffness is required. In that case they reduce to:

$$\frac{\dot{A}(t)}{A(t)} = -\frac{c}{2m} \quad \text{and} \quad \frac{\ddot{A}(t)}{A(t)} = \frac{c^2}{4m^2} \tag{25}$$

so that equation (23) is recovered [5].

3.2 System identification via multiple component signals

In those cases in which coupling of vibrating contributions must be retained the system as a whole should be considered and the above HT based identification procedure results no longer available and must be substituted by the simultaneous processing of all the components of the signal.

The nonlinearity detection reveals immediate due the following WD property [9]:

$$\omega_d(t) = \frac{\int_{\omega_1}^{\omega_2} \omega \mathcal{W}(t,\omega)d\omega}{\int_{\omega_1}^{\omega_2} \mathcal{W}(t,\omega)d\omega} \tag{26}$$

stating that the centroid of a time section of the WD graph is just the value attained, at the same time instant, by the sought instantaneous frequency $\omega_d(t)$, provided that the signal components are sufficently separated. Therefore, since the $\omega_d(t)$ curve proves typical for each nonlinearity type, it is a simple matter to qualify the restoring mechanism. Yet, unlike the HT case, since no "modal" mass can be defined, only an indirect quantitative assessment can be attempted by constructing reference tables that correlate the distorsion degree of a given family of $\omega_d(t)$ curves to some measure of the corresponding nonlinearity level. Moreover, misleading interpretations of the dissipating mechanism can be induced by possible amplitude decay alteration due to the energy transfer between the excited frequencies.

3.3 Numerical simulations

For illustrative purposes, the HT and WD based identification procedures discussed above, are hereafter applied to the response signals of nonstationary systems defined by the superpostion of linear damping forces $f_d = c\dot{x}$ plus nonlinear or time-variant restoring forces f_r. The following numerical simulations have been considered:

Example 1: single component signals - free vibrations - f_r: nonlinear or time variant

The selected test cases have a twofold aim: (i) to compare the resolution accuracy of either the HT or WD operator and (ii) to display possible aliasing effects if very different restoring mechanisms that give rise to similar global system behaviour are to be identified. To this end, a nonlinear and a time variant system have been simulated, see fig. **1**, and the same global behaviour has been obtained by assigning the same values attained at the initial and final recording time by the secant stiffness of the nonlinear system to the initial and residual stiffness of the time variant system. The identified frequency content, computed according to (22) and (26), is shown in fig. **1**. The comparison between the two cases allows to state that almost coincident results are obtained via either the HT or the WD operator. On the other hand, the results comparison within either case shows a strong "aliasing" effect; in fact the global frequency decrease towards asymptotic values would lead to conclude that a nonlinear hardening restoring force would be involved in both cases. Yet, a local inspection of the relevant graphs allows to distinguish between the two systems; in fact the revealing feature is represented by the perturbations absence in the time variant case, as indeed the case for an inherently linear, even if nonstationary, system.

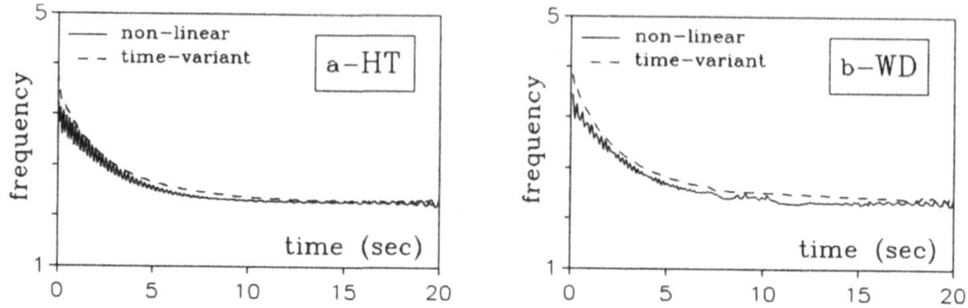

non-linear system : $f_r = k_0 x(1+\alpha x^2)$

time-variant system : $f_r = kx$; $k = k_0(1-\beta D)^\gamma$; $D = E_{diss}/E_{ref}$

Figure 1 System identification via single component response signals

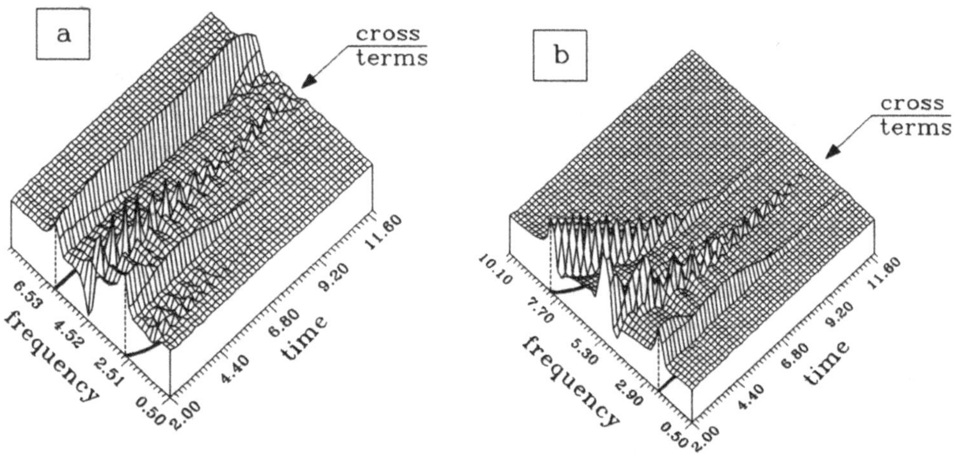

$f_r = k_0 x(1+\alpha x^2)$ $\qquad$ $f_{r1} = k_1 x$; $\quad f_{r2} = k_2 x(1+\alpha x^2)$

Figure 2 System identification via double component response signals

Example 2: double component signals - free or forced response - f_r: nonlinear
When analysing multiple component signals the WD only reveals applicable. Two typical cases have been investigated: (i) the forced response of a nonlinear single degree of freedom system and (ii) the free vibrations response of a double degree of freedom system having a nonlinear restoring term. The results are plotted according to the 3D representations of fig. **2** that show how the signal frequencies are isolated despite the cross-terms presence. It is then possible to detect the transient component (2 - 3 Hz) from the steady (harmonic) component (6 Hz), fig. **2a**, and also to detect the linear restoring mechanisms from the nonlinear one by the analysis of the frequency changes, fig. **2b**.

4 Test Case

The general formulation of the identification method previously presented is hereafter specialized to a case study aiming to illustrate how a damage severity assessment can be achieved in an actual structure. A viaduct with the statical scheme of a simply supported grillage was selected as a representative sample of a wide class of bridge structures. The monitoring was focused on the bridge deck whose transversal section is shown in fig. 3, together with the main features of the acquisition system and the probes arrangement. The measurement stations were strategically placed according to the peaks and nodes of the theoretical undamaged longitudinal beam mode shapes in order to give prominence both to the first lower vibration modes and to the support (neoprene pads) fixities.

The dynamic tests were performed by means of a truck running on the deck. The symmetric or antisymmetric (bending or torsional) deck vibration modes were alternatively excited by varying the running lanes from central to lateral, besides a step was alternated from midspan to the quarters to provide bumping action.

A typical recorded acceleration time history is plotted in fig. 4.

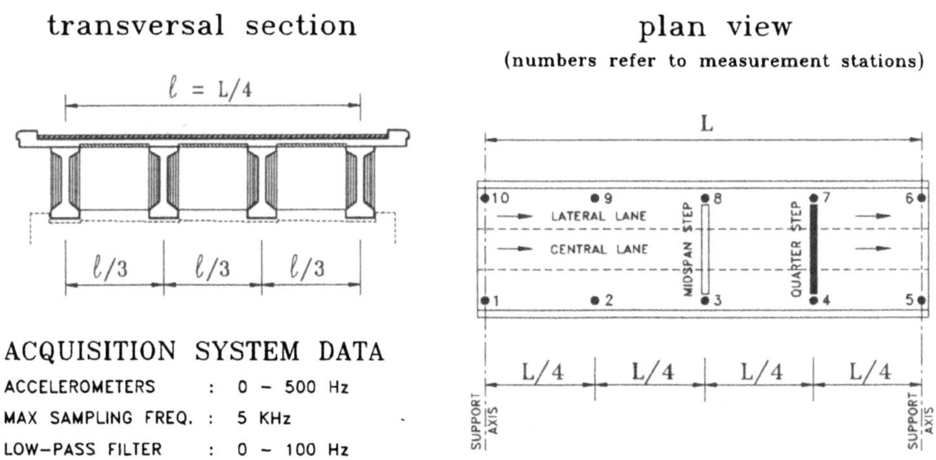

Figure 3 bridge deck and acquisition system layout

4.1 Identification via single component signals

Since the accelerations time histories present multiple components, see fig. 4, the HT based identification procedure must be preceeded by some form of modal filtering capable to produce out a new set of "uncoupled" single component signals. To this end, it has been laid down a problem oriented model updating technique allowing for a quick evaluation of the modal filter.

The recorded deck motion $\mathbf{w}(t)$ is approximated by the following discrete series

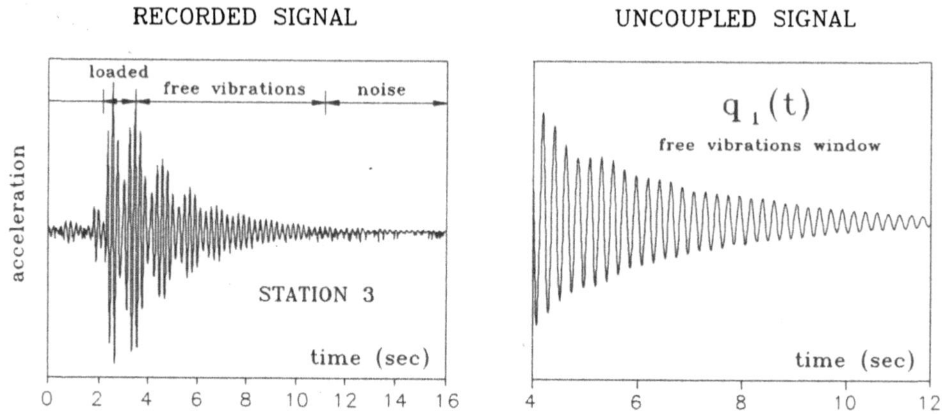

Figure 4 Typical recorded and uncoupled signals

written for the n_1 measurement stations and for the n_2 interested vibration modes u_{ij} scaled by the factors $q_j(t)$ (lagrangian coordinates):

$$w_i(t) = u_{ij}q_j(t) \quad (j = 1, 2, ...n_2) \quad (i = 1, 2, ...n_1) \tag{27}$$

Assuming $n_1 = n_2$, the sought new set of "uncoupled" signals to be processed is given by the inverse form of (27) that, in square matrix form, reads:

$$\mathbf{q}(t) = \mathbf{U}^{-1}\mathbf{w}(t) \tag{28}$$

From (28) it is apparent how the filter action is performed by the matrix **U** that collects the mode shapes of the deck. Therefore the true problem to be tackled in solving (28) is to find an expression for **U**. An iterative procedure has then been established according to the following steps:
- define the column $\mathbf{u}_j$ of **U** according to the conjectured damage;
- compute $\mathbf{q}(t)$ from (28);
- check if single frequency signals have been obtained;
- if not adjust **U** and repeat the above steps.

The mode shapes $\mathbf{u}_j$ are computed by means of a simplified analytical model of the deck developed according to the following assumptions: (i) damage uniformly distributed in the longitudinal beams and (ii) uncoupled behaviour between longitudinal and transversal deck modes. The former assumptions allow for the substitution of the tridimensional problem by the sequence of the two plane problems synthetized in fig. 5. The longitudinal model is used first to compute, for each k-th beam mode shape, the modal masses and stiffnesses whose values depend, in turn, on the possibly damaged material and sectional properties of the longitudinal beams. The computed values are then inserted into the transversal model and a sequence of eigenvalue problems is solved for k ranging from 1 to $n_2/2$; note in fact that each k-th mode of the longitudinal model gives rise to a pair of modes of the transversal model (k-th bending and torsional deck modes). The matrix **U** is then constructed

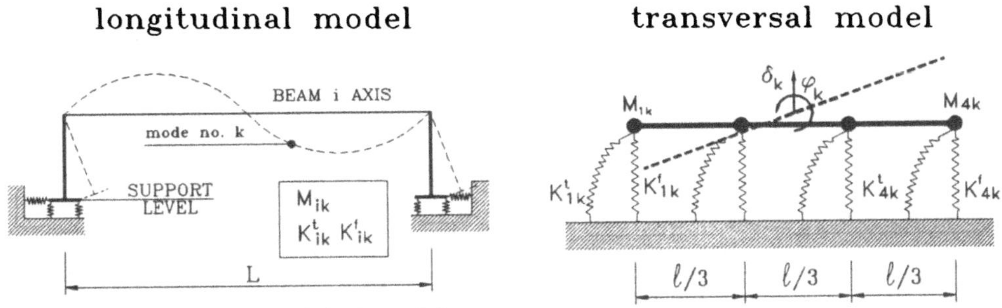

Figure 5 Analytical deck models

by column pairs by expanding each eigenvector pair $[\delta_k, \phi_k]_{1,2}$, first transversally according to the beam location and then longitudinally according to the k-th mode shape of the longitudinal model in order to match the i-th accelerometer location and hence the $\mathbf{u}_j$ components.

The initial temptative modal filter matrix $\mathbf{U}_o$ (no damage assumed) is listed in table 1 together with the final matrix $\mathbf{U}_d$ that realizes the best uncoupling features according to the supposed damage distribution that is 10% stiffness loss in only one lateral beam the others remaining undamaged. The modal filter action is shown in fig. 4 where only the $\mathbf{q}_1(t)$ signal, concerned to the first bending deck mode and windowed in the free oscillations time interval is plotted.

Table 1 – Modal filter matrices

$\mathbf{U}_o$

0.354	0.354	0.500	0.500	0.354	0.354
0.500	0.500	0.000	0.000	−0.500	−0.500
0.354	0.354	−0.500	−0.500	0.354	0.354
0.354	−0.354	0.500	−0.500	0.354	−0.354
0.500	−0.500	0.000	0.000	−0.500	0.500
0.354	−0.354	−0.500	0.500	0.354	−0.354

$\mathbf{U}_d$

0.405	0.322	0.653	0.379	0.482	0.235
0.573	0.455	0.000	0.000	−0.681	−0.332
0.405	0.322	−0.653	−0.379	0.482	0.235
0.293	−0.383	0.271	−0.590	0.134	−0.442
0.414	−0.542	0.000	0.000	−0.190	0.624
0.293	−0.383	−0.271	0.590	0.134	−0.442

Once the signal uncoupling has been obtained the HT based identification procedure can be started and carried out mode by mode. The frequency distribution in time, corresponding to the shown $\mathbf{q}_1(t)$ is given in fig. 6 together with the estimated value f_{b1} that reveals constant in time.

4.2 Identification via multiple component signals

The WD, as opposed to the HT, allows for the direct processing of the recorded accelerations regardless the forced components present into the analysed signals. A sample result, referred to the signal recorded at the measurement station no. 3, see

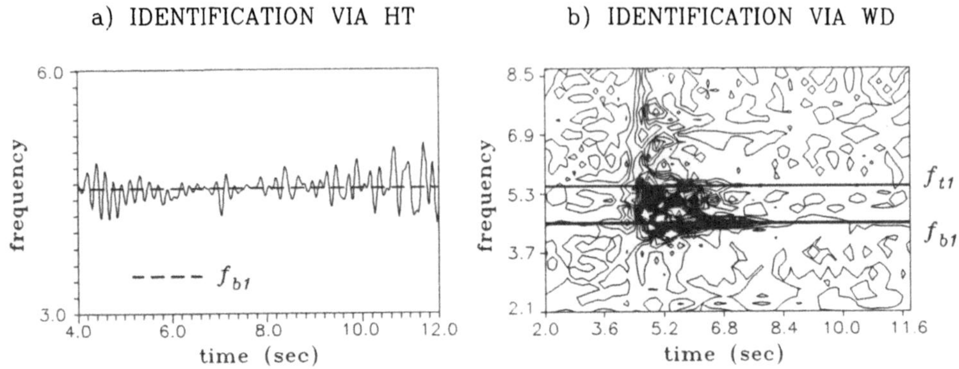

Figure 6 Frequency identification via HT or WD

fig. 4, is given in fig. 6 in terms of contour lines plot that allow to estimate the first bending f_{b1} and torsional f_{t1} deck frequencies.

Finally, for comparison purposes, the estimated and averaged (over the samples number) frequency values, as derived from the two identification approaches, are listed in fig. 7. Note that the shown results are limited to the two first bending f_{b1} and torsional f_{t1} mode shapes, and that no damping can be estimated by the WD.

4.3 Damage Assessment

In those cases in which the modal filter reveals applicable one could suppose that the knowledge of **U** alone would be sufficent to identify the damage. Unfortunately the output of the modal filter shows rather insensitive to small perturbations to $\mathbf{u}_j$, in fact even a raw estimate of $\mathbf{u}_j$ often reveals able to realize the modal uncoupling [10]. The matrix **U** remains thus foundamental for damage detection and localization, but must be integrated by the knowledge of the frequency changes for a quantitative assessment. To this end, the theoretical ratio f_{b1}/f_{t1}, that relates the frequency changes to the damage progress, has been derived by the simplified analytical models and has been assumed as reference, beacuse it does not depend on the geometric and material properties in the case of uniform damage. The computed curve is plotted in fig. 7 a from which a 30% reduction of the lateral beam stiffness, that overestimates the modal filter result, is found by inserting the estimated f_{b1}/f_{t1} ratio in the analytical curve.

4.4 Safety Assessment

Once the values of the damage indicators have been estimated, the second and final step to perform is to find out the corresponding values reached by the safety indicators. In the present case the key parameter in defining the reduced load carrying

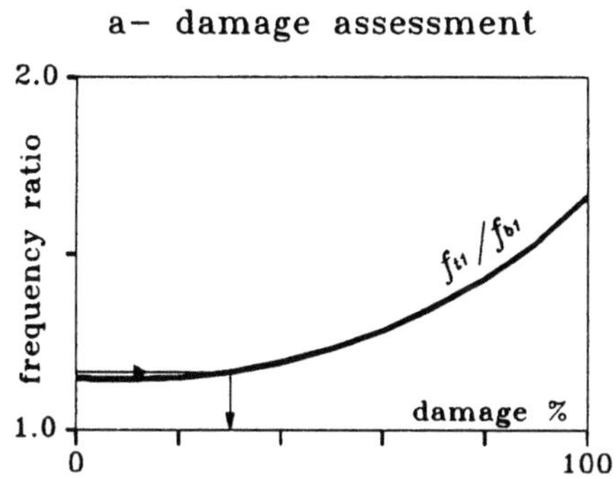

Estimated average values

	HT	WD
f_{b1}	4.54 Hz	4.51 Hz
f_{t1}	5.44 Hz	5.25 Hz

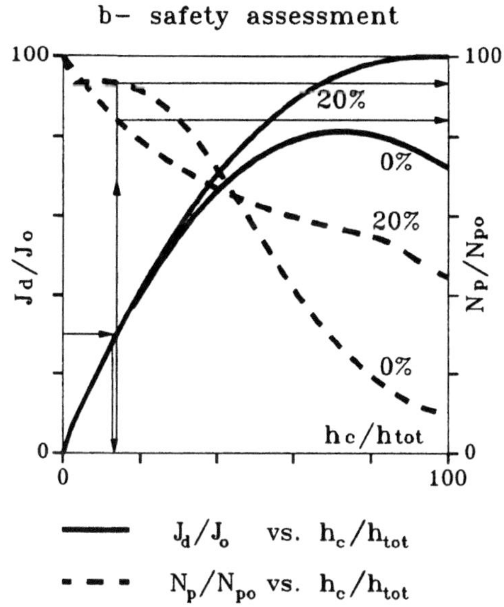

Figure 7 Damage and safety assessment

capacity of the deck is the residual prestressing force N_p in the longitudinal beams that can be estimated by means of conventional design methods provided that the reduced inertia moment J_d has been derived by the damage assessment.

Using the above estimated value for the residual stiffness of the lateral beam (30% loss), it has been computed the corresponding crack height $h_c = 12$ cm the prestressing force compatible with the crack picture; The 7% reduction of the N_p theoretical undamaged value has been found.

Anyway, in order to achieve some generality, it can be useful to discuss how typical correlation tables can be constructed and used as a first safety ascertainment. If it is assumed, as indeed the case for traffic induced vibrations, that the dynamic increment M_{dyn} is an almost constant fraction of the dead load bending moment M_o (say 20%) it is possible to relate J_d to the crack height h_c and then this latter quantity to the N_p necessary to balance $M_{dyn} + M_o$. For illustrative purposes, two temptative correlations have been plotted in fig. 7b where the concerned quantities are expressed in nondimensional form as a percent ratio of the initial undamaged value. The reported results refer to a T shaped beam and cover the practical application range according to two limit situations of the added steel reinforcement (0% and 20% of the gross cross section). The safety assessment can then be immediately achieved if one inserts the estimated value of the damage indicator J_d, see fig. 7a), into the $J_d/J_o - h_c/h_{tot}$ curve that returns a crack height $h_c = 15$ cm. If the h_c value is, in turn, inserted into the $h_c/h_{tot} - N_p/N_{po}$ curve, a a 10-15% loss for N_p is found, depending on the reinforcement amount, The comparison between the N_p reduction found by the help of fig. 7b and the "correct" result previously computed by means of the actual (recorded) values allows to conclude for the conservativeness of the introduced correlations.

Finally, it is worth to stress that the proposed procedure relies on the hypòtesis of uniform distributed damage in the longitudinal deck beams; therefore the estimated values for J_d, as well as that for h_c, key parameters for the computation of the safety indicator N_p, must be understood as equivalent and spread measures of the actual beam bending stiffness and cracking distribution that substitute the complex interaction between cracked and uncracked zones along the beam lenght; measures that are herein used for a quick, yet reliable safety assessment.

5 Conclusions

It has been presented an identification procedure to be used in a condition monitoring strategy for the assessment of the remaining load carrying capacity in progressively deteriorating structures. The posed problem has been solved by means of two subsequent analysis stages: the first pertains to the identification procedure itself and it is devoted to the estimate of the global damage effects by means of proper damage indicators; the second aims to link the global damage effects to the quantities responsible for the damage growth that act as safety indicators. It has shown that this latter stage can be simply accomplished using conventional design methods, whereas the damage detection and quantification reveals a critical step depending on the identification technique adopted. Due to the nonstationarity of the considered systems, the proposed identification procedure is centred on the analysis of the features displayed by a suitable joint time - frequency representations of the system response signals. To this end the theoretical framework of two signal processing methods, devoted to the identification of nearly arbitrary nonlinear and time variant systems, has been presented: the time domain Hilbert transform (HT) is used whenever some filtering process is applicable, on the contrary if vibrations coupling must be retained the Wigner distribution (WD) is substituted for it.

Finally, a sample case study pertaining to a wide class of bridge structures has been discussed in detail in order to check the effectiveness of the proposed identification procedure when applied to an actual structure. Both the HT and WD based representations have been considered and similar results have been found for the damage indicators, yet it has been recognised that better insight into the structural behaviour can be gained whenever a preliminar "modal" filter is applicable. Temptative correlations between the damage and safety indicators have also been provided aiming to illustrate how a safety assessment can be achieved.

As a concluding remark, it is worth to stress that the primary target of the proposed monitoring technique is to provide a synthetic index of the structural state. Therefore the damage, as herein identified, must be understood as an equivalent measure of the actual structural damage and if a detailed damage picture is required then specific and problem oriented methods must follow the one presented.

References

[1] Bridge Management, Eds. J.E. Harding, G.A.R. Parke and M.J. Ryall, 1st International Conference on Bridge Management, Univ. Surrey, Guildford, UK, Elsevier Applied Science, London & New York, March 1990.

[2] Bridge Rehabilitation, Eds. G. Konig, A.S. Nowak and B. Novsk, Proc. 3rd International Workshop on Bridge Rehabilitation, Univ. Darmstadt, Verlag fur Architektur und Technische Wissenschaften, Berlin, June 1992.

[3] M.P. Petrangeli, D. Spina, C. Valente, On the Use of the Hilbert Transform for the Identification of the Dynamic Response of Nonlinear Systems (in Italian), Research Report, DISG Dept. Univ. of Rome 'La Sapienza', March 1990.

[4] J.S. Bendat, The Hilbert Transform and Applications to Correlation Measurement, Bruèl and Kjaer Publications, 1985.

[5] F. Brancaleoni, D. Spina, C. Valente, A Free Oscillation Based Technique for the Identification of Nonlinear Dynamic Systems, 13th IMACS, Dublin, Ireland, Elseveir, July 1991.

[6] A. Agneni, L. Balis-Crema, Damping Measurement from Truncated Signals via Hilbert Transform, Jnl. of Mechanical System and Signal Proc., Vol. 3(1), 1989.

[7] F. Hlawatsch, Duality and Classification of Bilinear Time-Frequency Signal Representations, IEEE Transact. on Signal Proc., 39, 7, pp. 1564-1574, 1991.

[8] F. Brancaleoni, D. Spina, C. Valente, Analysis of the Nonstationary Structural Response in the Joint Time-Frequency Domain, (in Italian), XI AIMETA National Congress, Trento, Italy, October 1992.

[9] T.A.C.M. Claasen and W.F.G. Mecklenbrauker, The Wigner Distribution - A Tool for Time-Frequency Analysis, Part I-II-III, Philips J. Res., vol. 35, pp. 217-250, 276-300, 372-389, 1980.

[10] M.P. Petrangeli, D. Spina, C. Valente, The Nonlinear Hilbert Transform Technique for the Identification of the Dynamic Behaviour of Bridge Decks, Proc. 3rd International Workshop on Bridge Rehabilitation, Univ. Darmstadt, Verlag fur Architektur und Technische Wissenschaften, Berlin, June 1992.

Some Issues in the Identification of Structural Systems Through the Use of Neural Networks

S.F. Masri
University of Southern California
Department of Civil Engineering
Los Angeles, California, USA

A.G. Chassiakos
California State University
Long Beach, California, USA

1 INTRODUCTION

The identification of mathematical models of physical structures on the basis of experimental measurements is a problem that has been receiving increasing attention in the recent past. The main impetus for this phenomenon is the central role that appropriate mathematical models have in active structural control applications. For example, the success of precision active control of large space structures is based on the availability of accurate mathematical models of those structures. Since the size, nature, and configuration of these large, flexible structures preclude their identification on the ground, future space missions have an essential need for efficient on-orbit system identification procedures.

Similarly, the potential for using active control approaches to reduce the response of large *civil* structures under arbitrary dynamic environments, such as earthquakes or wind, has been drawing a considerable amount of interest worldwide. In fact, the first "U.S. National Workshop on Structural Control Research" (Housner and Masri, 1990) was convened to address the research needs of this rapidly emerging field. Among the key research topics identified by the attendees as needing much more investigation was the development of system identification approaches that can cope with the challenging nature of physical structures encountered in the structural mechanics field.

A related area in which there is also a rapidly increasing need for accurate mathematical models of large systems is the emerging field of "health monitoring" of structures (civil as well as aerospace types). Techniques are being developed for damage assessment of large structures on the basis of system identification of high-fidelity mathematical models (Natke and Yao, 1988).

Even though there are many studies dealing with all computational aspects of system identification, there is a paucity of publications in the open literature that address the use of parallel distributed processing methodologies (artificial neural networks) for identifying physical systems encountered in the applied mechanics field. It is this subject that the present paper explores.

2 NEURAL NETWORK BASED IDENTIFICATION

2.1 Artificial Neural Networks

An *Artificial Neural Network* (ANN) is a system with inputs and outputs, composed of a number of similar nonlinear processing elements. These processing elements operate in parallel and are arranged in patterns similar to the patterns found in biological neural nets. The processing elements or *nodes* are connected to each other by adjustable *weights* (Nguyen and Widrow, 1990; Lipmann, 1987). Changing these weights will change the input/output behavior of the network, hence the following is a natural goal for such a system: *Choose the weights of the net in such a way as to achieve a desired input/output relationship.* To achieve this goal, systematic ways of adjusting the weights have to be developed, which are named by the general term *training* or *learning* algorithms.

Artificial neural network models have several ***inherent properties*** which distinguish them from traditional computational models (Lippmann, 1987):

1. Neural network models do not perform sequential computations as a Von Neuman machine does, but are ***massively parallel*** nets of simpler computational elements. Analog VLSI implementation of these networks provides much higher computation rates than sequential machines because of the inherent parallelism.

2. Because neural networks contain many more processing elements than a sequential machine, and because these elements have primarily local connections, an artificial neural network has a higher degree of ***robustness*** or ***fault tolerance***. If a few of the many links in the network are severed, the overall performance will not downgrade dramatically.

3. Artificial neural networks are systems that have the property of ***adaptation or learning***. Given a system and a set of "training" patterns, the neural network can "learn" about the system by adapting its internal parameters to match the training examples.

The above properties (1)-(3) make artificial neural networks the ideal choice in cases in which real-time adaptation and fast processing of large amounts of data is required. For this reason a great deal of attention has been paid to artificial neural networks for high performance speech or image recognition.

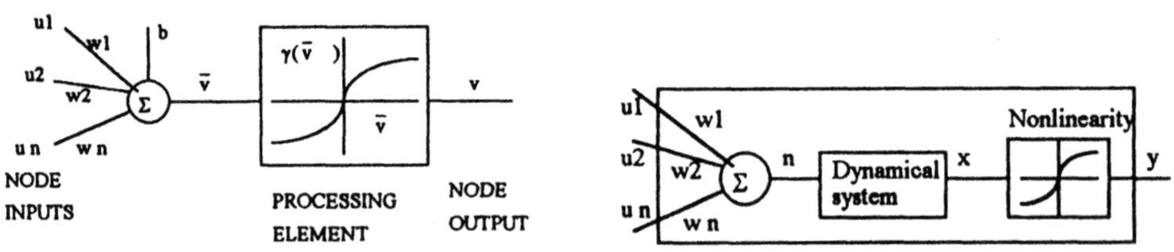

Figure 1 Static and dynamic neurons; (a) static; (b) dynamic

A neural net is characterized by the the ***processing elements***, the ***network topology***, and the ***learning or training algorithm***. Each of these network characteristics is described next.

A typical *processing element* (node) sums n weighted inputs $u_1, u_2, \ldots u_n$ and a bias term b and passes the result through a nonlinear function $\gamma(.)$ (Fig. **1**).

$$v = \gamma(\bar{v}) \quad ; \quad \bar{v} = \sum_{i=1}^{n} w_i u_i + b \tag{1}$$

Typical nonlinearities used in the nodes are: hard-limiters, threshold operations, the logistic function (sigmoid), and the hyperbolic tangent function.

The choice of the nonlinearity depends on the particular application for which the network is being used. In cases when it is required that the nonlinearity is differentiable, the logistic and the hyperbolic tangent functions are widely used.

The *network topology* depends on the way the nodes are connected to each other and to the input and output vectors. According to their topology, neural networks can be classified as: *single layer networks*, when only one layer of nodes is present, namely the output layer; or *multi-layer networks*, when the nodes are arranged in more than one layer. Moreover, a network can be characterized as: a *feedforward network*, if there is no feedback to previous layers from the output of subsequent layers or a *recurrent network*, if such a feedback connection exists.

2.2 Back Propagation

By the term *training* or *learning* algorithm we mean a systematic procedure for adjusting the weights in the network to achieve a desired input/output relationship. A general feature of "supervised" learning algorithms is that an error criterion $E = f(y, y_d)$ is evaluated, where y is the actual output vector and y_d is the desired output vector. The weights w_{ij} of the net are then adjusted in such a way as to reduce the value of the error criterion. The various learning algorithms reported in the literature differ in the ways this adjustment is being done. As an example, the *Hebbian rule* for a single layer net computes the weight change Δw_{ij} as proportional to the product of the node outputs

$$\Delta w_{ij} = k v_i v_j \tag{2}$$

where v_i and v_j are the outputs of nodes i and j.

Another example is the celebrated *Back Propagation* training algorithm, which is a gradient descent procedure based on a square error criterion. The back propagation algorithm is the one used for the applications demonstrated in Sections 3.1-3.2.

The following notation is used (as in Narendra and Parthasarathy, 1990):

- $u = [u_1, u_2, \ldots u_n]^T = n \times 1$ input vector
- $v = [v_1, v_2, \ldots v_p]^T = p \times 1$ vector, output of layer 1 (first hidden layer)
- $z = [z_1, z_2, \ldots z_q]^T = q \times 1$ vector, output of layer 2 (second hidden layer)
- $y = [y_1, y_2, \ldots y_m]^T = m \times 1$ vector, output of the net
- $\bar{v} = [\bar{v}_1, \bar{v}_2, \ldots \bar{v}_p]^T = p \times 1$ vector , intermediate output of layer 1 (first hidden layer)

- $\bar{z} = [\bar{z}_1, \bar{z}_2, \ldots \bar{z}_q]^T = q \times 1$ vector , intermediate output of layer 2 (second hidden layer)
- $\bar{y} = [\bar{y}_1, \bar{y}_2, \ldots \bar{y}_m] = m \times 1$ vector , intermediate output of layer 3 (output layer)
- $W^1 = \{w_{ij}^1\} = p \times n$ matrix of weights connecting the j-th input to the i-th node of layer 1
- $W^2 = \{w_{ij}^2\} = q \times p$ matrix of weights connecting the j-th node of layer 1 to the i-th node of layer 2
- $W^3 = \{w_{ij}^3\} = m \times q$ matrix of weights connecting the j-th node of layer 2 to the i-th node of layer 3
- $\gamma(x)$ is the differentiable nonlinearity $\gamma(x) = (1 - e^{-\alpha x})/(1 + e^{-\alpha x})$; $\alpha > 0$

The various inputs and outputs in each layer are related as follows:

- Layer 1: $\bar{v} = W^1 u$; $v_i = \gamma(\bar{v}_i)$, $i = 1, \ldots p$
- Layer 2: $\bar{z} = W^2 v$; $z_i = \gamma(\bar{z}_i)$, $i = 1, \ldots q$
- Layer 3: $\bar{y} = W^3 z$; $y_i = \gamma(\bar{y}_i)$, $i = 1, \ldots m$

In the case of "supervised" learning, the network is being presented a large number of input vector and desired output vector pairs (u^r, y_d^r), where the superscript r ranges over all pairs used to train the network and the subscript d stands for "desired" output vector. During training, the network learns to associate the input vector u^r with the output vector y_d^r. For a given set of weights, if the network is presented an input u^r it will produce an output y^r, which should be identical or very close to y_d^r if the training is successful.

In the back propagation algorithm, the performance criterion to be minimized by an appropriate choice of the weights is defined as

$$E = \frac{1}{2} \sum_r \sum_{k=1}^{m} (y_k^r - y_{dk}^r)^2 \tag{3}$$

where the index r ranges over all training patterns and the subscript k refers to the k-th component of the y^r and y_d^r vectors.

After the gradient has been computed, the weights are adjusted in the negative gradient direction, so that the error criterion E is reduced. A typical weight w_{ij} (which could belong to any layer) is adjusted from its old value w_{ij}^{old} to its new value w_{ij}^{new} according to

$$w_{ij}^{new} = w_{ij}^{old} - \eta \frac{\partial E}{\partial w_{ij}^{old}} \tag{4}$$

The stepsize η is termed the "learning rate". This η is usually chosen as constant during training, but it could also be adjustable, if this facilitates the training process.

2.3 Identification of Structural Systems

So far, very little work has been reported in the literature concerning the identification of *structural* systems by neural networks. In this section, the problem of identifying nonlinear structural systems by neural networks is formulated and the identification procedure is developed.

Specifically, let the multi-degree of freedom (MDOF) system to be identified obey the following equation of motion:

$$M\ddot{y}(t) + g(y(t), \dot{y}(t)) = u(t) \tag{5}$$

where y is the system displacement vector, $g(y, \dot{y})$ is the unknown vector of restoring forces, M is the known mass matrix and $u(t)$ is the system excitation. This formulation represents general classes of structural systems, such as systems with linear force/deflection characteristics, nonlinear systems having polynomial-form nonlinearities (such as the Duffing oscillator), as well as hysteretic systems.

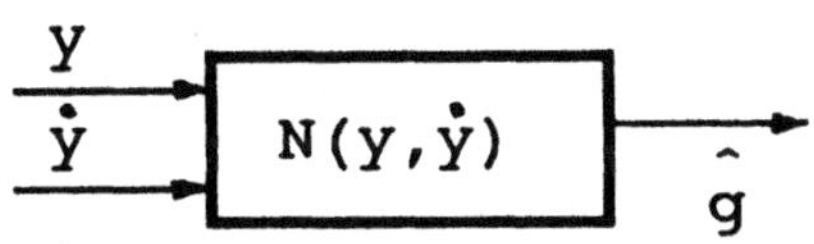

Figure 2 Block diagram for a neural identifier of a structural system

The neural identifier is shown in block diagram form in Fig.(**2**). The network, represented by N, is of undetermined size, topology and node nonlinearities, since these characteristics are application dependent. Presently no concrete analytical results exist and no specific guidelines are available concerning the criteria for the selection of the network characteristics. The selection of network size, topology and node characteristics is based largely on experience and extensive experimentation with different networks.

The inputs to the net are the displacement and velocity vectors $y, \dot{y}$. The output of the net is a nonlinear transformation of the inputs: $N(y, \dot{y})$. This output $N(y, \dot{y}) = \hat{g}$ should approximate the unknown function $g(y, \dot{y})$ of Eq. (5) if the network is a good identifier of the system.

Assume that the experimental measurements for $u(t)$ and $\ddot{y}(t)$ are available and that the corresponding system displacement $y(t)$ and velocity $\dot{y}(t)$ vectors can be found by direct measurements or through integration of $\ddot{y}(t)$. If the measurements are taken at discrete times t_k:

$$y_k = y(t_k); \quad \dot{y}_k = \dot{y}(t_k); \quad \ddot{y}_k = \ddot{y}(t_k); \quad u_k = u(t_k)$$

the values $g_k = g(y(t_k), \dot{y}(t_k))$ can then be computed from

$$g_k = u_k - M\ddot{y}_k$$

During the training phase, the network is presented with the sequence of input vectors $\{[y_k, \dot{y}_k]^T\}$ and the sequence $\{g_k\}$ of desired output vectors. Given a set of weights (which is initially chosen randomly), the input vector $\{[y_k, \dot{y}_k]^T\}$ is propagated forward through the net at time t_k and the network output $\hat{g}_k$ is calculated. This $\hat{g}_k$ is compared to the desired output g_k. The performance criterion is then evaluated and its gradient $\partial E/\partial\{w_{ij}\}$ is computed with respect to the network parameters. The weights are now adjusted according to Eq.(4), and the procedure is repeated until an acceptable value of the error criterion is obtained. Now, given an input vector $[y_k, \dot{y}_k]^T$, the network will produce an output $\hat{g}_k$ close to the desired output g_k. At this stage, the neural net has "learned" the system.

During the validation phase, the network is given other input vector sequences $\{[y_\alpha, \dot{y}_\alpha]^T\}$ not among those used for training. If the training was successful and the network is a good identifier, it should produce an output sequence $\{\hat{g}_\alpha\}$ very close to the actual system output $\{g_\alpha\} = \{g(y_\alpha, \dot{y}_\alpha)\}$.

Applications on identification of the internal forces of structural systems by neural networks are presented in the next section. Specific details of such applications are presented in Chassiakos and Masri (1991a, 1991b), and Masri et al. (1992a, 1992b).

2.4 Neural Network Based Identification of Dynamical Systems

Although the identification of dynamical systems is a subject that has attracted researchers for many years, it is only very recently that artificial neural networks have been used for this purpose. The nonlinear mappings between inputs and outputs of a neural network , and the ability of the net to adapt its parameters so as to minimize an error criterion, make the use of artificial neural networks particularly well suited for the identification of both nonlinear and linear dynamical systems. During the past two years a few papers have appeared in the literature dealing with dynamical system identification using neural networks (Werbos,(1990); Narendra and Parthasarathy, (1991)).

The neural networks described in sections 2.1-2.3, are nets with *static nonlinearities* in the nodes, since the function $\gamma(x)$ is a *static* function of its inputs (no dynamics involved). These networks work very well in modeling linear or nonlinear *static* functions, but they do not perform well in modeling *dynamic* systems. Experience with neural network architectures has shown that good modeling of systems with dynamics requires the introduction of some dynamics into the network. Different architectures and various training algorithms have been proposed in the literature for the solution of this problem (Williams and Zipser, (1989); Polycarpou and Ioannou, (1992)). Chassiakos et al. (1991d) and Kosmatopoulos et al. (1991) propose an architecture that gives rise to neural networks with *dynamic neurons*: each neuron contains a (not necessarily linear) dynamic system in addition to the nonlinear sigmoid function (Fig. **1**(b)). The neuron output y is related to the neuron inputs $u_1, \ldots u_N$ by the following equations (Cohen and Grossberg (1983), Pineda (1987)):

$$n = \sum_{j=1}^{N} w_{ij}^k u_j \quad ; \quad \dot{x} = -a(x)[b(x) + n] \quad ; \quad y = \gamma(x) \tag{6}$$

where x is the neuron *dynamic state*. The network is trained by a gradient technique termed *distributed dynamic back propagation*. The results obtained suggest that superior performance can be achieved by such an architecture, which is based on *distributing* the dynamics over the network nodes rather than lumping them into one dynamical system as is done in the "generalized neural networks" approach of Narendra and Parthasarathy (1990, 1991).

3 APPLICATIONS

3.1 Identification of the Internal Forces in a Linear Structural System

The identification procedure under discussion will be outlined in the context of a single-degree-of-freedom (SDOF) system in which the unknown restoring force acting on the system is $g(y, \dot{y})$. It is assumed that the excitation $u(t)$ and the acceleration $\ddot{y}(t)$ of the system are available from measurements, and that the mass m is known or easily estimated.

The time-history record for the SDOF system acceleration, velocity, displacement, and the internal force $g(y, \dot{y})$ corresponding to the example linear SDOF system under swept-sine excitation is shown in Fig.(**3**).

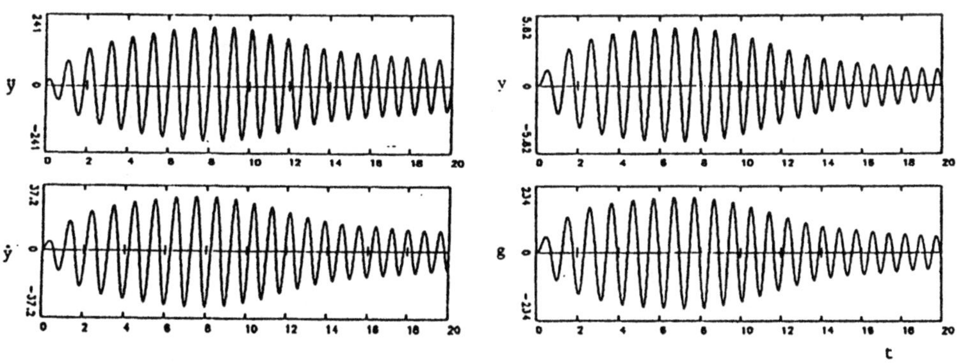

Figure 3 Response records used to train the neural network

The example system has a ratio of critical damping $\zeta = 0.10$. The plots in Fig.(**3**) correspond to a case where the lower and upper ranges of the swept-sine excitation, ω_l and ω_u, respectively, surround the natural frequency ω_n: $\omega_l/\omega_n = 0.9$, and $\omega_u/\omega_n = 1.1$. The amplitude of the excitation was $(2\pi)^2$, and the identification data set corresponded to a total excitation period of 20 natural periods.

The neural network chosen for the identification of the restoring forces is a three-layer feedforward net. Inputs to the net are the displacement and velocity measurements $y(t)$ and $\dot{y}(t)$ and the network output is the force $g(y, \dot{y})$. For the application at hand, it was decided that a net with 15 and 10 nodes in the first and second hidden layers, respectively, be tried. Hence, the total number of the network parameters to be adjusted was 216 (190 weights and 26 bias terms). This decision was made after experimentation with several other net sizes. Initially, the network parameters are assigned random values since nothing is known about the optimal parameter set (i.e. the set that will minimize the performance criterion). Training is performed by the back propagation algorithm with a hyperbolic tangent function as the node nonlinearity.

The example system presented here was trained over data covering 20.0 sec (equal to about 20 natural periods). The input/output data set was sampled every 0.2 sec, giving a total of 100 training patterns: $\{[y_1, \dot{y}_1]^T, \ldots [y_{100}, \dot{y}_{100}]^T\}$ and $\{g_1, \ldots g_{100}\}$. Moreover, since the neuron nonlinearity $\gamma(x)$ saturates for high values of $|x|$, some scaling of the training patterns is necessary to keep the network away from saturation. The scaling performed here was done so as to keep the ranges of variables in the training patterns approximately in the $[-1, 1]$ interval. The input sequence was scaled as $\{[y_k, \dot{y}_k]^T\} \rightarrow \{[y_k/5, \dot{y}_k/50]^T\}$ and the output sequence was preprocessed as $\{g_k\} \rightarrow \{g_k/500\}$. It is clear from the results that the size of the training set and the scaling chosen was adequate to give satisfactory performance.

By applying the identification procedure under discussion, a neural network model of the test system is obtained. The network topology is represented by the various arrays that relate the inputs and outputs in each of the three layers, as discussed in Section 2.1.

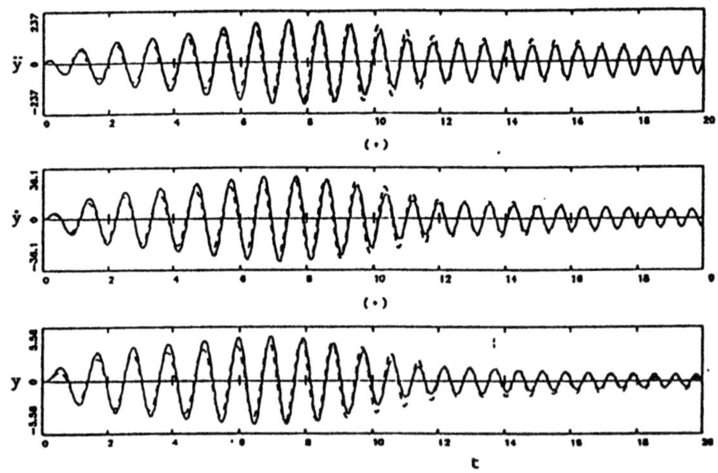

Figure 4 Comparison of the measured and identified time history response

In order to check the accuracy of the identification procedure, standard time-marching techniques were used to numerically solve the system differential equation of motion, as given in Eq.(5). The magnitude of the restoring force $g(y, \dot{y})$ was determined from the output of the neural network, which produced an estimate $\hat{g}(y, \dot{y})$. Good agreement was obtained between the "measured" response and the identified response.

Notice that no extra information about the physical system was given to the net, other than the input/output sequences, and that *no model of the system was assumed* during the learning phase.

Next the neural net is validated as a system identifier: given other input vector sequences $\{[y_\alpha, \dot{y}_\alpha]^T\}$ not among those used for training, the net will produce an output sequence $\{\hat{g}_\alpha\}$ which should be very close to the actual system output $\{g_\alpha\} = \{g(y_\alpha, \dot{y}_\alpha)\}$. To investigate the validity of the identification results, a different probing signal was used to predict the response of the identified system. The test signal used in this case was also a swept-sine excitation except that the sweep range satisfied the following condition: $\omega_l/\omega_n = 0.8$ and $\omega_u/\omega_n = 1.2$ and the excitation amplitude was kept the same as before.

A comparison between the "measured" response and identified response is shown in Fig.(**4**), in which the solid line refers to the exact measurements and the dashed line corresponds to the output of the neural network.

Even though the excitation used in this case differs from the probing signal used to "train" the neural network, examination of the results reveals a good agreement, both qualitatively as well as quantitatively, between the exact system internal force $g(y, \dot{y})$ and the estimated restoring force $\hat{g}(y, \dot{y})$. The small error between the measured and predicted response is attributed to the loose convergence criteria used in the learning procedure with the neural network.

3.2 Identification of the Internal Forces in a Nonlinear Structural System

The example of this section corresponds to the Duffing oscillator with a hardening nonlinearity governed by

$$m\ddot{y} + g(y, \dot{y}) = u(t). \tag{7}$$

The restoring force is

$$g(y, \dot{y}) = ay + by^3 + c\dot{y} \tag{8}$$

where the system parameters have the following values: $m = 1$, $a = 2\pi$ $b = 10$, and $c = 1.25$. The training was performed using the data of Fig.(**5**): the network inputs were y and $\dot{y}$ and the desired output was the set of values of $g = u - m\ddot{y}$ over the time period from 0.0 to 20.0 sec. After the training was over, the weights and biases were fixed to the values obtained by the back propagation algorithm. From this point on, the network is used to model the unknown restoring force nonlinearity. Several tests of the network's identifying ability were performed:

1. First, the actual data y and $\dot{y}$ were used as inputs to the net. The network output $\hat{g}(y, \dot{y})$ is virtually identical to the plot of $g(y, \dot{y})$ shown in the bottom of Fig.(**5**). This, of course, is expected since the network was trained on the basis of this data.

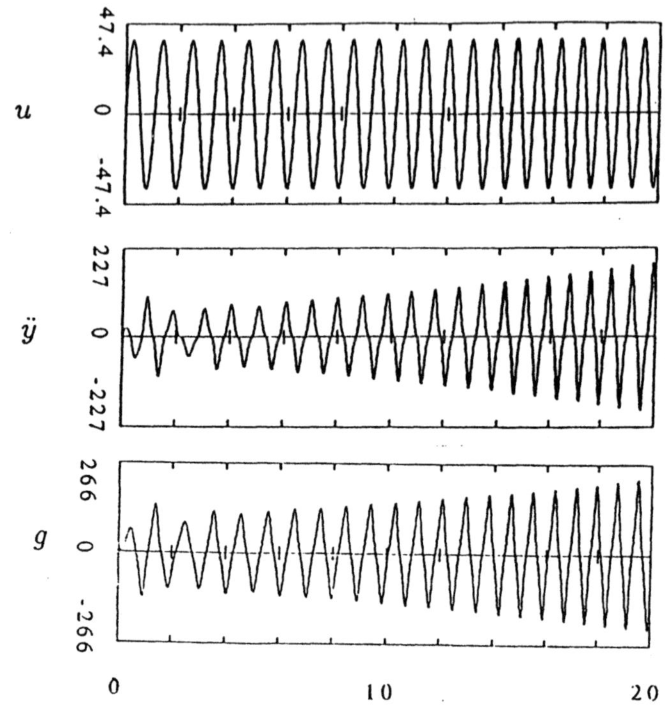

Figure 5 Time history records of the nonlinear system excitation and response under swept sine excitation used to train the neural network.: (a) excitation; (b) acceleration; (c) restoring force.

2. Next, the network was used as a model of the unknown nonlinearity for prediction of the system's responses to various excitations. The predicted responses $\hat{y}$ and $\hat{\dot{y}}$ are obtained by standard time-marching techniques as numerical solutions of the equation of motion

$$m\hat{\ddot{y}} + \hat{g}(\hat{y}, \hat{\dot{y}}) = u(t) \tag{9}$$

which is the same as the system Eq.(5), with $\hat{g}$ replacing the unknown function g.

The time plots of the predicted responses $\hat{y}(t), \hat{\dot{y}}(t)$ are virtually indistinguishable from the exact responses $y(t), \dot{y}(t)$. The state variable plots of the actual and predicted responses are presented in Fig.(**6**), in which the solid line refers to the exact measurements and the dashed line corresponds to the output of the neural network. The three plots shown in the upper half of Fig.(**6**) correspond to the exact data of Fig.(**5**), while the three plots shown in the lower half of Fig.(**6**) correspond to the predicted response $\hat{y}(t)$, $\hat{\dot{y}}(t)$ and $\hat{g}(\hat{y}, \hat{\dot{y}})$ numerically computed as shown above. Notice from Fig.(**6**b) that the $\hat{y}$ vs $\hat{g}(\hat{y}, \hat{\dot{y}})$ obtained from the neural network exhibits the same hardening characteristics, as well as the damping level of the exact measurements.

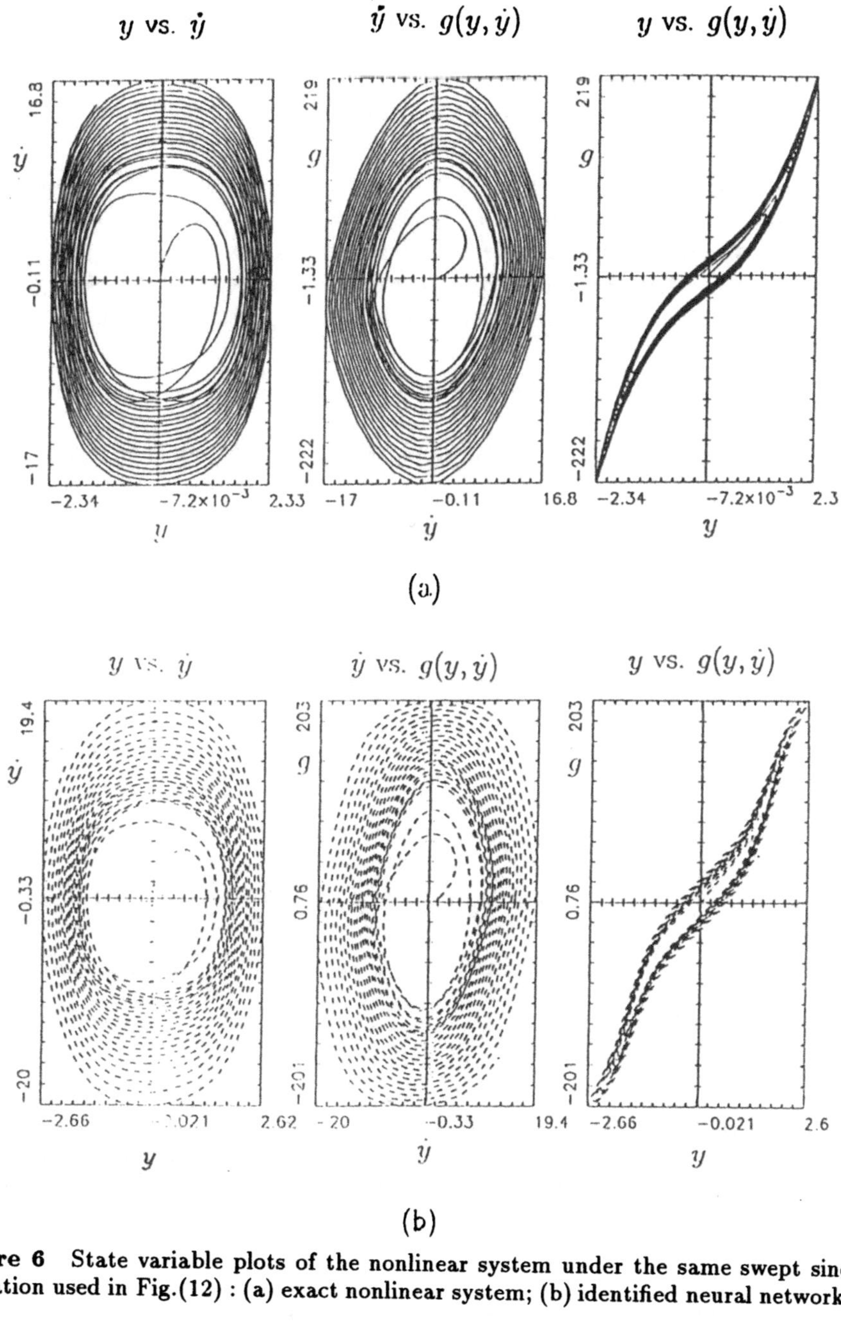

Figure 6 State variable plots of the nonlinear system under the same swept sine excitation used in Fig.(12) : (a) exact nonlinear system; (b) identified neural network.

3. Next, the network was tested under random excitation. This is a severe test of the network's modeling ability, since the training was performed under deterministic excitation of limited frequency content ($\pm 10\%$ of ω_n). In the present test, wide-band random excitation with Gaussian distribution of the

form $n(\mu, \sigma)$, where μ is the mean and σ is the standard deviation, was used as an excitation source:

$$u(t) = n(\mu, \sigma), \quad \mu = 0, \quad \sigma = 50 \tag{10}$$

A comparison in the time domain between the "measured" response and identified response is shown in Fig.(7) in which the solid line refers to the exact measurements and the dashed line corresponds to the output of the neural network. The accuracy of the neural network predictions is so high as to make the two distinct curves in each of the plots practically indistinguishable.

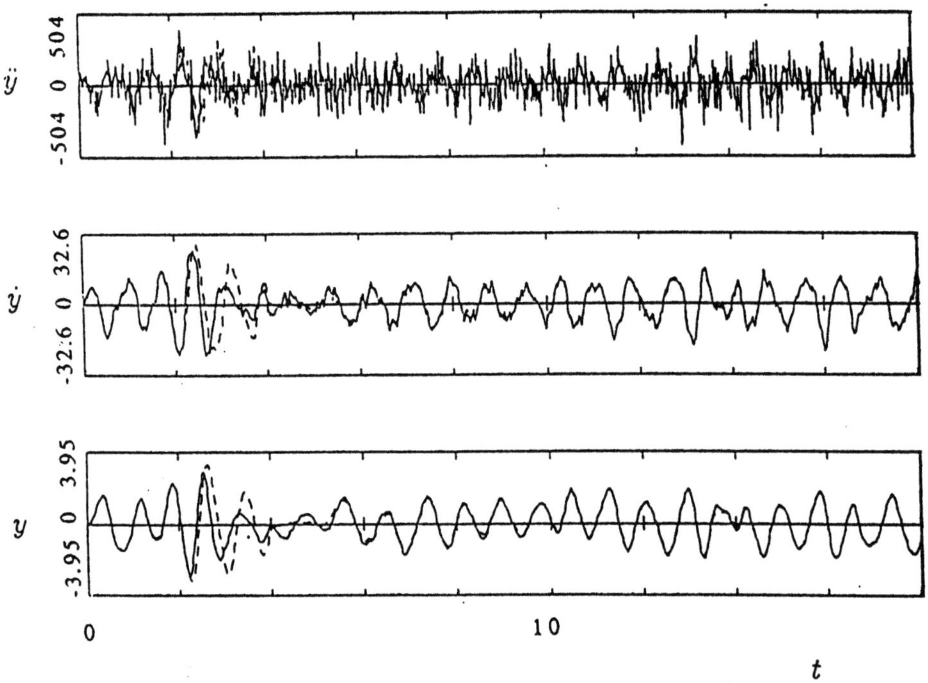

Figure 7 Comparison between the response time histories of the exact nonlinear system and the identified neural network under wide band random excitation. Solid line (—) = exact response; dashed line (- - -) = predictions of the neural net.

3.3 Identification of a Nonlinear Dynamic System by Nets with Dynamic Neurons

In this section an application of neural networks with *dynamic neurons* is presented (Kosmatopoulos et al. (1991), Chassiakos et al. (1991d)). The system to be identified is a two degree of freedom robotic manipulator, described by the following nonlinear vector differential equation

$$\tau(t) = M(\theta(t))\ddot{\theta}(t) + C(\theta(t), \dot{\theta}(t))\dot{\theta}(t) \tag{11}$$

where

- $\tau(t)$ is a 2×1 vector of joint torques
- $\theta(t)$ is a 2×1 vector containing the joint variables (joint angles)
- $M(\theta(t))$ is a 2×2 matrix, whose elements are nonlinear functions of $\theta(t)$, representing the contribution of the inertial forces to the dynamical equation
- $C(\theta(t), \dot{\theta}(t))\ddot{\theta}(t)$ represents the coriolis forces

The problem we are dealing with is that of modeling the *forward dynamics* of this system: given the vector $\tau(t)$ of input torques, reproduce the output vector $\theta(t)$ of joint angles. The training set consists of the time histories of input torques which are sinusoidal signals with time-varying frequencies, ranging from 0.001 rad/sec to 1.0 rad/sec, and the corresponding output joint angles $\theta(t)$. Initially a static network with only a sigmoid static function $\gamma(x)$ at the nodes was chosen and trained over the set of input/output data. The network chosen was a three layer net, with two inputs, two outputs and 10 nodes at each hidden layer. The results of the network identification are shown in Fig.(**8**a) for the angle $\theta_1(t)$. It is seen that the net

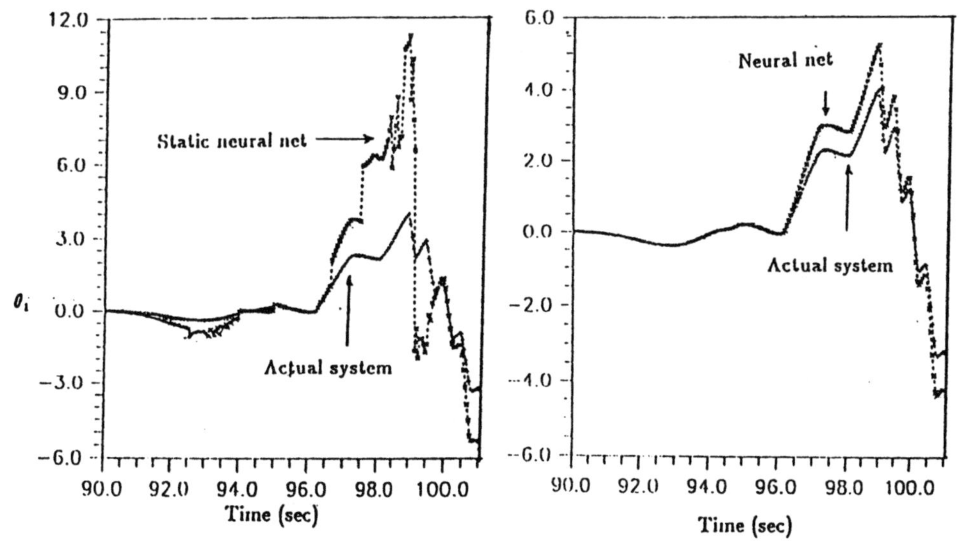

Figure 8 Modeling of a robotic system: (a) Using a static neural network; (b) Using a network with dynamic neurons

does not perform an accurate identification, since no dynamics were included in the network architecture. Next a network with *dynamic neurons* is used and trained by the *distributed dynamic back propagation* algorithm. The network chosen is of the same topology and size as the static neural net (three layered net, with two inputs, two outputs and 10 nodes at each hidden layer). A linear dynamical system of the form $\dot{x} = -ax + bn$ was employed at the nodes in addition to the static sigmoidal nonlinearity $\gamma(x)$. The network performs a very good modeling of the nonlinear robotic system, as can be seen from Fig.(8b), for the variable $\theta_1(t)$. Similar results were obtained for the second joint variable $\theta_2(t)$.

The authors' experience in networks with *dynamic neurons* provides strong indications that networks with distributed dynamics are a powerful tool for modeling unknown nonlinear dynamic systems.

4 SUMMARY AND CONCLUSIONS

This paper presents the development of a new method for identifying nonlinear structural systems typically encountered in the applied mechanics field. The method is based on the use of neural networks as system identifiers. The relevant background is discussed, the identification problem is formulated and an identification procedure which handles a general class of structural systems is developed.

Illustrative examples involving the use of multilayer, feed-forward neural networks to identify representative linear or nonlinear systems under a variety of deterministic as well as stochastic excitations are presented.

5 ACKNOWLEDGMENTS

This study was supported in part by a grant from the US National Science Foundation and by the Carpenters/Contractors Cooperation Committee, Inc.

6 REFERENCES

1. Chassiakos, A. G., and Masri, S. F., (1991a), "Identification of the Internal Forces of Structural Systems Using Neural Networks," *IEEE-INNS, International Joint Conf. on Neural Networks*, Seattle, WA., July 1991.

2. Chassiakos, A. G., and Masri, S. F., (1991b), "Identification of the Internal Forces of Structural Systems by Feedforward Multilayer Networks," *Computing Systems on Engineering*, Vol. 2, 1, pp.125-134.

3. Chassiakos, A. G. and Masri, S. F., (1992), "Neural Network Based Identification of Structural Systems," Parallel and Distributed Computing in Engineering Systems, Tzafestas, Borne and Gradinetti, eds., North Holland, 371-376

4. Chassiakos A.G., Kosmatopoulos E.B., Christodoulou M.A., (1992), "Identification of Robot Dynamics by Neural Networks with Dynamic Neurons," Neural Networks in Robotics, G. Bekey, ed. Kluwer Academic Publishers

5. Cohen, M.A., and Grossberg, S., (1983), "Absolute Stability of Global Pattern Formation and Parallel Memory Storage by Competitive Neural Networks," *IEEE Trans. on Systems, Man and Cybernetics*, Vol.13, 5, pp.815-826.

6. Housner, G.W. and Masri, S.F., (Editors), (1990), *Proceedings of the U.S. National Workshop on Structural Control Research*, Held at the University of Southern California, Los Angeles, CA, 25-26 October, USC Publication No. CE-9013.

7. Lippmann, R.P., (1987), "An Introduction to Computing with Neural Nets," *IEEE ASSP Mag.*, April, pp.4-22.

8. Masri S., Chassiakos A.G., Caughey T., (1992a), "Structure-unknown Nonlinear Dynamic Systems: Identification through Neural Networks," *Journal of Smart Materials and Structures, Institute of Physics*, 1, 1, March, pp. 45-56.

9. Masri S., Chassiakos A.G., Caughey T., (1992b), "Identification of Nonlinear Dynamic Systems Using Neural Networks," *Transactions of the ASME, Journal of Applied Mechanics*, to appear.

10. Narendra, K.S. and Parthasarathy, K., (1990), "Identification and Control of Dynamical Systems Using Neural Networks," *IEEE Transactions on Neural Networks*, Vol.1, 1, March, pp.4-27.

11. Narendra, K.S. and Parthasarathy, K., (1991), "Gradient methods for the optimization of dynamical systems containing neural networks", *IEEE Trans. on Neural Networks*, vol.2, no.2, March.

12. Natke, H.G. and Yao, J.T.P., (1988), *Structural Safety Evaluation Based on System Identification Approaches*, Frieder Vieweg & Sohn, Braunschweig.

13. Nguyen, D.H. and Widrow, B., (1990), "Neural Networks for Self Learning Control Systems," *IEEE Control Systems Magazine*, Vol.10,3, April, pp.18-23.

14. Pineda, F.J., (1987), "Generalization of Back Propagation to Recurrent Networks," *Phys. Rev. Lett.*, Vol. 59, 19, pp.2229-2232.

15. Polycarpou, M. and Ioannou, P. A., (1992), "Identification and Control of Nonlinear Systems Using Neural Network Models: Design and Stability Analysis," Neural Networks in Robotics, G. Bekey, ed. Kluwer Publishers.

16. Werbos, P., (1990), "Backpropagation Through Time: What it Does and How to do it," *Proceedings of the IEEE*, vol.78, no.10, Oct.

17. Williams, R.J. and Zipser, D., (1989), "A learning algorithm for continually running fully recurrent neural networks," *Neural Computation*, vol.1, no.2.

The Application of Neural Networks to Condition Monitoring of Non-linear Structures

P.H. Cowley
Rolls-Royce plc
Rolls-Royce Aerospace Group
Derby, U.K.

1 Introduction

Neural networks represent a powerful set of pattern recognition and pattern classification tools. They can identify the common features in a range of examples of a signal and then they can recognise future occurrence of the features as new signals are collected. A feature could be the presence of non-linear behaviour in a structural response, as described in Professor Tomlinson's studies (see Chapter ...) or it could be a particular stress pattern. Neural networks are being used for signal processing in many fields, from the recognition of patterns of pressure changes in the human body to geophysical measurements.

A short list of applications of neural networks relevant to structural assessment includes:

- feature recognition (signal processing and condition monitoring)
- fitting a response surface to a non linear process (to provide real time analysis or to speed up optimisation methods such as genetic algorithms)
- estimating the inverse of a structural-state to symptom model
- identification of key parameters affecting structural response

Neural networks are particularly useful when examples of measurements from structures are available and when there is incomplete understanding of the relationship between the structure and the measurements.

Some of the claims made for neural networks have been greeted with scepticism by traditional systems modellers, with good reason. There is justifiable concern about the use of a black box whose workings are hidden. Nevertheless, the practical results obtained with neural networks give an incentive to learn something about their workings. By making the black box a little more transparent, some insights into the processes may be obtained, allowing the outputs to be used with more confidence and helping with future modelling studies.

Because they learn to correlate inputs with outputs from a series of examples, neural networks are at a disadvantage when a comprehensive set of examples is hard to obtain.

This paper aims to communicate some practical experience which has helped to build confidence in the use of neural networks. Three factors have been identified which in combination increase the confidence in the output produced by a neural network. These are:

- Assurance that the signal being classified (the test signal) is adequately represented by the data used to train the network.
- Identification of which features of the input signal have been used to produce the signal classification.
- Provision of a comprehensive set of examples with which to train the network.

2 Network Verification

The ideal neural network will indicate when a test signal is not represented by the training data set.

The multi-layer perceptron, which is the most widely used form of neural network, calculates its outputs by summing weighted contributions from several intermediate (hidden layer) values.

The weighting factors are the network's knowledge base and are determined during the training phase of setting up a neural network. Each output of the network is a number, typically in the range 0-1 which can represent a scalar value or a binary classification.

One consequence of this network architecture is that if the input data are outside the range of the training data, the internal representation of the problem may be meaningless. Under these circumstances the network outputs will tend towards an intermediate value rather than one of the extremes. This is likely to lead to a "fail dangerous" condition if the intermediate value can be misinterpreted.

If there is a risk of this occurring, it is important for the network output to be chosen so as to avoid ambiguity. It must not attach particular significance to intermediate values.

2.1 Thermometer Encoding of Data

The thermometer code is a good general purpose representation of scalar data which avoids the ambiguity caused by indeterminate outputs. It uses several outputs to represent each scalar prediction. Most of the outputs will have values of 0 or 1 to indicate that the prediction is above or below the range covered by that output. These outputs will detect unrecognised input patterns by deviating from a clean binary output. One or two outputs at the transition between binary extremes will have intermediate values which allow the interpolation of accurate predicted values.

The number and range of values used for the encoding scheme needs to be chosen carefully. If the network is to interpolate between training data points, it must be trained with several examples in the range of each encoded output neuron, otherwise the network will not interpolate accurately. Instead the output will flip between the values presented during training. The use of several overlapping output ranges helps to avoid this problem.

Figure 1 illustrates a simple thermometer encoding scheme with 3 output ranges, with and without overlapping ranges.

2.2 Application example

The thermometer encoding scheme has been used successfully to locate the source of acoustic emission (AE) in composite structures. A neural network is used for this application because the speed of sound varies with direction in the structure, and there are multiple sound paths which make analytical location of an acoustic emission event difficult and not very accurate.

A neural network which uses the thermometer encoding scheme detects the small percentage of false indications caused by ringing, multiple overlapping events or sensor faults.

The acoustic emission source location and identification facility at the Rolls-Royce Applied Science Laboratory (RRASL) is based upon an automated "touch and learn" facility. AE is induced by artificial means at known locations in the component under test. The signals are detected by an array of broad band sensors. The data are pre-processed, either conventionally or by neural network, to identify the time of arrival of the AE signal at each sensor. These data may be supplemented by rise time and amplitude information to allow the neural network to correct for dispersion and attenuation of the signal. In subsequent use, the position of stress induced AE is identified by relating the signals to the reference data.

The use of overlapping range encoding makes two measures of the quality of the signal interpretation available, the difference between the multiple estimates of position, and the approach of "out of range" data to the threshold values. For this application, the number of "in range" classifications was checked, allowing location estimates with the wrong number of values to be rejected.

Figure 2 shows predicted and actual locations for one early AE source location trial, using data provided by Dr Noel Kerr of RRASL. The missing prediction, associated with the point at (165, 90), was eliminated on the basis described above. Examination of the raw signal showed that one sensor was faulty during this measurement.

The mean error of the estimated locations was 11mm which is significantly better than could be achieved by more conventional methods.

2.3 Input Verification

With any pattern recognition or classification scheme there is a risk of incorrect classification due to the coincidence of part of a noisy signal with features used for classification. A neural network classification is usually more noise resistant than a simple feature discriminator, because the network uses the combination of a number of features to arrive at a classification. It is still desirable to guard against a noise spike which coincides with one of the features.

If the removal of noise by pre-processing is not possible, and the network cannot be trained to disregard typical noise signals (say due to a lack of example data) then the network should be interrogated to determine which input features are used to make a classification.

There are two methods that can confirm that the features used to classify a test signal are the same as the features used to classify the original network training data. The first method is indirect, but can always be used. It involves injecting a small pseudo error at the network outputs which correspond with the network output. The pseudo error is propagated back through the network to give an inferred error at each input. This is an estimate of the change in each input needed to cancel the pseudo error. The inferred input errors indicate the relative importance of each input in arriving at that classification. To have confidence in the classification of the test signal one would expect several features in the signal to coincide with the inputs identified by the pseudo error.

The second method can only be used when there is one to one mapping between the inputs and outputs of the neural network. In this case it is possible to construct an inverse network which can reconstruct the input signal from the classification. A simple statistical comparison of the actual input signal with the reconstructed signal indicates the deviation between the test signal and the signal which the network training has led it to associate with that diagnosis.

With both methods there is a need to calibrate the significance of the difference between the test signal and the expected signal. For many cases the deviation can be compared with the deviation calculated for known good data.

3 Modelling Studies

A side effect of network verification is that the information can be useful in modelling the system. The important input features are clearly those which should be incorporated into a systems model. If one of the network inputs is set to the output of the system model, the network will rank the other input parameters in line with their ability to reduce the residual error. This process is closely related to an analysis of residual errors, but it remains effective when the relationship between the inputs and the residuals is non linear.

The use of neural networks and the quantity of training data required can be helped by incorporating any known relationships between the input data in the network inputs. This may involve giving the network simple model outputs, simple combinations of variables (such as $yieldstress/modulus$ or $mass * velocity^2$) or may recognise a strong correlation between inputs.

Figure 3 shows a Coherent Anti-Stokes Raman Spectrum (CARS) measured by Dr John Black from Rolls-Royce Applied Science Laboratories. These spectra may be used for remote temperature sensing. Figure 4 shows the relative importance of each part of the spectrum in performing a neural network classification. The spectral intensity of adjacent points is strongly correlated. With a small set of training data (12 points) and 200 inputs, the neural network is unable to determine that the large importance attributed to adjacent inputs should be cancelled as they have opposite signs. The temperature predictions made by this network are quite accurate with good data, but with noisy data they are prone to error. Averaging groups of input data help a little (Figure 5) but the sensitivity to the input spectral intensities is still erratic.

Having identified the potential problem from the input sensitivities, it is possible to retrain the network, constraining it to recognise the correlation between adjacent points. The noise is substantially reduced. The important features of the spectrum become clear, and further simplification of the network becomes possible. The end result is illustrated in Figure 6. It has been possible to reduce the number of inputs to 50 and the temperature predictions are substantially more accurate and noise tolerant.

4 Synthesising Training Data

The largest flaw in the use of neural networks for condition monitoring large or unique structures is the lack of data to use to train the network. This problem also exists, albeit to a lesser extent, for analytical localisation methods. There is a need to calibrate the size of the abnormality in the structural response to the size of the engineering problem.

For some structures and signal sources it is possible to model the behaviour completely. In this rare case the model may provide the required diagnosis directly. More commonly the model will only map from cause to symptoms. In this case the model output may be used as the input for a neural network, and the network can be trained to predict the model inputs, relating symptoms to diagnosis.

For many structures, it seems unlikely that we will be able to model the whole structure due to a lack of data or sufficiently powerful methods. It is much more likely that small parts of the structure can be modelled.

If a defect can be modelled in part of a structure, the model should allow a physical "pseudo defect" to be constructed. When attached to the structure, the effect of a known "defect" on the dynamic response of the structure can be measured experimentally.

The concept for introducing pseudo defects involves comparing the structural response of a member of a structure with the expected response of a defected member. The difference between the two responses constitutes an error term which can be used to drive a force transducer (electromagnetic at RRA or piezoceramic at Manchester) which modifies the behaviour of the structural member in response to a measured excitation. If the effect of the defect on the member is well understood theoretically or empirically, then the member can be made to respond exactly as though it has a defect. Its effect on the rest of the structure can be monitored, either to provide training data for a neural network, or to provide calibration data for analytical structural dynamics methods.

Figure 7 illustrates the pseudo defect concept. It shows a structure made up of beam elements. The element which has the simulated defect has a transducer to detect the exciting forces and a pair of force transducers.

If the structure is statically loaded and the defect to be simulated is a transverse crack, then a strain gauge could be used to measure the exciting force. For forces acting while the simulated crack is closed, the force transducer would be inactive, but once the conditions would open a crack the force transducer becomes active, acting in sympathy with the exciting force to increase the deflection of the beam.

The same principal applies for a dynamically loaded structure. However in this case the control system has to be more sophisticated in order to maintain the correct phase relationship between the measured exciting forces and the force transducer signal. The figure shows an open loop control system. Replacing the non-defected structure model by the measured structural response would effectively close this control loop at the expense of increasing the complexity of the adaptive controller.

5 Conclusions

Neural networks can provide a useful tool for analysing and interpreting signals for the condition monitoring of structures. There are shortcomings which result from the hidden nature of the neural network interpretation, but there are several routes available to help to overcome these shortcomings. In finding out more about how a neural network has interpreted a set of results, the user may find out more about the process itself.

The problem with virtually all attempts to characterise non linear structures, whether by neural network or by theoretical analysis, is that the defect indication needs to be calibrated. Without calibration, the investigator will be able to deduce that a new non-linear mode has been generated and he may be able to relate this to a particular type of defect. However the structural engineer is usually less interested in the presence of a defect than in sizing the defect and then estimating the consequence of this on structural integrity.

No general methods are known for relating the non-linear characteristics of a defect in a complex structure to the size of the defect. For common structures there will be field experience of the structural performance. For inexpensive structures destructive testing can be used. This leaves unique, expensive structures; for these structures the method for generating pseudo defects appears to be potentially valuable.

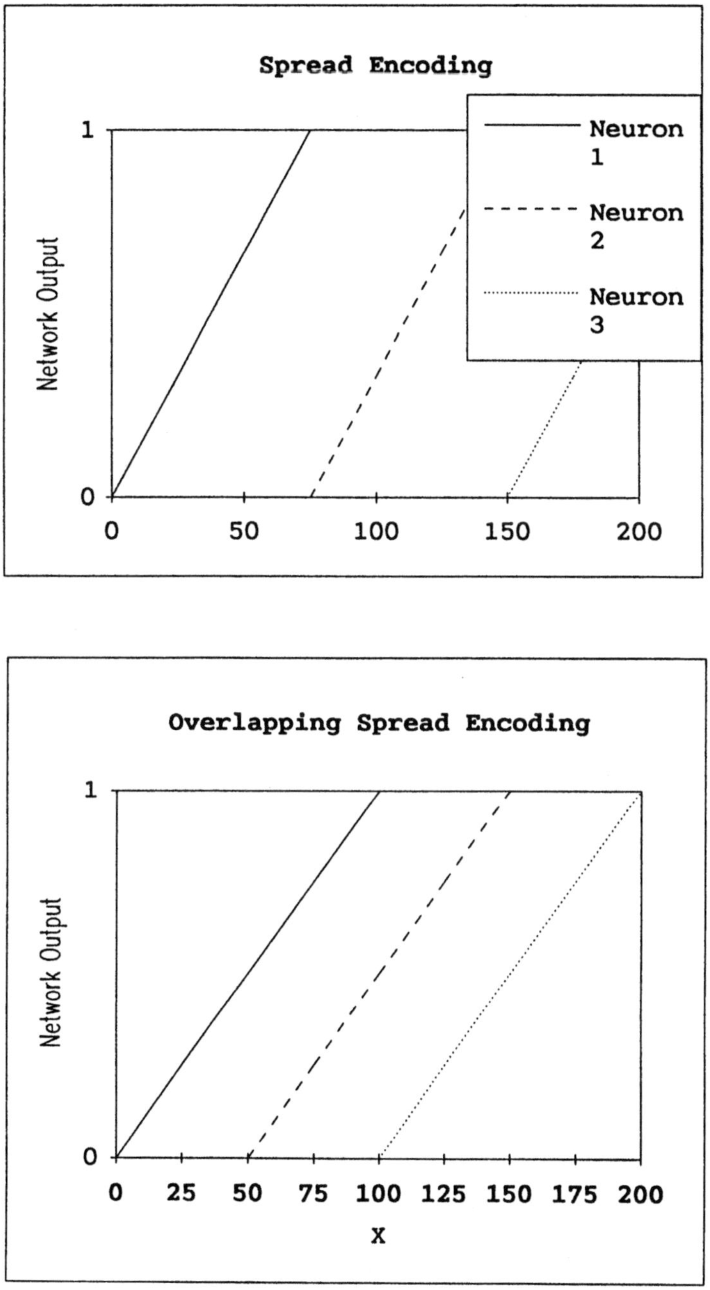
Spread Encoding
Neuron 1
Neuron 2
Neuron 3
Network Output
1
0
0
50
100
150
200
Overlapping Spread Encoding
Network Output
1
0
0
25
50
75
100
125
150
175
200
X

Figure 1

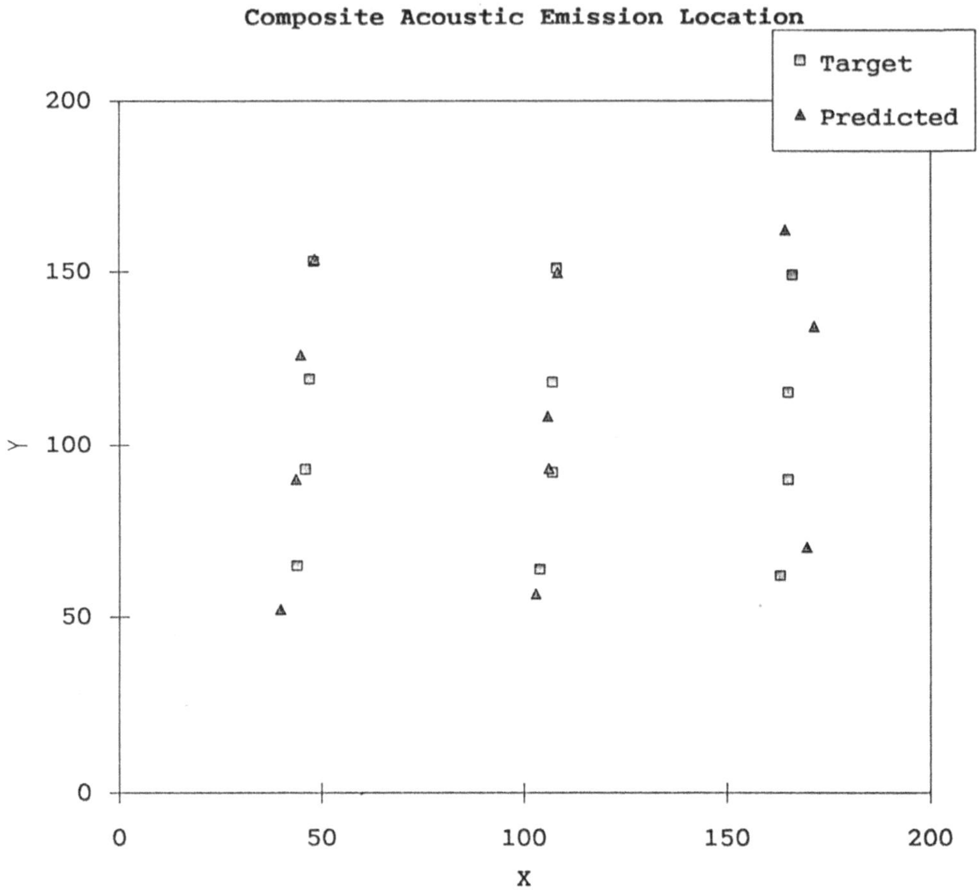
Composite Acoustic Emission Location
Target
Predicted
200
150
100
50
0
Y
0
50
100
150
200
X

Figure 2

CARS spectrum

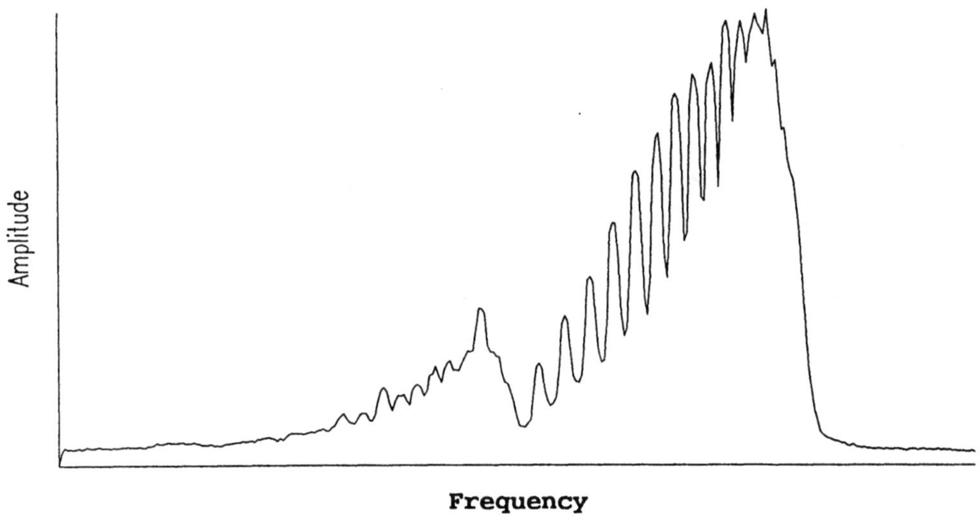

Figure 3

Raw data
Standard algorithm

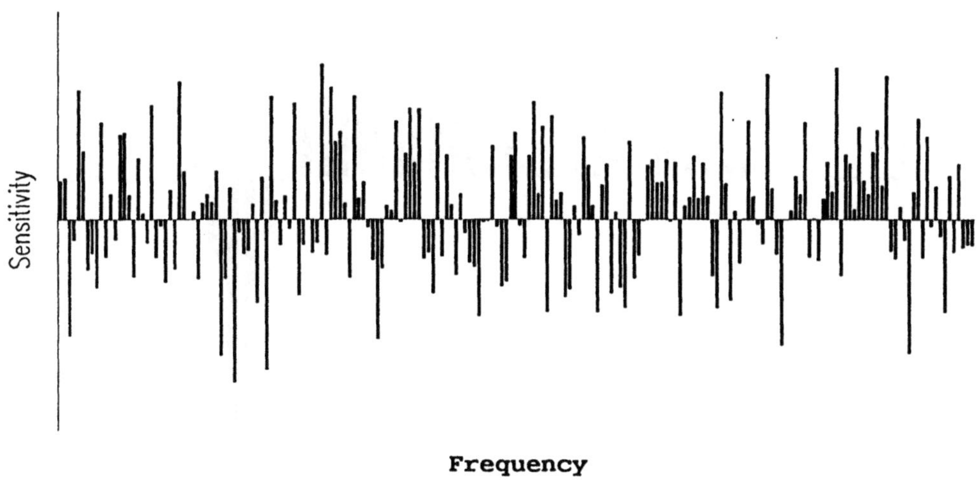

Figure 4

Averaged data
Standard algorithm

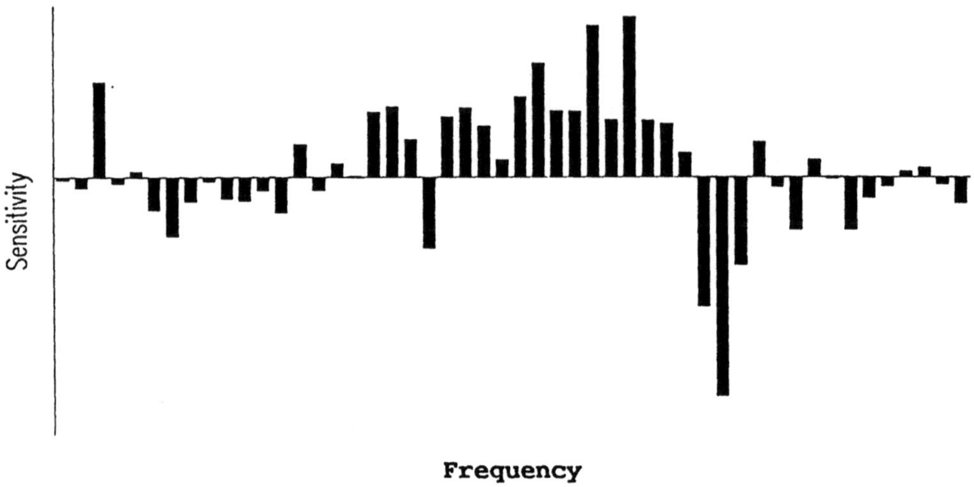

Figure 5

Averaged data
Correlated weights

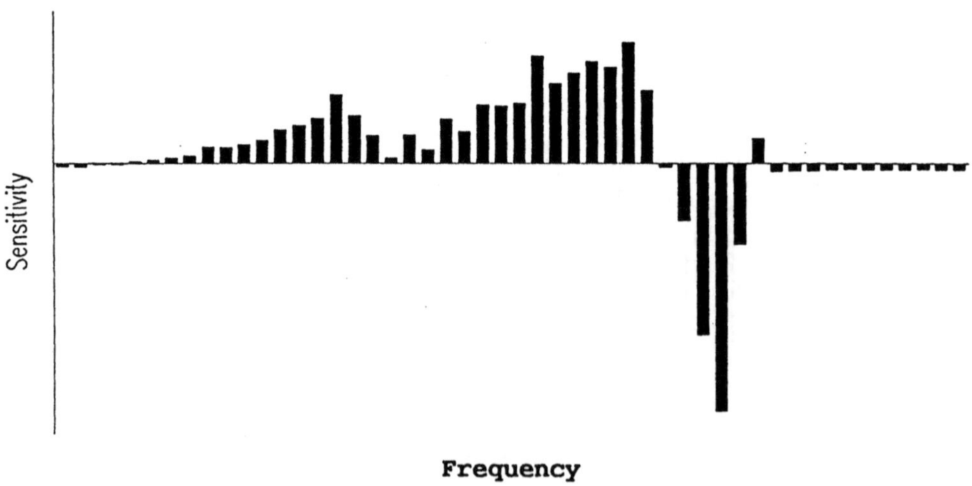

Figure 6

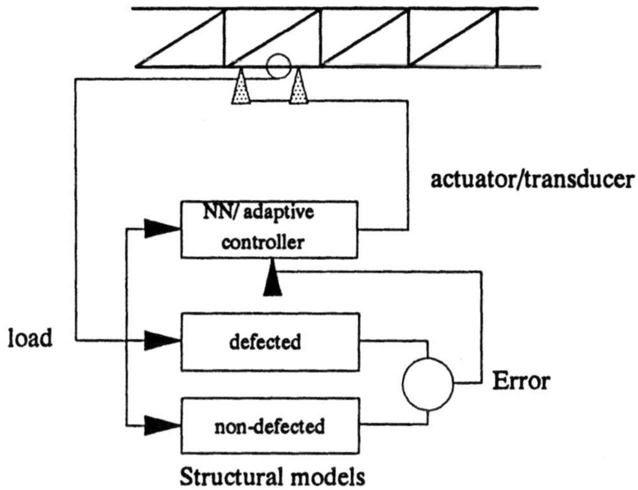
actuator/transducer
NN/ adaptive controller
load
defected
Error
non-defected
Structural models

Figure 7

Fuzzy Logic, Neural Networks and Soft Computing

Lotfi A. Zadeh
University of California
College of Engineering
Department of Electrical Engineering
and Computer Sciences
Berkley, CA, USA

(Paper not submitted for publication)

Prof. Zadeh presented a comprehensive lecture on fuzzy logic, neural networks, and soft computing. In addition, he lead a spirited discussion of how these relatively new techniques may be applied to safety evaluation of time variant and nonlinear structures based on identification approaches. The abstract of his lecture is given as follows.

The past few years have witnessed a rapid growth of interest in novel modes of computation which are collectively referred to as *soft computing.*

The distinguishing characteristic of soft computing is that its primary aim is to exploit the tolerance for imprecision and uncertainty to achieve tractability, robustness and low cost. Thus, in soft computing what is usually sought is an approximate solution to a precisely formulated problem or, more typically, an approximate solution to an imprecisely formulated problem. A simple case in point is the problem of parking a car. Generally, a car can be parked rather easily because the final position of the car is not specified exactly. If it were specified to within, say, a fraction of a millimeter and a few seconds of arc, it would take hours or days of maneuvering and precise measurements of distance and angular position to solve the problem. What this simple example points to is the fact that, in general, high precision carries a high cost. The challenge, then, is to exploit the tolerance for imprecision by devising methods of computation which lead to an acceptable solution at low cost. This, in essence, is the guiding principle of soft computing.

At this juncture, the major components of soft computing are fuzzy logic (FL), neural network theory (NN) and probabilistic reasoning techniques (PR), including genetic algorithms, chaos theory and parts of learning theory. It may be argued that it is soft computing — rather than hard computing — that should be viewed is the foundation for artificial intelligence. In the years ahead, this may well become a widely held position. Our talk presents a rationale for this view. Possible applications for the identification of complex and highly nonlinear systems will be discussed.

SUMMARY AND CONCLUSIONS

1 OVERVIEW

This workshop was organized with the following objectives

1. to focus attention on the problem of damage assessment and safety evaluation of time-varying and nonlinear systems using identification approaches, and

2. to exchange information among a group of invited participants with expertise in a variety of subject areas ranging from mathematics via acoustics to engineering applications of fuzzy logic and neural networks.

Four general subject areas were addressed in this workshop: The utilisation of new measurement techniques and of new sensors for damage detection in time-variant and nonlinear systems, damage detection and localisation methods including sensitivity investigations and regularisation methods, the identification of nonlinear systems, and damage evaluation.

Signal processing is essential in fault detection, and in addition to the classical methods techniques such as wavelets, the time Hilbert transform and the Wigner distribution were discussed.

Guided stress waves, propagating energy measurements and intensities as directional quantities can be used to localise sources and sinks (damage). However, methods for identifying nonlinear structures are under further research and development.

Sensoring is important in monitoring and inspection. Local contacting and non-contacting sensors are in common use, but spatially continuous (field) measuring devices (distributed sensors) represent a relatively new topic. It can be distinguished between fully, segmented, shaped and convolved designs of distributed sensors. Applications are in active structures (sensing and control) and for high-precision systems.

Many papers deal with damage detection and localisation based on system identification methods. Several useful approaches exist for specific problems to be solved. System identification methods may apply a forward strategy and an inverse problem formulation which is ill-posed. Ill-posed problems requires regularisation for its stable solution. Dimensionality plays a role. The regularisation theory for linear (operator) equations is almost complete from a mathematical point of view, which is not true for nonlinear problems including parameter identification. Identification methods exist for special problems. The principal missing solution is that of (model) structure identification. This is substituted in many cases by parametrisation through the use of polynomials. The first steps in applying neural networks are accomplished, neural network algorithms and training seems to be well established.

Model-based fault diagnosis with parameter estimation and knowledge-based inference is applied for electrical and structural processes and systems. The mathematical model used must be, of course, validated. Dynamic mathematical models can serve, in general, for fault detection and localisation. However, if the damage assessment is based on (local) maximum stresses, then a more detailed (static) model is needed.

In addition, damage measures and damage evolution are important and their further development is being studied. This is a step in the direction to include environmental effects. This research is closely connected with damage measures, transformation into symptoms and the reverse including sensitivity to damage.

Specific damage detection and localisation schemes were applied for nonlinear composites using stationary and progressive waves, and for electrical towers using vibration signatures. Problems of a more general nature to be solved describe faults and their causes. Finite element calculations were used to explain the effect of various types of damage on the response of nonlinear systems.

2 IMPACT OF THE WORKSHOP

Because of the diverse backgrounds of participants, each learned a lot from others. Meanwhile, much interaction and group dynamics enabled cross fertilisation and leading to potential collaboration among several participants.

During the workshop, several new methods for instrumentation, parameter identification, decision-making and damage assessment were discussed and explored. There seem to exist variability in several new methods for studying nonlinear and time-varying systems. In most studies, simulations were used to catagorise the effects of faults and there is a lack of application to real data. In fact, there exists a huge communication gap between researchers and practitioners. A significant result of this workshop was to sensitise all participants as to the need for improved communication between researchers and practicing engineers.

Work has been done to find procedures not requiring parametrization and into which a variety of engineering damage models can be utilized. This is a positive step toward the development of a practical and implementable damage assessment procedure.

3 FUTURE DEVELOPMENT AND TRENDS: Model-based Detection, Diagnosis and Assessment

Due to the diversity of nonlinear phenomena there is no universal and unique approach for safety evaluation of such systems. Detection and localisation of nonlinearities in systems, for example, due to damage/faults, needs more detailed investigation for each particular application (symptoms, patterns, features). Guided stress waves, intensity measurements may be developed for detection and localisation as possible methods for practical implementation. Active systems should be included which will impact the investigation by new possibilities, for example, by using sensor/actuator signals.

Improved simulations with *validated* models are necessary for studying possible faults to summarise them in a catalogue/diagram as patterns for diagnosis purposes. Precision in itself is not and cannot be the aim, however, accuracy in the meaning of uncertainty considerations (uncertainty modelling) and error estimation is important and should be emphasized.

From a methodological point of view inverse formulations have to be applied carefully due to their ill-posedness. Regularisation methods for nonlinear problems, sensitivity measures for parameter identification are needed and should include convergence considerations. Alternatively other procedures are under investigation such as using neural nets and fuzzy sets.

Structure (of the mathematical model) identification is an unsolved problem. Patterns of nonlinear behaviour obtained from special transformations and from phase-plane diagrams should be used more intensively. Parametrical but physically not interpretable models are the first step within the procedures discussed, however, they must be developed in a direction embodying a physical understanding.

In addition to model-based assessment, diagnosis and assessment based on fuzzy logic and neural nets deserves further attention and development. Neural network has been used by several investigators dealing with nonlinear and time-varying systems with promising results. It is expected that more studies will be conducted using this relatively new tool for the identification.

There exist Type 1 and Type 2 errors (e.g., failure to replace the faulty system when it needs to be replaced and replacement of the system when it is not necessary). A bigger problem is to find the optimal policy which requires the evaluation of probabilities. There are too many factors involved in this problem to produce precise analysis for making the optimal policy. High precision is incompatible with high complexity (the principle of incompatibility). Damage is a matter of degree (e.g., slight damage or moderate damage) which may involve fuzziness.

It is clear that in order to combine identification procedures for nonlinear and time-variant systems with safety evaluation formulations some attention must be given to the diversity of the techniques and approaches used, as evident from the workshop. An accepted basis for evaluating the effectiveness and choice of a given methodology is needed which would provide a sub-grouping and hence a first step in the reduction of the diversity problem.

Generally, a holistic investigation which includes environmental effects as causes of damage as well as damage evolution is one of todays requirements.

GPSR Compliance

The European Union's (EU) General Product Safety Regulation (GPSR) is a set of rules that requires consumer products to be safe and our obligations to ensure this.

If you have any concerns about our products, you can contact us on ProductSafety@springernature.com

In case Publisher is established outside the EU, the EU authorized representative is:

Springer Nature Customer Service Center GmbH
Europaplatz 3
69115 Heidelberg, Germany

Batch number: 10577878

Printed by Printforce, the Netherlands